Metal Clusters and Their Reactivity

Zhixun Luo · Shiv N. Khanna

Metal Clusters and Their Reactivity

Springer

Zhixun Luo
State Key Laboratory for Structural
Chemistry of Unstable and Stable Species
Institute of Chemistry
Chinese Academy of Sciences
Beijing, China

Shiv N. Khanna
Department of Physics
Virginia Commonwealth University
Richmond, VA, USA

ISBN 978-981-15-9703-9 ISBN 978-981-15-9704-6 (eBook)
https://doi.org/10.1007/978-981-15-9704-6

This Springer imprint is published by the registered company Springer Nature Singapore Pte Ltd.
The registered company address is: 152 Beach Road, #21-01/04 Gateway East, Singapore 189721, Singapore

Dedicated to A. W. Castleman, Jr.

Preface

Clusters are found almost everywhere, known as a very tiny collection of atoms or molecules which form relatively stable microscopic and sub-microscopic aggregates through chemical bonding or physical force. Bridging the gap between atoms and macroscopic matter, clusters have become a scientific topic bearing increased great research interest in view of the development of precise chemistry. Clusters are ideal systems for observing the quantum effect and studying the initial state in forming macroscopic matter. Cluster researches are associated with many chemical reactions and material-changing processes, not only catalysis, but also combustion, crystal nucleation and growth, solidification and phase transformation, sol-gel, sputtering film formation, etc. The unique symmetry, stability, magic number of electrons/atoms, and high catalytic performance of clusters enable the exploration of stable and unstable species to give rise to consistent discovery of new materials. It has been recognized that one of the greatest triumphs of the last century is the development of cluster science, just as *R. Feynman* predicted. The blooming publications in the last decade concur with this prediction.

> What would the properties of materials be if we could really arrange the atoms the way we want them...I can't see exactly what would happen, but I can hardly doubt that when we have some control of the arrangement of things on a small scale we will get an enormously greater range of possible properties that substances can have, and of different things that we can do.—Richard P. Feynman (Presented at American Physical Society Meeting, 1959)

With the technical development of cluster experiments, nowadays researchers have been able to devise methods to fabricate structures which are so small that the energy levels of these systems present a discrete spectrum where the stability and reactivity are determined by the nature of electronic levels and the degree to which they are filled. Uprising research interest has been stimulated in recent years to generate free, supported, and embedded clusters (known as ligand-stabilized nanoclusters, i.e., NCs, or monolayer-protected metal clusters, i.e., MPCs) with controlled size and composition, nanoscale particles containing up to several million atoms, nano-composites, and nano-crystalline materials. Note that, the metal-metal bonds are generally weaker than ionic bonds and covalent bonds, and the valence electrons of metal often occupy the higher energy levels of the MPCs, showing a determining role in the cluster structure evolution, stability, and electronic transition of frontier orbitals. Meanwhile, various theoretical methods including global structure search and intelligent machine learning enable to provide a fundamental understanding of the properties of nanomaterials and to guide scientific thinking in the future.

> In many cases, people have worked from the top-down; that is, subdividing matter to get it smaller and smaller. We're trying to work with atoms and molecules and put them together--working our way from the bottom up. If we can retain the properties of aggregates, as we put them together, perhaps we will be able to construct new nanoscale materials.—A. W. Castleman, Jr. (Evan Pugh Professor of Chemistry and Physics and the Eberly Family Distinguished Chair in Science at Penn State University).

The current interest in the field of cluster science continues on a rapidly expanding trajectory in large measure due to two considerations. First, novel behaviors are found to emerge as the cluster size is reduced to the sub-nanometer scale. The electronic, chemical, and optical properties are all found to change with size, and in many cases, clusters of nonmagnetic solids are found to become magnetic. The second consideration is the connection to the field of nanoscale science where clusters offer the exciting prospect of serving as building blocks for new materials whose desired properties may be tailored through the selection of size and composition. Indeed, the fundamental research activity is inspired by the joint work of *Castleman, Khanna, Luo,* and others as a major goal of acquiring the underpinning knowledge for undertaking the formation of new materials "from the bottom up". This contrasts with the more conventional "top-down method" which typically involves the subdivision of matter of bulk dimensions. Particularly interesting and significant are systems whose properties vary dramatically with size and composition, one atom at a time, and don't simply scale with size or surface area directly. Most appealing among these are clusters that display interesting behaviors, whose composition can be selectively chosen and individual characteristics could be retained when assembled into an extended material. As *Castleman, Khanna, Luo* et al. demonstrated, some stable metallic clusters mimic the chemical behavior of elements in the periodic table and hence can be regarded as "superatoms" providing an unprecedented ability to design novel materials.

The focus of the current book is the reactivity of clusters. The chemical properties of matter depend on its energy levels which are greatly influenced by boundaries which restrict sizes. At the mesoscale, a large fraction of atoms are near the surface and are not interacting through bonding with as many neighboring atoms as in the case of the bulk substance. Frequently, they have different bonding characteristics, and hence their chemical properties are different. It is important to elucidate the details of the behavior and reactivity of metal cluster species in reaction cells or flow tube reactors, to apply the level of understanding obtainable for gas-phase species to the systems of practical interest in condensed phase chemistry.

The field of cluster science has developed along a few directions, including gas-phase clusters, monolayer-protected clusters (or termed as ligand-protected metal clusters), and surface-supported clusters. Around these fields, abundant efforts have been devoted to the investigations on carbon clusters, metal and semiconductor cluster species, and studies on rare gas and related van der Waals systems, as well as those comprised of hydrogen-bonded molecules. Such a book involving both gas-phase and wet-synthetic metal clusters was initially proposed by Prof. *A. W. Castleman, Jr.* who passed away in Feb. 2017. To continue *Will's* spirit, now we have completed the collection of a diverse set of topics in this book and highlight the stability and reactivity of metal clusters at reduced sizes. There has been a delay and *Will* is not there to see its publication. Accordingly, this book is dedicated to him for his honorable contributions in cluster science. Sincerely, we hope that readers will find the limited material interesting and helpful, although the present version may still not cover all aspects of metal clusters and their reactivity.

Beijing, China Zhixun Luo
Richmond, USA Shiv N. Khanna
June 2020

Contents

Chapter 1
An Overview of Metal Clusters and Their Reactivity

Atomic clusters containing a few to a few thousand atoms are emerging as a frontier area in science. The properties in this size regime are controlled by the discrete quantum conditions associated with reduced size as opposed to bulk where the properties are insensitive to the boundaries. Consequently, the physical, electronic, magnetic, and chemical properties are all found to change with size and in many cases, the properties at small sizes are different from those of bulk. Due to quantum effects, the addition of a single atom or an electron can lead to a completely different behavior. Consider the case of gold. Bulk gold is a noble element that is resistant to corrosion. Yet, small gold clusters are found to be excellent catalysts where the activity is highly dependent on the size. Bulk rhodium is non-magnetic while small rhodium clusters are found to exhibit large spin magnetic moments. The novel behaviors have led to the expectation that it should be possible to design materials with controlled properties by assembling size-selected clusters as the building blocks.

While the field of clusters can be traced to the formation of C_{60} clusters as existed in intersteller matter, the recent interest is inspired by the development in experimental techniques over the past four decades that have enabled synthesis and characterization of atomic clusters of finite size and any composition. In this book, we will primarily focus on clusters of metallic elements. These clusters bridge several interdisciplines by combining ideas within atomic, molecular and condensed matter with nuclear physics, chemistry, and biology [1–6]. Several amongest research topics in metal clusters, including structure and stability, size-dependent evolution and electronic behavior, thermal property, catalysis and reactivity etc., have been widely investigated [7–14]. In particular, techniques such as flow tube reactors coupled with quadrupole mass spectrometers, and a combination of quadrupole and octupole fields in a guided ion beam arrangement have been used in studies of cluster reactions, which are retrospect to about 40 years ago [15–25]. In recent years, extensive research interest has been stimulated by ligand-protected metal clusters [26–33], showing novel structural diversity and potential applications in catalysis, photochemistry, chemo-sensing, and

Z. Luo and S. N. Khanna, *Metal Clusters and Their Reactivity*,
https://doi.org/10.1007/978-981-15-9704-6_1

bioimaging, etc. [32–36]. Abundant metal nanoclusters (NCs) with precise atomic-level structures have been synthesized [37–55], and the high thermodynamic, electronic and chemical stabilities of these nanoclusters are often found to be rooted in bonding nature [56], degenerated electronic states and high symmetry of geometric structure of the metallic core [57, 58]. In particular, abundant MPCs prefer core-shell structures and similar building blocks of the kernel are frequently observed [59, 60], such as tetrahedral M_4 [61, 62], octahedral M_6 [63], bi-tetrahedral M_8 [64], hollow-cage M_{12} [65–67], and particularly icosahedral M_{13} [68]. For example, a vast of MPCs including Au_{18} [51], Au_{20} [49, 50], Au_{25} [45], Au_{30} [42, 43], Au_{36} [41, 69], Au_{38} [40], Au_{55} [39, 70, 71], Au_{60} [38], and Au_{102} [37] etc., all have a 13-atom icosahedral inner core [72, 73]. More and more stable metal clusters that were ascertained in gas phase have been successfully synthesized via wet chemistry approaches [74], revealing the correspondences of gas-phase stability/reactivity and condense-phase properties [48]. It is notable that the nature of the metallic core plays a determining role in the cluster structure evolution [75], stability and electronic transition between frontier orbitals.

Metal cluster reactivity in gas phase has been widely recognized of importance to determine the correspondence between activity and electronic/geometric structure stability. In general, as indicated by the HOMO-LUMO gaps, binding energies, ionization energies, etc., the metal cluster reactivity undergoes odd-even alternation effect especially for the charge-transfer dominated reactions. Also, certain reactions of the metal clusters could exhibit strong size-dependence and charge-state variation; some reactions are ADE (adiabatic detachment energy)-dependent with likely long-range charge transfer pertaining to a harpoon mechanism. Besides, the multiple valence states could account for the redox reactions of metal clusters with oxygenic chemicals [76–78]. Typical gas-phase reactivity has been unveiled for metal clusters reacting with water due to the importance of hydrogen evolution, where a mechanism of complementary active sites (Lewis acid/base) was demonstrated to account for irregular charge distribution on the cluster surface and thus the size-dependent reactivity of aluminum clusters with water. Similar hydrogen evolution reactions have also been found for aluminum clusters reacting with alcohols [79], allowing a competition and likely self-catalysis in the presence of multiple –OH reactants. Besides, the reactivity of metal clusters provides a variety of insightful information regarding the C–H bond activation [80–92], C–C bond cleavage [89, 93] C–N and C–S bond activation [94, 95], as well as C–C cross coupling reactions [96–98].

As quantum confinement effects often govern the behavior of matter in tiny size regime [2], studying the cluster reactivity provides fundamental insights into the interplay of atomic structure, geometry, and electronic property, enabling to manipulate the chemical behaviors. The studies of metal cluster reactivity can help develop tunable materials with possible catalytic or energetic qualities. This potential has been brought to fruition through recent advances. For example, an insight into the gas-phase reactivity experiments on aluminum clusters reacting with molecular oxygen have led to the discovery of specific clusters mimicking elements of the periodic table. This major finding, known as the superatom concept [9, 99–103], originated from the experimental study of the reactivity of aluminum clusters by noting a dramatic size dependence of the reactivity where cluster anions containing a certain number of Al atoms were unreactive towards oxygen while the other species were etched away [104]. Following that, ongoing efforts are devoted to studying metal cluster stability by measuring the relative mass abundances of clusters formed in comparable conditions using mass spectrometry, by further reacting with oxygen or other appropriate reactants, or by evaluating the size-dependent reactivity of mass-selected clusters [105, 106]. Based on near-free electron gas (NFEG) theory, a "jellium model" was introduced to account for the enhanced stability of these specific metal clusters [107–112]. Within this model, metal clusters with closed electronic shells often exhibit a large HOMO–LUMO gap hence enhanced chemical stability and reduced reactivity. However, not all metal clusters are subject to the same fundamental constraints, and those of favorable geometry within Mackay icosahedrons often find prominent stability [113]. In general, magically stable metal clusters are anticipated to be associated with both geometric and/or electronic shell closures [101, 102]. In comparison, superatoms are expected to retain their integrity when they undergo reactions or are used as building blocks for macro materials, as an atom retains its integrity when it bounds into a molecule. Therefore, superatoms do not have to possess closed electronic shell; instead, fruitful superatomic species may ultimately enable to establish a 3D periodic table of elements [2, 9, 101, 114]. Investigations of superatomic metal clusters could lead to the finding of new stable species with unique chemical activity and highly tunable properties that pave the way to design aimed catalysts, develop new materials with promising applications [5, 68, 115, 116].

In addition to the application in elucidating molecular details of condensed matter, studies on the kinetics of association reactions are important in the subject of phase transitions where progress is impeded because of rare fundamental data for comparison with general theories. Clusters comprised predominantly of free-electron metals often demonstrate different properties and reactivity with the addition or removal of a single atom, distinguishing them from bulk materials in this regard. The studies of metal cluster reactivity under a variety of experimental conditions have provided a wealth of information concerning the evolution of solid state with respect to that of clusters in the gas phase [117]. There is a fact that the metal-metal bonds are weaker than ionic bonds and covalent bonds, and the valence electrons of metal often occupy the higher energy levels of the cluster. It is unequivocally significant for insightful studies of naked metal clusters and their reactivities to bridge the knowledge obtained from gas-phase and soft-landing deposition on diverse supports, pertaining to practical applications in metal catalysis and genetic materials [68] which consist of metal clusters instead of metal atoms as building blocks [118].

In order to determine the binding energies of electrons in a substance, researchers performed energy measurement of electrons emitted from solids, gases or liquids by the photoelectric effect, known as photoelectron spectroscopy (PES) or photoemission spectroscopy. As bulk metals generally exhibit essentially free electrons, the metal clusters in a vast size regime between a single atom and the condensed phase offer opportunities for better understanding the electronic properties of metal aggregates generally. Among the metal clusters studied via photoelectron spectroscopy, alkali metal and coinage metal clusters offer relative simplicity as they are hydrogen-like and each bears a valence electron (s^1). Coinage metals typically have fully occupied d-orbital while half-filled s-orbitals, rendering a variety of tuneable metal-metal and metal-nonmetal bonding patterns. Extensive interest has also been devoted to the PES studies of metal clusters pertaining to both theoretical and experimental aspects [119–133]. Meanwhile, several other spectroscopies, including the VUV + IR two-color photoionization spectroscopy [134], infrared multiphoton dissociation (IRMPD) action spectroscopy [35, 135–146], are available for the identification of metal cluster structure. The IR experimental results are verified by theoretical calculation results, shedding light on the structure chemistry of gas-phase metal clusters within jellium model, under superatom characteristics [105, 147], and crystal-field-like splitting of orbitals [148, 149], and likely relativistic effect being involved [150].

This book surveys the advances that have emerged from investigations of metal cluster reactivity using mass spectrometry. Our focus is first directed toward the metal cluster reactions relating to etching effect, oxidation/reduction, halogenation, hydrogen evolution reactions (HER) and energetic reaction with hydrocarbons, collision-induced dissociation (CID) [151–158], photodissociation [159–163], cluster-size dependence and charge-state variation, charge-transfer dependence [164], shedding light on the odd-even alternation effect (or, spin effect) [149, 165–167], complementary-active-sites (CAS) [168] mechanisms, and the harpoon mechanism [169]. Continuing with the cluster reactivity surrounding with metallocarbohedrenes (Met-Cars) [170–172], some interesting aspects such as cluster catalysis and cluster-assembly materials are also summarized. Also involved, are correlative theoretical investigations based on the first-principles calculations which have been applied to depict the reaction coordinates [173]. Advances in cluster science have also stimulated interest in exploring superatoms and superatom complexes [101–103], shedding light on the stability and reactivity of both naked metal clusters [116], and ligand-protected metal nanoclusters [34, 36, 66, 174–187], enabling to prepare new materials with the characteristics of cluster genes being inherited [68].

References

1. A.W. Castleman Jr., R.G. Keesee, Acc. Chem. Res. **19**, 413–419 (1986)
2. S.N. Khanna, A.W. Castleman Jr., *Quantum Phenomena in Clusters and Nanostructures* (Springer, New York, 2003)
3. A.W. Castleman Jr., Environ. Sci. Technol. **22**, 1265–1267 (1988)
4. A.W. Castleman Jr., P. Jena, Proc. Natl. Acad. Sci. U. S. A. **103**, 10554–10559 (2006)
5. S.A. Claridge, A.W. Castleman, S.N. Khanna, C.B. Murray, A. Sen, P.S. Weiss, ACS Nano **3**, 244–255 (2009)
6. A.W. Castleman, Catal. Lett. **141**, 1243–1253 (2011)
7. P. Jena, S. Behera, *Clusters and Nanostructured Materials* (Nova Scientific Publishers, New York, 1996)
8. K.A. Zemski, D.R. Justes, A.W. Castleman Jr., J. Phys. Chem. B **106**, 6136–6148 (2002)
9. A.W. Castleman Jr., S.N. Khanna, J. Phys. Chem. C **113**, 2664–2675 (2009)

10. G.E. Johnson, R. Mitrić, E.C. Tyo, V. Bonačić-Koutecký, A.W. Castleman Jr., J. Am. Chem. Soc. **130**, 13912–13920 (2008)
11. A.E.E.W.H. Schnockel, Nature **387**, 379–381 (1997)
12. R. Burgert, H. Schnöckel, A. Grubisic, X. Li, S.T. Stokes, Science **319**, 438–442 (2008)
13. V. Bonacic-Koutecky, T.M. Bernhardt, Phys. Chem. Chem. Phys. **14**, 9252–9254 (2012)
14. H. Schwarz, Angew. Chem. Int. Ed. **50**, 10096–10115 (2011)
15. I.N. Tang, A.W. Castleman Jr., J. Chem. Phys. **60**, 3981–3986 (1974)
16. A.W. Castleman Jr., R.E. Davis, H.R. Munkelwitz, I.N. Tang, W.P. Wood, Int. J. Chem. Kinet. **1**, 629–640 (1975)
17. I.N. Tang, M.S. Lian, A.W. Castleman Jr., J. Chem. Phys. **65**, 4022–4027 (1976)
18. A.W. Castleman Jr., I.N. Tang, J. Photochem. **6**, 349–354 (1977)
19. R. Hofmannsievert, A.W. Castleman Jr., J. Phys. Chem. **88**, 3329–3333 (1984)
20. A.W. Castleman, Jr., R.G. Keesee, Z. Phys. D At. Mol. Clusters **3**, 167–176 (1986)
21. S. Morgan, A.W. Castleman Jr., J. Am. Chem. Soc. **109**, 2867–2870 (1987)
22. R. Passarella, R.J. Shul, R.G. Keesee, A.W. Castleman Jr., Int. J. Mass Spectrom. Ion Processes **81**, 227–233 (1987)
23. B.L. Upschulte, R.J. Shul, R. Passarella, R.G. Keesee, A.W. Castleman Jr., Int. J. Mass Spectrom. Ion Processes **75**, 27–45 (1987)
24. J.J. Breen, K. Kilgore, S. Wei, W.B. Tzeng, R.G. Keesee, A.W. Castleman Jr., J. Phys. Chem. **93**, 7703–7707 (1989)
25. J.J. Breen, W.B. Tzeng, K. Kilgore, R.G. Keesee, A.W. Castleman Jr., J. Chem. Phys. **90**, 19–24 (1989)
26. M.A. Banares, J.H. Cardoso, G.J. Hutchings, J.M.C. Bueno, J.L.G. Fierro, Catal. Lett. **56**, 149–153 (1998)
27. M. Ziolek, Catal. Today **78**, 47–64 (2003)
28. X. Solans-Monfort, V. Branchadell, M. Sodupe, M. Sierka, J. Sauer, J. Chem. Phys. **121**, 6034–6041 (2004)
29. J. Guzman, B.C. Gates, J. Am. Chem. Soc. **126**, 2672–2673 (2004)
30. C.K. Siu, S.J. Reitmeier, I. Balteanu, V.E. Bondybey, M.K. Beyer, Eur. Phys. J. D **43**, 189–192 (2007)
31. S. Arndt, G. Laugel, S. Levchenko, R. Horn, M. Baerns, M. Scheffler, R. Schlögl, R. Schomäcker, Catal. Rev. **53**, 424–514 (2011)
32. Z. Gan, N. Xia, Z. Wu, Acc. Chem. Res. **51**, 2774–2783 (2018)
33. S. Wang, Q. Li, X. Kang, M. Zhu, Acc. Chem. Res. **51**, 2784–2792 (2018)
34. T. Higaki, Q. Li, M. Zhou, S. Zhao, Y. Li, S. Li, R. Jin, Acc. Chem. Res. **51**, 2764–2773 (2018)
35. B. Nieto-Ortega, T. Burgi, Acc. Chem. Res. **51**, 2811–2819 (2018)
36. Q. Tang, G. Hu, V. Fung, D.E. Jiang, Acc. Chem. Res. **51**, 2793–2802 (2018)
37. P.D. Jadzinsky, G. Calero, C.J. Ackerson, D.A. Bushnell, R.D. Kornberg, Science **318**, 430–433 (2007)
38. Y. Song, F. Fu, J. Zhang, J. Chai, X. Kang, P. Li, S. Li, H. Zhou, M. Zhu, Angew. Chem. Int. Ed. **54**, 8430–8434 (2015)
39. C.M. Chang, C. Cheng, C.M. Wei, J. Chem. Phys. **128**, 124710 (2008)
40. H. Qian, W.T. Eckenhoff, Y. Zhu, T. Pintauer, R. Jin, J. Am. Chem. Soc. **132**, 8280–8281 (2010)
41. C.J. Zeng, H.F. Qian, T. Li, G. Li, N.L. Rosi, B. Yoon, R.N. Barnett, R.L. Whetten, U. Landman, R.C. Jin, Angew. Chem. Int. Ed. **51**, 13114–13118 (2012)
42. D. Crasto, S. Malola, G. Brosofsky, A. Dass, H. Häkkinen, J. Am. Chem. Soc. **136**, 5000–5005 (2014)
43. H.Y. Yang, Y. Wang, A.J. Edwards, J.Z. Yan, N.F. Zheng, Chem. Commun. **50**, 14325–14327 (2014)
44. M.W. Heaven, A. Dass, P.S. White, K.M. Holt, R.W. Murray, J. Am. Chem. Soc. **130**, 3754–3755 (2008)

45. M. Zhu, C.M. Aikens, F.J. Hollander, G.C. Schatz, R. Jin, J. Am. Chem. Soc. **130**, 5883–5885 (2008)
46. M.Z. Zhu, W.T. Eckenhoff, T. Pintauer, R.C. Jin, J. Phys. Chem. C **112**, 14221–14224 (2008)
47. Z. Luo, C.J. Grover, A.C. Reber, S.N. Khanna, A.W. Castleman, J. Am. Chem. Soc. **135**, 4307–4313 (2013)
48. Z. Luo, A.C. Reber, M. Jia, W.H. Blades, S.N. Khanna, A.W. Castleman, Chem. Sci. **7**, 3067–3074 (2016)
49. X.K. Wan, Z.W. Lin, Q.M. Wang, J. Am. Chem. Soc. **134**, 14750–14752 (2012)
50. C.J. Zeng, C. Liu, Y.X. Chen, N.L. Rosi, R.C. Jin, J. Am. Chem. Soc. **136**, 11922–11925 (2014)
51. Y. Chen, C. Zeng, C. Liu, K. Kirschbaum, C. Gayathri, R.R. Gil, N.L. Rosi, R. Jin, J. Am. Chem. Soc. **137**, 10076–10079 (2015)
52. C.P. Joshi, M.S. Bootharaju, M.J. Alhilaly, O.M. Bakr, J. Am. Chem. Soc. **137**, 11578–11581 (2015)
53. L.G. AbdulHalim, M.S. Bootharaju, Q. Tang, S. Del Gobbo, R.G. AbdulHalim, M. Eddaoudi, D.E. Jiang, O.M. Bakr, J. Am. Chem. Soc. **137**, 11970–11975 (2015)
54. A. Desireddy, B.E. Conn, J. Guo, B. Yoon, R.N. Barnett, B.M. Monahan, K. Kirschbaum, W.P. Griffith, R.L. Whetten, U. Landman, T.P. Bigioni, Nature **501**, 399–402 (2013)
55. M.J. Alhilaly, M.S. Bootharaju, C.P. Joshi, T.M. Besong, A.H. Emwas, R. Juarez-Mosqueda, S. Kaappa, S. Malola, K. Adil, A. Shkurenko, H. Hakkinen, M. Eddaoudi, O.M. Bakr, J. Am. Chem. Soc. **138**, 14727–14732 (2016)
56. G. Frenking, N. Fröhlich, Chem. Rev. **100**, 717–774 (2000)
57. N. Haruta, T. Tsukamoto, A. Kuzume, T. Kambe, K. Yamamoto, Nat. Commun. **9**, 3758 (2018)
58. B. Stephanidis, S. Adichtchev, S. Etienne, S. Migot, E. Duval, A. Mermet, Phys. Rev. B **76**, 121404 (2007)
59. D.-E. Jiang, M.L. Tiago, W. Luo, S. Dai, J. Am. Chem. Soc. **130**, 2777–2779 (2008)
60. G. Hu, R. Jin, D.-E. Jiang, Nanoscale **8**, 20103–20110 (2016)
61. D.M. Chevrier, C. Zeng, R. Jin, A. Chatt, P. Zhang, J. Phys. Chem. C **119**, 1217–1223 (2014)
62. A. Tlahuice-Flores, J. Phys. Chem. C **123**, 10831–10841 (2019)
63. H. Yang, J. Lei, B. Wu, Y. Wang, M. Zhou, A. Xia, L. Zheng, N. Zheng, Chem. Commun. **49**, 300–302 (2013)
64. Y. Song, S. Wang, J. Zhang, X. Kang, S. Chen, P. Li, H. Sheng, M. Zhu, J. Am. Chem. Soc. **136**, 2963–2965 (2014)
65. K.M. Harkness, Y. Tang, A. Dass, J. Pan, N. Kothalawala, V.J. Reddy, D.E. Cliffel, B. Demeler, F. Stellacci, O.M. Bakr, Nanoscale **4**, 4269–4274 (2012)
66. H. Yang, Y. Wang, H. Huang, L. Gell, L. Lehtovaara, S. Malola, H. Hakkinen, N. Zheng, Nat. Commun. **4**, 2422 (2013)
67. S.-F. Yuan, C.-Q. Xu, J. Li, Q.-M. Wang, Angew. Chem. Int. Ed. **58**, 5906–5909 (2019)
68. Y. Jia, Z. Luo, Coord. Chem. Rev. **400**, 213053 (2019)
69. A. Das, C. Liu, C.J. Zeng, G. Li, T. Li, N.L. Rosi, R.C. Jin, J. Phys. Chem. A **118**, 8264–8269 (2014)
70. H.G. Boyen, G. Kastle, F. Weigl, B. Koslowski, C. Dietrich, P. Ziemann, J.P. Spatz, S. Riethmuller, C. Hartmann, M. Moller, G. Schmid, M.G. Garnier, P. Oelhafen, Science **297**, 1533–1536 (2002)
71. M. Turner, V.B. Golovko, O.P.H. Vaughan, P. Abdulkin, A. Berenguer-Murcia, M.S. Tikhov, B.F.G. Johnson, R.M. Lambert, Nature **454**, 981–983 (2008)
72. G. Schmid, Chem. Soc. Rev. **37**, 1909–1930 (2008)
73. Y. Shichibu, K. Suzuki, K. Konishi, Nanoscale **4**, 4125–4129 (2012)
74. T. Kambe, N. Haruta, T. Imaoka, K. Yamamoto, Nat. Commun. **8**, 2046 (2017)
75. H. Zhang, A.C. Reber, L. Geng, D. Rabayda, H. Wu, Z. Luo, J. Yao, S.N. Khanna, CCS Chem. **1**, 571–581 (2019)
76. H. Zhang, H. Wu, Y. Jia, L. Geng, Z. Luo, H. Fu, J. Yao, Rev. Sci. Instrum. **90**, 073101 (2019)

77. A. Armstrong, H. Zhang, A.C. Reber, Y. Jia, H. Wu, Z. Luo, S.N. Khanna, J. Phys. Chem. A **123**, 7463–7469 (2019)
78. H. Zhang, H. Wu, L. Geng, Y. Jia, M. Yang, Z. Luo, Phys. Chem. Chem. Phys. **21**, 11234–11241 (2019)
79. Z. Luo, J.C. Smith, W.H. Woodward, A.W. Castleman Jr., J. Phys. Chem. Lett. **3**, 3818–3821 (2012)
80. Q.Z. Zheng, N. Jiao, Chem. Soc. Rev. **45**, 4590–4627 (2016)
81. D. Schröder, H. Schwarz, Angew. Chem. Int. Ed. **34**, 1973–1995 (1995)
82. X.-L. Ding, X.-N. Wu, Y.-X. Zhao, S.-G. He, Acc. Chem. Res. **45**, 382–390 (2012)
83. D. Balcells, E. Clot, O. Eisenstein, Chem. Rev. **110**, 749–823 (2010)
84. C.L. Sun, B.J. Li, Z.J. Shi, Chem. Rev. **111**, 1293–1314 (2011)
85. K.R. Campos, Chem. Soc. Rev. **36**, 1069–1084 (2007)
86. T.C. Boorman, I. Larrosa, Chem. Soc. Rev. **40**, 1910–1925 (2011)
87. R. Shang, L. Ilies, E. Nakamura, Chem. Rev. **117**, 9086–9139 (2017)
88. S. Gaillard, C.S.J. Cazin, S.P. Nolan, Acc. Chem. Res. **45**, 778–787 (2012)
89. H. Schwarz, Acc. Chem. Res. **22**, 282–287 (1989)
90. K. Liao, S. Negretti, D.G. Musaev, J. Bacsa, H.M. Davies, Nature **533**, 230–234 (2016)
91. K. Chen, A. Eschenmoser, P.S. Baran, Angew. Chem. Int. Ed. **48**, 9705–9708 (2009)
92. J.A. Labinger, J.E. Bercaw, Nature **417**, 507–514 (2002)
93. J.-B. Ma, Z. Yuan, J.-H. Meng, Q.-Y. Liu, S.-G. He, ChemPhysChem **15**, 4117–4125 (2014)
94. Z.X. Luo, J.C. Smith, C. Berkdemir, A.W. Castleman Jr., Chem. Phys. Lett. **590**, 63–68 (2013)
95. Z. Luo, G.U. Gamboa, M. Jia, A.C. Reber, S.N. Khanna, A.W. Castleman Jr., J. Phys. Chem. A **118**, 8345–8350 (2014)
96. K.S. Yoo, C.H. Yoon, K.W. Jung, J. Am. Chem. Soc. **128**, 16384–16393 (2006)
97. K.S. Yoo, C.H. Yoon, R.K. Mishra, Y.C. Jung, S.W. Yi, K.W. Jung, J. Am. Chem. Soc. **129**, 3771–3771 (2007)
98. J.C. Lo, J. Gui, Y. Yabe, C.-M. Pan, P.S. Baran, Nature **516**, 343–348 (2014)
99. S.N. Khanna, P. Jena, Phys. Rev. B **51**, 13705–13716 (1995)
100. D.E. Bergeron, A.W. Castleman Jr., T. Morisato, S.N. Khanna, Science **304**, 84–87 (2004)
101. Z. Luo, A.W. Castleman, Acc. Chem. Res. **47**, 2931–2940 (2014)
102. A.C. Reber, S.N. Khanna, Acc. Chem. Res. **50**, 255–263 (2017)
103. P. Jena, Q. Sun, Chem. Rev. **118**, 5755–5780 (2018)
104. R.E. Leuchtner, A.C. Harms, A.W. Castleman Jr., J. Chem. Phys. **91**, 2753–2754 (1989)
105. W.A. De Heer, Rev. Mod. Phys. **65**, 611–676 (1993)
106. M. Brack, Rev. Mod. Phys. **65**, 677–732 (1993)
107. C.Y. Cha, G. Ganteför, W. Eberhardt, J. Chem. Phys. **99**, 6308–6312 (1993)
108. R.B. King, J. Zhao, Chem. Commun. 4204–4205 (2006)
109. C.E. Jones Jr., P.A. Clayborne, J.U. Reveles, J.J. Melko, U. Gupta, S.N. Khanna, A.W. Castleman, J. Phys. Chem. A **112**, 13316–13325 (2008)
110. R.B. King, I. Silaghi-Dumitrescu, Dalton Trans. 6083–6088 (2008)
111. R.G. Polozkov, V.K. Ivanov, A.V. Verkhovtsev, A.V. Solov'yov, Phys. Rev. A **79** (2009)
112. J.J. Melko, P.A. Clayborne, C.E. Jones, J.U. Reveles, U. Gupta, S.N. Khanna, A. W. Castleman, J. Phys. Chem. A **114**, 2045–2052 (2010)
113. K.H. Kuo, Struct. Chem. **13**, 221–222 (2002)
114. A.W. Castleman Jr., J. Phys. Chem. Lett. **2**, 1062–1069 (2011)
115. J.U. Reveles, P.A. Clayborne, A.C. Reber, S.N. Khanna, K. Pradhan, P. Sen, M.R. Pederson, Nat. Chem. **1**, 310–315 (2009)
116. C.-N. Cui, H.-Y. Zhang, Z.-X. Luo, F. Pan, Chin. J. Struct. Chem. **39**, 989–998 (2020)
117. F. Trager, G.z. Putlitz, *Metal Clusters* (Springer, Heidelberg, 1986)
118. T.E. Dermota, Q. Zhong, A.W. Castleman Jr., Chem. Rev. **104**, 1861–1886 (2004)
119. D.G. Leopold, J. Ho, W.C. Lineberger, J. Chem. Phys. **86**, 1715–1726 (1987)
120. W.C. Lineberger, K.M. Ervin, J. Ho, Abstr. Pap. Am. Chem. Soc. **196**, 21-PHYS (1988)
121. W.C. Lineberger, J. Ho, K.M. Ervin, Abstr. Pap. Am. Chem. Soc. **199**, 123-COLL (1990)

122. J.G. Eaton, L.H. Kidder, H.W. Sarkas, K.M. McHugh, K.H. Bowen, Abstr. Pap. Am. Chem. Soc. **204**, 222-PHYS (1992)
123. K.S. Busenlener, M.L. Polak, J. Ho, W.C. Lineberger, Abstr. Pap. Am. Chem. Soc. **205**, 146-PHYS (1993)
124. S.J. DixonWarren, R.F. Gunion, W.C. Lineberger, J. Chem. Phys. **104**, 4902–4910 (1996)
125. A. Nakajima, T. Taguwa, K. Nakao, K. Hoshino, S. Iwata, K. Kaya, Surf. Rev. Lett. **3**, 417–421 (1996)
126. A. Nakajima, H. Kawamata, T. Hayase, Y. Negishi, K. Kaya, Z. Phys. D At. Mol. Clusters **40**, 17–21 (1997)
127. S.R. Liu, H.J. Zhai, L.S. Wang, Abstr. Pap. Am. Chem. Soc. **222**, U194–U194 (2001)
128. A. Pramann, Y. Nakamura, A. Nakajima, K. Kaya, J. Phys. Chem. A **105**, 7534–7540 (2001)
129. C.C. Jarrold, Abstr. Pap. Am. Chem. Soc. **227**, U276–U276 (2004)
130. K. Koyasu, Y. Naono, M. Akutsu, M. Mitsui, A. Nakajima, Chem. Phys. Lett. **422**, 62–66 (2006)
131. K. Koyasu, J. Atobe, S. Furuse, A. Nakajima, J. Chem. Phys. **129**, 214301 (2008)
132. D.G. Leopold, S.R. Miller, P.W. Villalta, T.P. Marcy, E.L. Millam, Abstr. Pap. Am. Chem. Soc. **236**, 561-PHYS (2008)
133. J. Atobe, K. Koyasu, S. Furuse, A. Nakajima, Phys. Chem. Chem. Phys. **14**, 9403–9410 (2012)
134. J.R. Woodward, H. Watanabe, S. Ishiuchi, M. Fujii, Rev. Sci. Instrum. **83**, 014102 (2012)
135. K.R. Asmis, G. Santambrogio, M. Brümmer, J. Sauer, Angew. Chem. Int. Ed. **44**, 3122–3125 (2005)
136. D.R. Zahn, G.N. Gavrila, G. Salvan, Chem. Rev. **107**, 1161–1232 (2007)
137. R. Zenobi, Chem. Rev. **117**, 4943–4944 (2017)
138. G. Santambrogio, E. Janssens, S. Li, T. Siebert, G. Meijer, K.R. Asmis, J. Dobler, M. Sierka, J. Sauer, J. Am. Chem. Soc. **130**, 15143–15149 (2008)
139. H.B. Fu, Y.J. Hu, E.R. Bernstein, J. Chem. Phys. **124**, 024302 (2006)
140. A. Engdahl, B. Nelander, Science **295**, 482–483 (2002)
141. J.M. Headrick, E.G. Diken, R.S. Walters, N.I. Hammer, R.A. Christie, J. Cui, E.M. Myshakin, M.A. Duncan, M.A. Johnson, K.D. Jordan, Science **308**, 1765–1769 (2005)
142. A.C. Luntz, Science **302**, 70–71 (2003)
143. M. Miyazaki, A. Fujii, T. Ebata, N. Mikami, Science **304**, 1134–1137 (2004)
144. K. Nauta, D.T. Moore, P.L. Stiles, R.E. Miller, Science **292**, 481–484 (2001)
145. P. Gruene, D.M. Rayner, B. Redlich, A.F.G. van der Meer, J.T. Lyon, G. Meijer, A. Fielicke, Science **321**, 674–676 (2008)
146. K.R. Asmis, Phys. Chem. Chem. Phys. **14**, 9270–9281 (2012)
147. V.M. Medel, J.U. Reveles, S.N. Khanna, V. Chauhan, P. Sen, A.W. Castleman, Proc. Natl. Acad. Sci. U. S. A. **108**, 10062–10066 (2011)
148. J.M. Pettibone, J.W. Hudgens, J. Phys. Chem. Lett. **1**, 2536–2540 (2010)
149. Z. Luo, G.U. Gamboa, J.C. Smith, A.C. Reber, J.U. Reveles, S.N. Khanna, A.W. Castleman Jr., J. Am. Chem. Soc. **134**, 18973–18978 (2012)
150. H. Hakkinen, M. Moseler, U. Landman, Phys. Rev. Lett. **89**, 033401 (2002)
151. B. Xu, Y.-X. Zhao, X.-L. Ding, Q.-Y. Liu, S.-G. He, J. Phys. Chem. A **117**, 2961–2970 (2013)
152. X.N. Wu, J.B. Ma, B. Xu, Y.X. Zhao, X.L. Ding, S.G. He, J. Phys. Chem. A **115**, 5238–5246 (2011)
153. M.T. Rodgers, P.B. Armentrout, J. Phys. Chem. A **101**, 1238–1249 (1997)
154. H. Koizumi, X.-G. Zhang, P.B. Armentrout, J. Phys. Chem. A **105**, 2444–2452 (2001)
155. N. Aristov, P.B. Armentrout, J. Phys. Chem. **90**, 5135–5140 (1986)
156. J. Ree, Y.H. Kim, H.K. Shin, J. Chem. Phys. **124**, 074307 (2006)
157. M. Li, S.R. Liu, P.B. Armentrout, J. Chem. Phys. **131**, 144310 (2009)
158. R.B. Huang, Z.Y. Liu, H.F. Liu, L.H. Chen, Q. Zhang, C.R. Wang, L.S. Zheng, F.Y. Liu, S.Q. Yu, X.X. Ma, Int. J. Mass Spectrom. Ion Processes **151**, 55–62 (1995)
159. K. Mizuse, A. Fujii, Phys. Chem. Chem. Phys. **13**, 7129–7135 (2011)
160. G. Wang, C. Chi, J. Cui, X. Xing, M. Zhou, J. Phys. Chem. A **116**, 2484–2489 (2012)

161. R.S. Walters, N.R. Brinkmann, H.F. Schaefer, M.A. Duncan, J. Phys. Chem. A **107**, 7396–7405 (2003)
162. Z.A. Reed, M.A. Duncan, J. Phys. Chem. A **112**, 5354–5362 (2008)
163. K.S. Molek, C. Anfuso-Cleary, M.A. Duncan, J. Phys. Chem. A **112**, 9238–9247 (2008)
164. D.H. Ess, R.J. Nielsen, W.A. Goddard, R.A. Periana, J. Am. Chem. Soc. **131**, 11686–11688 (2009)
165. S. Álvarez-Barcia, J.R. Flores, J. Phys. Chem. A **116**, 8040–8050 (2012)
166. R. Burgert, H. Schnockel, A. Grubisic, X. Li, S.T. Stokes, K.H. Bowen, G.F. Gantefor, B. Kiran, P. Jena, Science **319**, 438–442 (2008)
167. A.C. Reber, S.N. Khanna, P.J. Roach, W.H. Woodward, A.W. Castleman Jr., J. Am. Chem. Soc. **129**, 16098–16101 (2007)
168. P.J. Roach, W.H. Woodward, A.W. Castleman Jr., A.C. Reber, S.N. Khanna, Science **323**, 492–495 (2009)
169. Z.X. Luo, C. Berkdemir, J.C. Smith, A.W. Castleman Jr., Chem. Phys. Lett. **582**, 24–30 (2013)
170. L. Gao, M.E. Lyn, D.E. Bergeron, A.W. Castleman, Int. J. Mass Spectrom. **229**, 11–17 (2003)
171. H. Sakurai, A.W. Castleman Jr., J. Phys. Chem. A **101**, 7695–7698 (1997)
172. S.F. Cartier, Z.Y. Chen, G.J. Walder, C.R. Sleppy, A.W. Castleman Jr., Science **260**, 195–196 (1993)
173. Z. Luo, A.W. Castleman Jr., S.N. Khanna, Chem. Rev. **116**, 14456–14492 (2016)
174. F. Baletto, R. Ferrando, Rev. Mod. Phys. **77**, 371–423 (2005)
175. Y. Tao, M. Li, J. Rena, X. Qu, Chem. Soc. Rev. **44**, 8636–8663 (2015)
176. R.S. Dhayal, W.E. van Zyl, C.W. Liu, Acc. Chem. Res. **49**, 86–95 (2016)
177. O. Fuhr, S. Dehnen, D. Fenske, Chem. Soc. Rev. **42**, 1871–1906 (2013)
178. R.R. Nasaruddin, T.K. Chen, N. Yan, J.P. Xie, Coord. Chem. Rev. **368**, 60–79 (2018)
179. K.R. Krishnadas, A. Baksi, A. Ghosh, G. Natarajan, A. Som, T. Pradeep, Acc. Chem. Res. **50**, 1988–1996 (2017)
180. S. Yamazoe, K. Koyasu, T. Tsukuda, Acc. Chem. Res. **47**, 816–824 (2014)
181. Z. Han, X.-Y. Dong, P. Luo, S. Li, Z.-Y. Wang, S.-Q. Zang, T.C.W. Mak, Sci. Adv. **6**, eaay0107 (2020)
182. R. Jin, C. Zeng, M. Zhou, Y. Chen, Chem. Rev. **116**, 10346–10413 (2016)
183. G. Li, R. Jin, Acc. Chem. Res. **46**, 1749–1758 (2013)
184. H. Qian, M. Zhu, Z. Wu, R. Jin, Acc. Chem. Res. **45**, 1470–1479 (2012)
185. J.Q. Wang, Z.J. Guan, W.D. Liu, Y. Yang, Q.M. Wang, J. Am. Chem. Soc. **141**, 2384–2390 (2019)
186. J. Yan, B.K. Teo, N. Zheng, Acc. Chem. Res. **51**, 3084–3093 (2018)
187. X. Kang, M. Zhu, Chem. Soc. Rev. **48**, 2422–2457 (2019)

Chapter 2
Instrumentation for Cluster Science

There are several critical concerns regarding different strategies for producing gaseous metal atoms/clusters and studying their gas-phase reactivity. In this chapter we endeavour to summarize all the experimental methods used for cluster preparation and cluster reaction studies. Stress will be laid on a few representative aspects including cluster sources, cluster growth principles, cluster reaction apparatus and cluster spectroscopic techniques.

2.1 Cluster Sources

While the utility of molecular beams for studying atomic/molecular collisions has been well-known for many years, the application of molecule beam technology to metal clusters requires suitable sources that produce metal atoms/clusters in the gas phase. A simple way to produce clusters is retrospected to the use of a Knudsen cell which however was found to be inefficient [1]. With the development of science and technology, there are many sources used today for an efficient production of free metal clusters. According to the differences in cluster forming process, the various cluster sources can be catalogued as:

(a) *Supersonic jets*, where clusters traverse a skimmer and are ionized by electron or photon impact and then separated via mass spectrometer;
(b) *Gas aggregation sources,* which are favoured to produce large clusters;
(c) *Surface erosion sources*, including those by laser ablation or sputtering;
(d) *Pick-up sources*, known as the pulsed arc-discharge ion source, etc. [2, 3].

On the other hand, on a basis of the different techniques being involved, the various cluster sources generally include the following aspects: (i) thermal heated oven sources; (ii) electrospray ionization; (iii) laser vaporization cluster sources; (iv)

Z. Luo and S. N. Khanna, *Metal Clusters and Their Reactivity*,
https://doi.org/10.1007/978-981-15-9704-6_2

sputtering sources including arc-discharge sputtering and magnetron sputtering. The working principles of these sources are described in detail below.

2.1.1 Thermal Heated Oven Sources

Metal cluster sources have been used as an integral part of molecular beam experiments for many years. Among them, thermal heated oven sources (also known as Knudsen ovens) were widely used in the early investigations [4–14]. Afterwards Larsen et al. [4] extended this kind of sources and applied to the study of alkali metal atom/cluster systems by using a supersonic nozzle and "seeding" technique. Their design of the oven, as displayed in Fig. 2.1, allowed the beam gas of metal atoms/clusters mixing with a light diluent gas, and hence the beam species accelerated to a velocity equal to that of the diluent gas. The seeded beams were used to measure cross sections for electronic excitation in collisions of fast alkali atoms, utilizing thermal energy beams of the alkali atoms and mercury atoms, as well as various diatomic and polyatomic molecules [4].

Similar sources were also built with thermionic emission from heated metal salt cathodes, and decomposition of volatile organometallics [15]. For example, Draves et al. [16] developed a thermionic and molecular-beam source and firstly applied it to the studies upon $Cs(CH_3OH)_n^+$ cluster systems. The thermionic metal-ion source was combined with an electrostatic lens system, a quadrupole mass filter, and an electron multiplier for signal detection. The apparatus consistsed of two chambers, a source chamber and a detector chamber which were pumped by diffusion pumps, respectively. During the operation, the metal-ion source was placed away from the nozzle at a linear distance of ~5 mm. An electric current of ~5 A from a floating power supply (±100 V) was passed through the filament. The metal-ion emission can be enhanced by biasing the filament relative to ground [16].

2.1.2 Electron Impact and Electrospray Ionization (ESI)

Electron impact (EI) and electrospray ionization (ESI) sources [17–19], as another kind of free-jet themes, are known of great value applied in mass spectrometry [17–19]. Utilizing the ESI technique, protonated water clusters and series of alkali metal halide clusters were successfully prepared and investigated [20–22]. Up to now, both for practical purposes and to discern the mechanisms of ESI itself, various salt cluster ions have been studied by the ESI mass spectrometry (ESI-MS) [18, 23–27]. Figure 2.2 depicts a sketch drawing of a typical ESI source [28], where the nebuliser (N_2 gas) makes the ionised liquid to spray through a taylor-cone nozzle, resulting to aerosol plume and forming gas-phase cluster ions. There are two mechanisms for the ESI process: the ion evaporation model proposed by Iribarne and Thomson [29, 30], and the charge residue model originated by Dole et al. [17] and developed

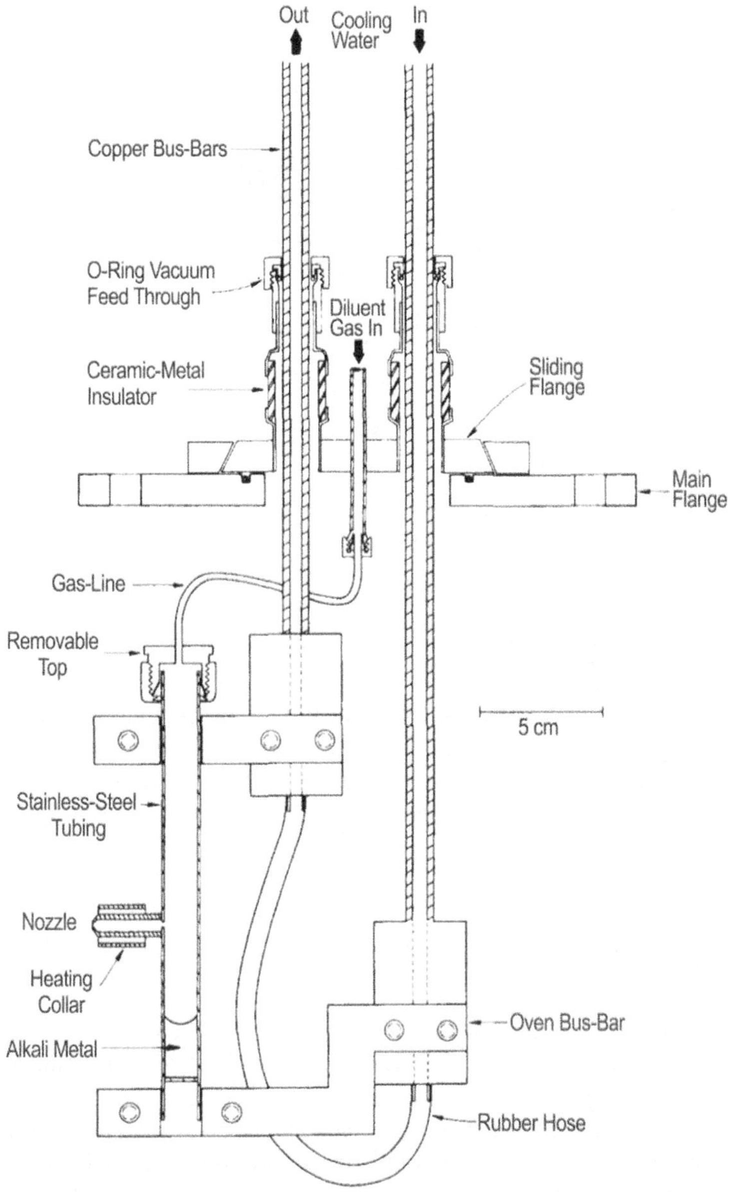

Fig. 2.1 Plan view of the supersonic beam source, showing the arrangement for heating the nozzle and supply chambers separately. The source is suspended from a sliding flange by water-cooled bus-bars which are electrically isolated by ceramic-to-metal insulators. The flange can be scanned transverse to the beam path to align or flag the beam. Adapted with permission from Ref. [4]. Copyright 1974 by the American Physical Society

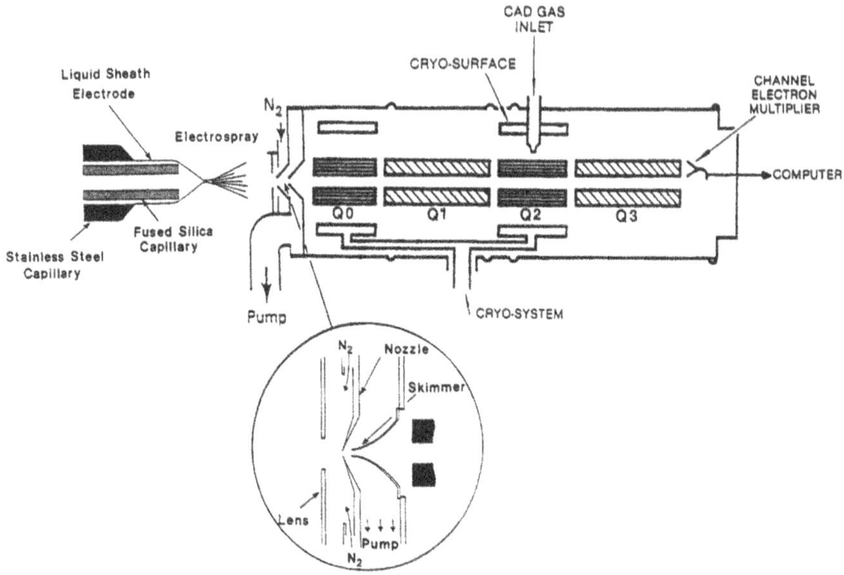

Fig. 2.2 A sketch of the electrospray ionisation (ESI) and ion source, reproduced with the permission from Ref. [28]. Copyright 1990 American Chemistry Society

by Röllgen and co-workers [31, 32]. In general, for ionic compound solutions, ion evaporation model could dominate the ESI process, while for certain organic or biological compounds, the charge residue model is believed to play an vital role. ESI-MS has become a proven technique and various commercial ESI instruments are available now days.

2.1.3 Laser Vaporization Cluster Sources

Laser vaporization (LaVa) sources, originally known as "Smalley sources", were based on the setup by Smalley and coworkers in the early 1980s where this method was used to ablate graphite leading to the discovery of buckminster fullerene [33]. It has been developed into an effective method for creating clusters of high-boiling point metals [34, 35]. In particular, LaVa sources are widely applied in the studies of metal clusters and metal-containing molecules and complexes. Together with the development and reliability of modern lasers, the versatility regarding the source materials makes the LaVa source an attractive option in cluster researches.

The construction of a LaVa source is in many ways an art form [4, 34–56]. It has been 30 years since the original report, but it is clear that LaVa sources continue to dominate the approaches for gas-phase chemistry of metal clusters [2, 57–65]. An overall design of the LaVa source was greatly improved by de Heer (Fig. 2.3a) [12]. The general principle behind the LaVa source is to use a focused high-power

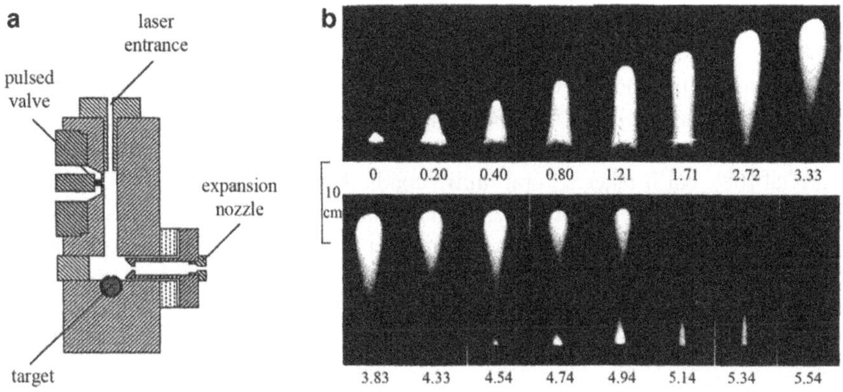

Fig. 2.3 **a** A sketch map of the standard laser vaporization source. Focused laser is used to ablate a target, which is typically a rotating/translating rod. Pulsed gas then carries the nascent clusters through the expansion nozzle into the vacuum instrument. **b** High-speed photographic images of a high-power laser pulse striking aluminum. Units of time are in ms; exposure time is 6 μs. Adapted from Refs. [67, 68]. Copyright 1997 and 1990 by the American Physical Society

laser beam to ablate a target and create a gaseous form of the target material. The target is either a rod or a disk which is rotated (and/or movable) to increase material availability [50, 66]. The gas-phase atoms, ions and small clusters of the ablated material then condense into aim clusters with a certain size distribution through an expansion nozzle, and then are carried into the vacuum apparatus via a backing/carrier gas such as helium.

The applications for fundamental studies of metal clusters typically include those on aluminium and other main group metals [34, 64, 69, 70], as well as transition metals [35, 43, 45, 46, 50]. The target can also be semiconductors such as silicon [71, 72], germanium, etc. [72, 73]. Care was taken, experiments were also extended to cluster investigations of non-metal systems, such as carbon materials, etc. [33, 71, 74–76]. In addition, there are extension of the LaVa source applied to study those composed of more than one element in order to create bi-elemental species [77, 78], especially metal alloy clusters and their mixtures [77, 79–81]. To attain metal alloy clusters, one method is to directly utilize an alloy metal target; while on the other hand, especially for those metal mixtures not available as alloys, volatile organometallics (which decomposed in the laser plasma) can be added to the gas flow leading to mixed metal clusters [82]. Researchers may also employ vapour or electrochemical deposition to make composite samples with a thin film of one metal coated onto a solid sample of another [15, 83–86]. This approach on LaVa source has led to the discovery of metal carbide clusters, i.e., "Met-Cars" species, which were produced by laser vaporization of a metal target (such as titanium, vanadium or zirconium) with a hydrocarbon gas (such as methane or acetylene) being added to the gas flow [87–99].

Although the use of a continuous-wave laser is not forbidden, LaVa sources usually employ a pulse laser system because of the following reasons: (i) it is easier to achieve the powers necessarily enough to evaporate metals with a Q-switched laser; (ii) as the buffer gas used to carry the burgeoning clusters out of the source must be rapidly pumped out of the vacuum instrument, it is much easier to manage when the laser and thus the gas is pulsed; (iii) the formation of plasma around the target area may briefly shield the material as tailing light is absorbed by the hot and opaque blowoff material [68], but it takes several milliseconds (Fig. 2.3b) for the plasma-shielded material to cool down when the plasma fully dissociates, so it is not necessary to use too long laser pulse. Actually, if the laser faster than 200 Hz (including continuous-wave lasers), stronger than 10^8 W/cm^2 and with a pulse duration longer than 10^{-4} s, the amount of material evaporated will be reduced and hence direct influence to the intensity of the clusters produced [100].

2.1.4 Sputtering Sources

(a) Arc-discharge sputtering

The arc-discharge metal cluster source is to employ a DC discharge at a metal cathode in a helium/argon flow [101–103]. Cluster ions are produced from sputtering of the cathodic metal target by ionized Ar^+ and followed by further clustering in the discharge plasma. Typically, Fig. 2.4a displays one of such cluster source used by Ho et al. [103]. The use of carrier gas flowing in around the cathode feedthrough helped to prevent the plating of metals onto the glass insulator. The cathode was negatively biased (e.g., 3–5 kV) with respect to the grounded flow tube, where the parameters of gas composition, flow rate and DC-voltage were adjustable to optimize the cluster anion yields. In order to minimize the arcing from the cathode to ground, they simply employed electrical ballast consisting of a 100 kΩ resistor and a 4H inductor connected in series with the DC power supply, resulting in a discharge with

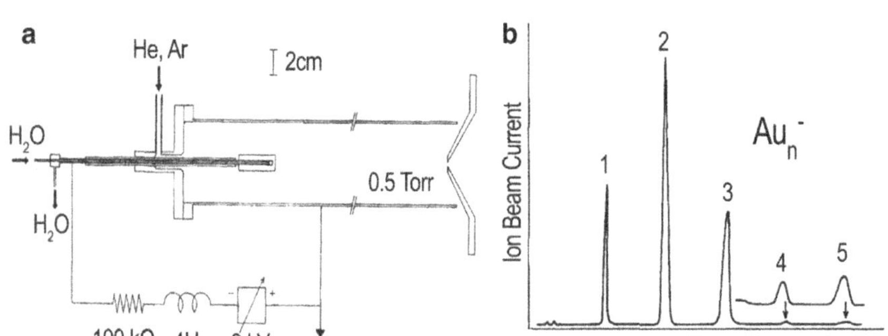

Fig. 2.4 **a** Schematic diagram of the flowing afterglow system with the metal cathode discharge ion source; **b** Mass spectra of Au_n^- clusters produced in the metal cathode discharge ion source. Adapted from Ref. [103]. Copyright 1994 by the American Physical Society

an electric current of ~10–30 mA to ground. Such sputtering sources were found to take advantages for producing noble metal clusters (Fig. 2.4b).

(b) Magnetron sputtering (MagS)

The discovery of magnetron sputtering (MagS) that a magnetic field could confine plasma to the target surface and amplify the rate of evaporation has attracted reasonable interest in thin-film deposition and related materials [104]. Compared with usual chemical vapor deposition (CVD) which has existed over a century, MagS bears advantages in that it requires less amounts of energy to evaporate the target medium especially those metals. In particular, MagS source is controllable of the correlated parameters such as voltage and working gas, which enables to attain tunable size distribution of clusters/particles [105–107]. Advances in this field has made the round planar MagS-sources being widely used [107–110]. The outstanding sputtering efficiency of MagS source has been applied to produce metal clusters and correlative investigations, as shown in Fig. 2.5A [107, 109, 110].

Briefly, a MagS-source works in the following way (Fig. 2.5B) [111]. When a negative voltage is applied to the target surface, the inert gas (typically argon) which

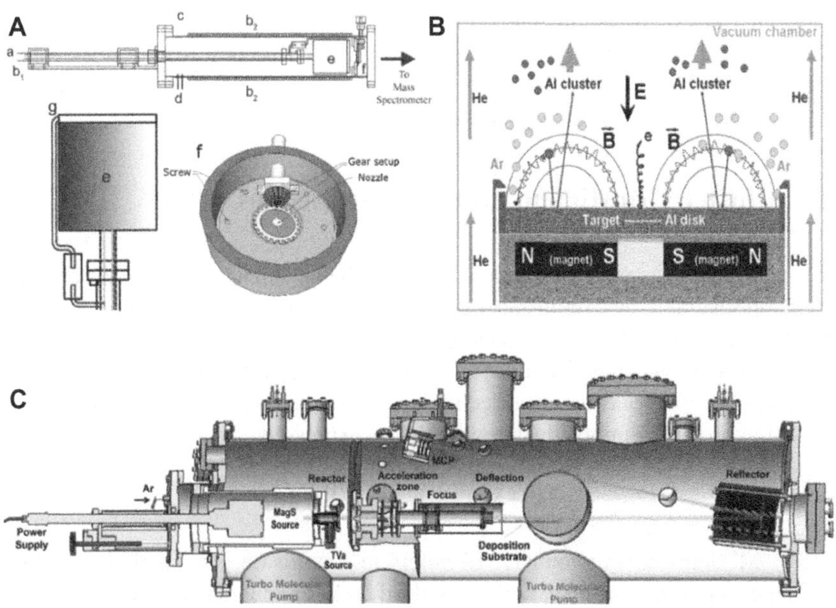

Fig. 2.5 **A** Sketch of the experimental set-up for production and analysis of the size-controlled clusters. Some of the components are labeled: **a** inlet of Ar and electric wire; **b** inlet of cooling water; **c** magnetron axial mount and the shell of vacuum chamber with cooling water; **d** inlet of He; **e** magnetron head with target; **f** nozzle; **g** outlet of Ar; **B** Mechanism sketch of the MagS-source, corresponding to the enlarged part of the magnetron head. **C** A sketch of the customized reflection time-of-flight mass spectrometer combined with a MagS source and a compact reaction cell which is connected with a portable thermal evaporation setup

flows over the surface will be ionized due to quantum effects or background radiation. Consequently, the ejected electrons display helical motion under the magnetic field restriction; simultaneously, the positively charged Ar^+ suddenly feels a large coulombic attraction towards the cathode target. As the Ar^+ ions impact the surface at high speed, enough energy is transferred to eject some atoms, electrons, and/or nascent clusters off the target. The ejected target clusters then expand outward and accompany with the buffer gas (e.g., He) to get through the nozzle. The electrons ejected from the target surface could collide with additional argon atoms, causing further ionization and hence intensifying the sputtering process [107].

Similar to other cluster sources, the plasma of atoms/clusters created near the magnetron head can also cause aggregation before they exit from the source via an adjustable iris. Analogous to Eq. 2.1, the sizes of clusters formed by such a MagS source are determined by the aggregation time and the speed of the initial aggregation. In general, the aggregation time can be adjusted by increasing or decreasing the flow inside the source, including the backing gas helium (also working gas Ar), the distance between the magnetron head and the exit iris, as well as the hole size of the iris [107, 112]. It is worth mentioning that an RF power supply instead of the DC power supply could make the MagS source applicable to non-conductive target maerials [113] Recently, there has been combined applications of the MagS-source with a thermal vaporization source (TVa-source), as shown in Fig. 2.5C, enabling to syntheisize metal-organic complexes and clusters directly from the gas phase reactions [111].

2.2 Cluster Growth and Statistical Principles

The growth of clusters is an important research scheme, but it is still open to academic debate on several issues, for instances, whether the monomers can be simply considered as single atoms or initially small clusters/molecules, and to what extent cluster may be meticulously tuned and dissociated [114, 115]. In general, cluster growth is initated by atoms, followed by three-body collisions that grow dimers/trimers in a sequential process. Simply, this process can be expressed as,

$$\mathrm{M} + \mathrm{M} + \mathrm{L} \rightarrow \mathrm{M}_2 + \mathrm{L} \tag{2.1}$$

where "M" aims at a monomer; "L" refers to a collision gas, single or multiple molecules, providing a higher density of third bodies and it promotes condensation. This is why the condensation efficiency depends on collision gas pressures and it can be optimized by using a proper nozzle to adjust the density in the condensation process.

Besides the promotion of condensation, the collision gas can take place in other excitation and relaxation processes in the plasma, where collision-induced energy transfer or quenching removes translation, electronic, vibrational and rotational energy from the metal atoms or growing clusters, expressed as:

$$M_n^* + L \rightarrow M_n + L \quad (2.2)$$

Evidently, the efficiency of this process also varies with the collision gas itself and the pressures. The cluster growth can continue via cluster-cluster collisions when the ratio of monomers to large clusters reaches a certain point, resulting in a twin-peaked mass distribution [112, 115].

The growth mechanism to form a dimer based on three-body collisions may also be assisted by a heavy inert carrier gas (e.g. argon instead of helium), but this is not recommended for a LaVa source to avoid possible formation of argon additions on the metal clusters [2, 116]. It is worth mentioning that rare gas atoms in plasma may also become ionized or electronically excited, and then their collisions with metal atoms or clusters may cause electronic excitation [15, 117].

$$M_n + L^* \rightarrow M_n^* + L \quad (2.3)$$

In general, as the rare gas bear relatively higher ionization energies (He: 24.6 eV; Ne: 21.6 eV; Ar: 15.8 eV) than metals (typically 5–9 eV) [118], there are little chances for direct ionization of these rare gases. But the energies at metastable excited states could be comparable to metal ionization potentials, which allow penning ionization being efficient and exothermic enough to cause fragmentation of the growing clusters [15].

$$M_n + L^* \rightarrow M_{n-X}^* + x\,M + L \quad (2.4)$$

In addition to the nascent neutral atoms/clusters, experimental results have demonstrated that both positive and negative ion clusters are produced by variation of the laser plasma conditions. Moreover, the electron attachment to neutral clusters contributes to producing anion clusters, which can be promoted in case of the addition of an electron source [15, 55].

$$M_n + e^- \rightarrow M_n^- \quad (2.5)$$

Following these origins in forming clusters in a LaVa source, further insights into the LaVa sources shed light on the statistical principles. In general, the evaporation of atoms/clusters into vacuum is dependent on the element studied, but can be broadly represented by the following equation where clusters with a diameter smaller than d^* will evaporate while the larger will grow:

$$d^* = \frac{4\sigma m}{kT\varrho} \cdot \frac{1}{\ln(\Phi_k)} \quad (2.6)$$

In this equation, σ is the surface tension of the liquid cluster, m is the mass of the monomer, k is Boltzmann's constant. T and ϱ are the temperature and density of the cluster respectively. Φ_k is the supersaturation level of the liquid/vapor interface, represented by the ratio of the partial pressure of the monomer over the vapor pressure of the cluster, P_k/P_∞. This equation helps explain why a cluster distribution is usually seen with a normal or an inverse gamma distribution while opposed to a logarithmic decay (i.e. the monomer is more intense than the dimer which is also more intense than the trimer, etc.). This equation also indicates interpretion why metals with higher boiling points tend to form smaller clusters even though under the same experimental conditions (due to the reduced partial pressure of the monomer, and then Φ_k).

Also Eq. 2.6 stands to reason its time independence: once the cluster has exited from the source it will no longer grow. The time allowed for cluster formation is also called "aggregation time" [119] which is dependent on the parameters of the source and the speed/pressure of the backing gas [120]. In experimental design, the "aggregation time" can be qualitatively ascertained by the flow rate of the backing gas and the space immediately between the target material and the exit of the source, called "waiting room". A practical means of increasing the "aggregation time" (i.e., gas pressures in the "waiting room") is by simply enlarging the flow of the gas. Nevertheless, the operation to simply increase the flow of the buffer gas could be insufficient to adjust the "aggregation time", as the gas is expanding into vacuum and therefore small changes in flow will do rare changes to the pressure in the source [121]. In comparison, adjusting the exit nozzle is a more efficient solution. The conductance "C" of a tube is inversely proportional to its length "L" and directly proportional to the fourth power of its diameter "D" according to fluid mechanics ($C \propto L^{-1}$; $C \propto D^4$) [121].

2.3 Cluster Reaction Apparatus

The investigations into cluster reactivity have been accomplished by several approaches [122–145], for example, through the study of the ensuing dynamics of product evolution of excited species formed on the excitation of the neutral cluster parents [146]. Alternatively, it is accomplished via a direct production of cluster ions through supersonic coexpansion of cluster and molecule constituents. For these approaches, the selected cluster species may also undergo studies of metastability, photoexcitation and dissociation, or collision induced dissociation (CID) processes, etc. [147]. Besides, advances in the technology of flow-tube reactors (such as the ones in *Castleman* and *Schwarz groups*) [148–150] have opened up a new realm of investigations in the past decades [147, 151, 152]. This will be introduced in detail following. Functioning as mini-sized flow tube, tandem reaction cells (also known as collision cells) have become popular and convenient along with pulsed buffer gas, such as those used in *Lievens* group [153–155], *Andersson* group [156], *Bowen group* [157], and *Fielicke* group [158], allowing extensive studies of cluster reacvity along with multiple-photon-dissociation (MPD) spectroscopy and photoelectron spectroscopy of in situ synthetic clusters.

2.3.1 *Ion Traps and Tandem Quadrupole/Hexapole Reactors*

Three-dimensional and two-dimensional linear quadrupole ion traps (developed by *Wolfgang Paul*) [159–161] generate a RF quadrupole field to store ions within defined boundaries, combined with mass spectrometry, enabling to probe metal cluster reactions up to approximate 10 Pa pressures. Similar to ion traps, tandem quadrupole or hexapole reactors are also a kind of customized cluster systems that use dynamic RF/DC electric fields to host and guide charged clusters allowing for gas-phase reactions at low pressures [162–164]. Typically a RF-octupole ion trap is combined with a time-of-flight (TOF) mass spectrometer [165], such as those built in *Bernhardt* group [163] and *He* group [166]. When running reactions, small cluster ions are concentrated in a helium-filled ion trap with tunable experimental conditions allowing for reactions for ~0.1 s or even longer [163]. The ions of both reactants and products are then released from the ion trap for mass analysis [167–171].

2.3.2 *Selected Ion Flow Tube (SIFT)*

In the past, the flowing afterglow technique [172] and other related flow reactors [173, 174], have provided a wealth of information on general ion-molecule reactions although limited attention was paid to cluster systems [175, 176]. Selected ion flow tube (SIFT) is a technique termed by *Adams* and *Smith* in 1976 [161] as shown in Fig. 2.6, where a mass-selected positive ion beam derived from a certain source is injected into a flowing gas. A combination of SIFT with mass spectrometry, also abbreviated as SIFT-MS, has been known as a convincing quantitative mass spectrometric technique for trace gas analysis, which involves the chemical ionization of trace volatile compounds by selected positive precursor ions during a well-defined time period along a flow tube. Reasonable interest is attracted on the subsequent reaction of the ions with neutral molecules which are introduced into the carrier gas downstream of the injection point. In addition to the extensive studies to explore ion-molecule reaction kinetics, crucial advances had been made on its application to ionospheric and interstellar ion chemistry [147], SIFT-MS was also used in human breath analysis and showed promise as a non-invasive tool for physiological monitoring and correlated disease diagnosis [152].

Among others, a typical SIFT system in *Schwarz* group allow to study reactions of ions produced via one of the three sources, i.e., electron or chemical ionization (EI/CI), electrospray ionization (ESI) or glow-discharge ionization (GDI) [148, 149]. Before going through the flow tube, the resulting ions from these optional sources are mass-analyzed or selected using the first quadrupole mass filter (Q_1). Next, there are options to choose either all ions or just a single-mass ion to be directed towards the flow tube. A pressure of ~0.5 mbar with buffer gas helium in the flow tube is kept under SIFT conditions for ion-molecular reactions. The ion products extracted from the flow tube are then directed towards the second quadrupole analyzer Q_2, pass an octupole

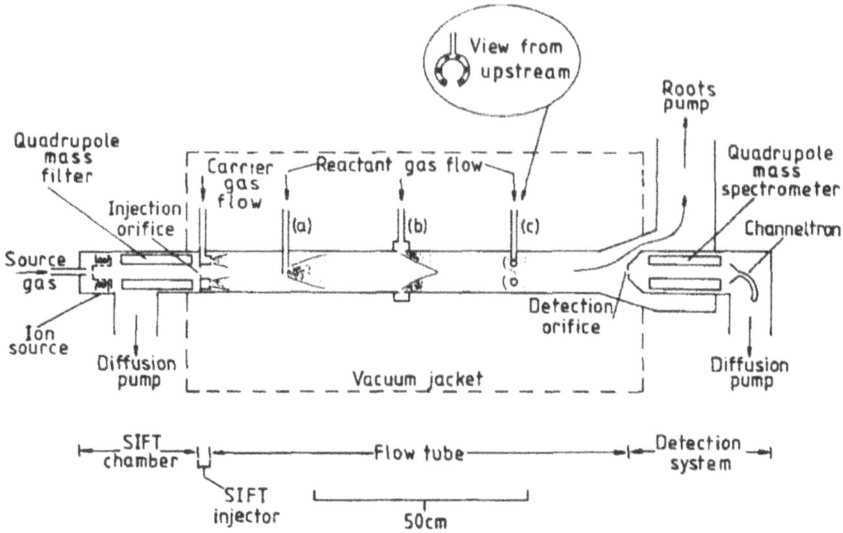

Fig. 2.6 A schematic diagram of the SIFT apparatus illustrating its major features. The SIFT chamber, the flow tube (typically 100 cm long and 8 cm in diameter). The profiles of reactant gas flows into the carrier gas stream are illustrated for a simple axial port (**a**), a radial port (**b**), and a "ring" port (**c**). Reproduced with permission from Ref. [161]. Copyright 1976 Elsevier

collision cell O and then the third quadrupole Q_3. Abundant investigations on gas-phase cluster reactivity have been undertaken on such an instrument [148, 149].

2.3.3 *Multiple-Ion Laminar Flow Tube (MIFT)*

Figure 2.7 shows a sketch map of the multiple-ion laminar flow tube reactor in tandem with a triple quadrupole mass spectrometer (MIFT-TQMS) in Luo's group [177]. Simply, metal clusters are created in a MagS source, transported through a reaction vessel where a reactant gas is introduced, and the reactants/products are sampled through a series of ion optics that culminate in a quadrupole mass spectrometer. Note that, for clusters created in the MagS source, large amounts of carrier gas helium are expanded into the flow tube enabling the multiple-collisions reaction studies. This requires additional differential pumping in a system where the source is coupled directly to the high-vacuum part of the instrument. With the development of cluster sources which take advantage of the flow tube reactors and enable the exploration of interactions of cluster ions with a wide variety of molecules under thermal reaction conditions (i.e., well-defined temperatures and collision conditions). In particular, understanding cluster reactivity can help develop tunable materials with possible catalytic or energetic qualities.

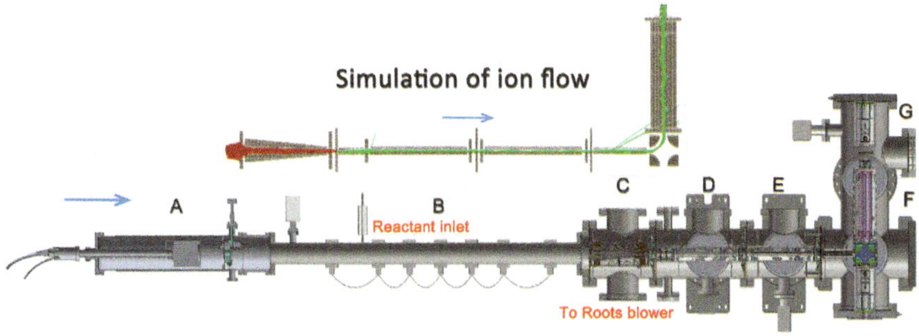

Fig. 2.7 A sketch map of the multiple-ion laminar flow tube reactor in tandem with a triple quadrupole mass spectrometer (MIFT-TQMS) built in *Luo's* group [177]. From left to right are the magnetron cluster source chamber (**A**), flow tube reactor (**B**), ion guide chambers consisting of a conical octupole ion focuser (**C**) and two linear octupole ion guides (**D, E**), and ion detection chamber with a quadrupole mass analyser (**F**) and a counting electron multiplier (**G**)

The importance of laminar flow [172], which was developed in 1960s, has been recogznied for the determination of ion-molecule reaction rate constants and activation energies, especially for reactions where the constraints of a vessel would normally compromise the reactants due to their being extremely labile under normal circumstances. The laminar flow in a fast-flow tube is defined as a static state of flow wherein velocity is represented by a parabolic distribution of layers, or streamlines. In general, the viscosity of an ideal gas can be represented by the following equation [178]:

$$\mu = \frac{2}{3\pi^{2/3}} \frac{\sqrt{mkT}}{d^2} \tag{2.7}$$

where m is the mass of the cluster, k is Boltzmann's constant, T is the temperature of the cluster, d is the diameter of the cluster/molecule. Further, the radial-dependent velocity of the gas in the flow tube is represented by:

$$v_z(r) = \frac{(P_0 - P_L)R^2}{4\mu L}\left[1 - \left(\frac{r}{R}\right)^2\right] \tag{2.8}$$

and then the laminar flow is represented by:

$$Q = \frac{\pi(P_0 - P_L)R^4}{8\mu L} \tag{2.9}$$

Here P_0 and P_L are the pressures at the beginning and end of the flow region of length L ($P_0 > P_L$), and R is the radius of the tube. Note that systems with transient or turbulent flow will not strictly follow this simple equation, nor will imperfect shapes due to their being eddies along the tube at a defect site. Considering the Reynolds number of a system ($\mathrm{Re} = 2\mathrm{R} < \mathrm{v_z} > \rho/\mu$), it is estimated that the force of the gas flow on an object at a given radial position along a laminar flow tube follows [172]:

$$F_z(r) = -v_z \cdot S \cdot \rho = -\frac{\pi R^2 \rho}{2}\left[1 - \left(\frac{r}{R}\right)^2\right]\left[\frac{(P_0 - P_L)R^2}{4\mu L}\right]^2$$
$$= -\mathrm{Re}\frac{\pi R v_z \mu}{4}\left[1 - \left(\frac{r}{R}\right)^2\right] \tag{2.10}$$

Providing a laminar flow condition, the introduction of any reactant gas will be regarded as reacting with clusters at a well-defined temperature, thus allowing for kinetics studies, such as determining activation energy and the Arrhenius prefactor. For example, simply considering Ferguson's original analysis of laminar flow reaction vessels, where an model for reaction kinetics for a basic reaction such as "A + B→C" can be derived,

$$k(T) = \frac{\pi \mathrm{R}^2 v_z^2}{Q_\mathrm{B}\mathrm{L}} \ln\left(\frac{[\mathrm{A}]}{[\mathrm{A}]_0}\right) \tag{2.11}$$

where Q_B is the rate of introduction of reactant B, in units of flow [172]. With the laminar flow reaction vessel also comes the ability to control temperature. A temperature-dependent study will yield both the Arrhenious prefactor and the activation energy of the reaction.

Although it is available to use laminar flow to check out the reaction kinetics, as the initial clusters are not mass-selected before reaction in these experiments, it is not always possible to ascertain the detailed product channel branching ratios for a given sized clusters especially when metal-metal bond breaking channels are significant. Nevertheless, by taking large numbers of scans of the ionic products as a function of reagent concentration in the flow tube, overall trends can be followed. Anyway, such MIFT systems have advantages for probing and identifying the most stable product species in view of the thermalizing collision conditions (~ 10^4–10^5 collisions) which are intended to quench the initial vibrational and electronic excitation of the parent clusters before their reaching the reaction zone; also the high pressures (c.a., 0.7 torr) in the flow tube mean sufficient gas-phase collisions allowing for the cluster-reactant interaction and chemical reactions.

2.3.4 Compact Flow-Tube Reactor and Collisional Cell

Compact flow-tube reactors and tandem reaction cells (also known as collision cells) have also been often used conveniently for gas-phase cluster reactions, such as those used in *Andersson* group [156], *Lievens* group [154–156], and *Fielicke* group [158] where multiple-photon-dissociation (MPD) spectroscopy is combined to obtain vibrational spectral information on clusters in the gas phase. Figure 2.8 shows such an instruments in *Luo's* group [179], where a 80-mm long flow tube is used. Among

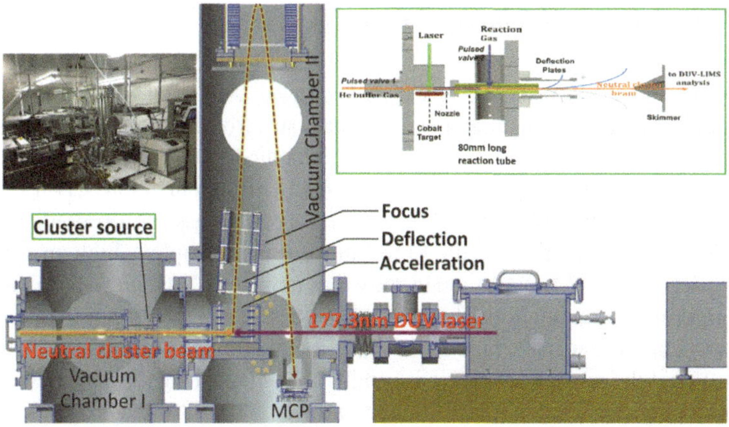

Fig. 2.8 A sketch showing the showing the compact flow-tube reactor combined with a home-made reflection time-of-flight mass spectrometer (Re-TOFMS) in *Luo's* group

others, customized reaction cells have also been applied by a few other research groups such as *Bowen's* [157], allowing studies of photoelectron spectroscopy of the in situ synthetic clusters.

2.4 Detection and Characterization

2.4.1 Cluster Mass Spectrometry

As the detection of a slow neutral cluster seems to be impossible, the clusters have to be ionised for an efficient mass-selective detection. Simply by measuring the ions abundance relating to mass-to-charge ratios, mass spectrometry is the unique tool for identifying the quantity and type of clusters and their reaction products. According to the mass analysis techniques, the most popular candidates for mass spectrometry include quadrupole mass spectrometer (QMS), ion-trap mass spectrometer (ITMS), and time-of-flight mass spectrometer (TOFMS). Among others, tandem quadrupole/ion-trap mass spectrometer and Fourier transform ion cyclotron resonance (FT-ICR) mass spectrometer have also been used in cluster reaction investigations [180–182]. In tandem mass spectrometry, the cluster ions successively lose energy due to the nonreactive collisions with background gasses and the adiabaticity of the multipole [55, 56]. Temperature-controlled tandem mass spectrometers are available but the technique is not as precise as the laminar flow tube and generally can only be controlled when lower than room temperature [183]. FT-ICR mass spectrometry adds resonance excitation to the trapped species hence enabling analysis of internal energy distributions [184–186].

2.4.2 *Ionization of Cluster Neutrals*

Detection of neutral clusters is associated with an additional ionization process. According to different ionization methods, mass spectrometry techniques are generally classified into hard ionization (i.e., EI method) [187] and soft ionization which includes photoionization (PI) [188–194], chemical ionization (CI) [195, 196], electrospray ionization (ESI) [197], and matrix-assisted laser desorption/ionization (MALDI) [198]. Among them, the EI technique usually employs high-energy electron impacts (c.a., ~70 eV) to attain maximal ionization efficiency but inevitably brings rigorous fragmentation. This is not only questionable for organic compounds, but also metal clusters of which metal-metal bond strengths are relatively smaller than 5 eV and the ionization energy is smaller than 10 eV (Fig. 2.9). Soft ionization techniques (e.g., ESI and MALDI, brought into being in 1980s) have been known as a milestone in the history of mass spectrometry, stimulating extensive applications in chemistry, biology, and material science.

History keeps moving forward. Recently researchers find higher requirements for mass spectrometry in view of the development of nanomaterials and the tendency of precise chemistry. Considering that most compounds have absorption in the deep ultraviolet (DUV) region ($\lambda < 200$ nm), effective DUV light sources are crucial for their ionization. DUV sources were usually obtained by using nonlinear frequency conversion of the radiation of lasers [199], DUV lamps [200, 201], gas discharges [202], or electron synchrotrons [203, 204]. Recently the technique of all-solid-state 177.3 nm DUV laser by harmonic generation has been developed [199], by frequency-doubling of a ps-pulsed 355 nm laser in going through a KBBF-CaF_2 prism coupled device [199, 205]. This ultrafast DUV laser has shown unique advantages of high photon flux, narrow bandwidth, good beam quality and coherence [206–208], and also relatively convenient comparing with a synchrotron radiation facility. Among others, a tunable vacuum ultraviolet free electron laser (VUV-FEL) facility provides

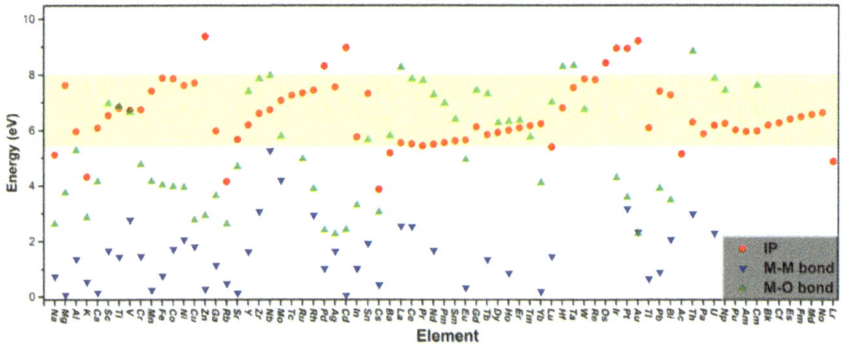

Fig. 2.9 Scatter plots of the metal ionization potentials (IPs), metal-metal bond strengths, and metal-oxygen bond strengths for the metal elements

a powerful tool for efficient size-selective soft ionization of neutral clusters with even larger ionization energies [209].

2.4.3 Cluster Storage and Deposition

In addition to sampling the masss of the cluster species, there are extensive investigations endevouring to possess the ability to strore ions and clusters via vacuum equipments such as cryogenic storage ring [210–213], and to host clusters onto solid surfaces via soft-landing deposition [214–250]. This is important as it leads to cluster assembly materials and also allowing for spectroscopic investigations. Supported metal catalysts enable practical applications in industrial processes with excellent catalystic performance [251, 252], providing an precise strategy to exploit the interface interactions, to improve the catalytic reaction selectivity and catalyst efficiency [253–259]. For example, Vajda and co-workers [260–263], Laskin and Johnson et al. [264], also Cooks [265], Nakajima [219], and Heiz et al. [266–268] conducted extensive investigations on the size effect of supported clusters for catalytic reactions [269–272], revealing the size dependence of metal cluster catalysis and opening up a new window for catalyst design.

Figure 2.9 presents an example of the cluster depositoin for matrix-isolation cavity ringdown spectroscopy, allowing for convenient disassembly and installation so as to realize technology combination with other vacuum instrument such as STM characterization [246]. A sketch of these experimental approaches is shown in Fig. 2.10 where the part A refers to a setup for the air-sensitive sample transport, prepared in a sample holder that is attached to the linear translator of the suitcase via a fixed screw thread. The suitcase is attached to the vacuum instrument via two gate valves, which protects the vacuum at the connection and removal of the suitcase-related chamber (i.e., deposition sampling part) in which vacuum is maintained by a battery-powered ion pump. After deposition sample is ready, the holder is translated into the suitcase where the gate valve can be closed and the entire suitcase can be detached, transported, and connected to the other vacuum apparatus (e.g., STM), the sample kept in a certain vacuum environment can be securely removed by a wobble stick with modified jaws [246].

2.4.4 Cluster Spectroscopy

Along with uprising research interest towards cluster chemistry, cluster spectroscopy has been known as a challenging and innovative field for the past decades [273]. Firstly, cluster photoelectron spectroscopy, also known as photoemission spectroscopy (PES), simply by measuring energies of electrons emitted from clusters by the photoelectric effect, has been widely utilized in serval research groups to help determine the binding energies of electrons in a cluster [274–323]. Cavity ringdown

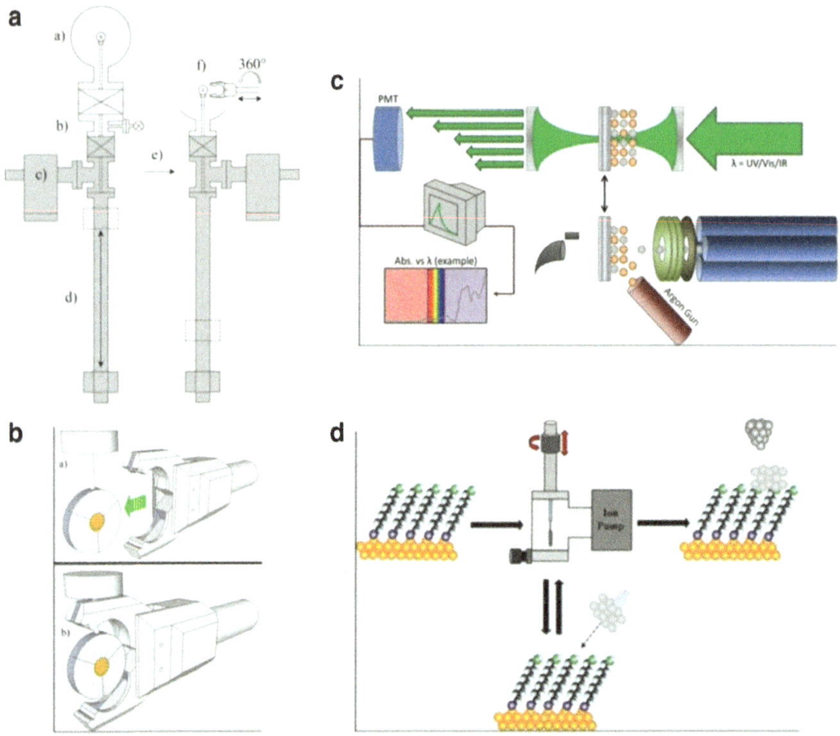

Fig. 2.10 **a** Diagram of the custom-built vacuum suitcase. **b** Diagram of sample transfer. **c/d** A sketch showing the soft-landing deposition of Al clusters on a hydroxyl terminated self-assembled monolayer

spectroscopy (CRDS) is known to take good advantages of relatively simple experimental setup, when compared with other highly sensitive spectroscopic techniques [324–329]. CRDS allows the detection of clusters that are deposited in a reasonable amount of time, so as to perform absorption spectroscopic measurements on clusters with eight or more orders of magnitude less material than is required for standard single-pass absorption spectroscopy. In addition to instrument simplicity, CRDS is also convenient because determining an absorbance spectrum is independent of several variables that are difficult to control, such as shot-to-shot fluctuations in laser intensity. In recent decade, it has been widely recognized that mass spectrometry combined with infrared multiphoton dissociation (IRMPD) active spectroscopy could be a versatile technology available for studying the electronic properties and identifying cluster structures [330–334]. Among other spectroscpies avalible for clusters, a few groups developed energy- and mass- resolved spectroscopies, such as IR + VUV two-color photoionization spectroscopy [335, 336], and IR + UV double resonance spectroscopy [337–342].

References

1. K. Hilpert, *Chemistry of Inorganic Vapors* (Springer, Berlin, Heidelberg, 1990)
2. H. Haberland, *Clusters of Atoms and Molecules: Theory, Experiment, and Clusters of Atoms* (Springer, 1994)
3. N. Satoh, K. Kimura, J. Am. Chem. Soc. **112**, 4688–4692 (1990)
4. R.A. Larsen, S.K. Neoh, D.R. Herschbach, Rev. Sci. Instrum. **45**, 1511–1516 (1974)
5. D.R. Preuss, S.A. Pace, J.L. Gole, J. Chem. Phys. **71**, 3553–3560 (1979)
6. K. Sattler, J. Muhlbach, E. Recknagel, Phys. Rev. Lett. **45**, 821–824 (1980)
7. S.J. Riley, E.K. Parks, C.R. Mao, L.G. Pobo, S. Wexler, J. Phys. Chem. **86**, 3911–3913 (1982)
8. W.D. Knight, K. Clemenger, W.A. Deheer, W.A. Saunders, M.Y. Chou, M.L. Cohen, Phys. Rev. Lett. **52**, 2141–2143 (1984)
9. K.M. McHugh, H.W. Sarkas, J.G. Eaton, C.R. Westgate, K.H. Bowen, Z. Phys. D At. Mol. Clusters **12**, 3–6 (1989)
10. T.P. Martin, T. Bergmann, H. Gohlich, T. Lange, J. Phys. Chem. **95**, 6421–6429 (1991)
11. I.M. Goldby, B. vonIssendorff, L. Kuipers, R.E. Palmer, Rev. Sci. Instrum. **68**, 3327–3334 (1997)
12. B. Meyer, *Low Temperature Spectroscopy* (Elsevier, New York, 1971)
13. R. Dunkin, *Matrix Isolation Techniques* (Oxford University Press, Oxford, 1998)
14. K.-L. Han, R.-C. Lu, H. Lin, E.B. Gallogy, W.M. Jackson, Chem. Phys. Lett. **243**, 29–35 (1995)
15. M.A. Duncan, Rev. Sci. Instrum. **83**, 041101 (2012)
16. J.A. Draves, Z. Lutheyschulten, W.L. Liu, J.M. Lisy, J. Chem. Phys. **93**, 4589–4602 (1990)
17. M. Dole, L.L. Mack, R.L. Hines, R.C. Mobley, L.D. Ferguson, M.B. Alice, J. Chem. Phys. **49**, 2240 (1968)
18. M. Yamashita, J.B. Fenn, J. Phys. Chem. **88**, 4451–4459 (1984)
19. C. Whitehouse, R. Dreyer, M. Yamashita, J.B. Fenn, Anal. Chem. **57**, 675 (1985)
20. R.G. Cooks, A.L. Rockwood, Rapid Commun. Mass Spectrom. **5**, 93 (1991)
21. P. Kebarle, M. Peschke, Anal. Chim. Acta **406**, 11 (2000)
22. G. Wang, R.B. Cole, Anal. Chim. Acta **406**, 53 (2000)
23. A.T. Blades, P. Jayaweera, M.G. Ikonomou, P. Kebarle, Int. J. Mass Spectrom. Ion Processes **102**, 251–267 (1990)
24. M.F. Bush, R.J. Saykally, E.R. Williams, ChemPhysChem **8**, 2245–2253 (2007)
25. J.T. O'Brien, E.R. Williams, J. Phys. Chem. A **112**, 5893–5901 (2008)
26. T.E. Cooper, D.R. Carl, P.B. Armentrout, J. Phys. Chem. A **113**, 13727–13741 (2009)
27. P. Kebarle, U.H. Verkerk, Mass Spectrom. Rev. **28**, 898–917 (2009)
28. R.D. Smith, J.A. Loo, C.G. Edmonds, C.J. Barinaga, H.R. Udseth, Anal. Chem. **62**, 882–899 (1990)
29. J.V. Iribarne, B.A. Thomson, J. Chem. Phys. **64**, 2287 (1976)
30. B.A. Thomson, J.V. Iribarne, J. Chem. Phys. **71**, 4451 (1979)
31. F.W. Röllgen, E. Bramer-Wegner, L. Buttering, J. Phys. Colloq. **45**, C9 (1984)
32. G. Schmelzeisen-Redeker, L. Buttering, F.W. Röllgen, Int. J. Mass Spectrom. Ion Proc. **90**, 139 (1989)
33. H.W. Kroto, J.R. Heath, S.C. Obrien, R.F. Curl, R.E. Smalley, Nature **318**, 162–163 (1985)
34. T.G. Dietz, M.A. Duncan, D.E. Powers, R.E. Smalley, J. Chem. Phys. **74**, 6511–6512 (1981)
35. D.E. Powers, S.G. Hansen, M.E. Geusic, A.C. Puiu, J.B. Hopkins, T.G. Dietz, M.A. Duncan, P.R.R. Langridgesmith, R.E. Smalley, J. Phys. Chem. **86**, 2556–2560 (1982)
36. W.C. Wiley, I.H. McLaren, Rev. Sci. Instrum. **26**, 1150–1157 (1955)
37. J.F. Friichte, Rev. Sci. Instrum. **45**, 51–56 (1974)
38. R.A. Bingham, P.L. Salter, Anal. Chem. **48**, 1735–1740 (1976)
39. N.G. Utterback, S.P. Tang, J.F. Friichtenicht, Phys. Fluids **19**, 900–905 (1976)
40. D.K. Anderson, D.W. Jones, J.D. McCullen, Rev. Sci. Instrum. **48**, 1381–1385 (1977)
41. B.G. Wicke, N.G. Utterback, S.P. Tang, J.F. Friichtenicht, Rev. Sci. Instrum. **51**, 151–153 (1980)

42. V.E. Bondybey, J.H. English, J. Chem. Phys. **74**, 6978–6979 (1981)
43. D.E. Powers, S.G. Hansen, M.E. Geusic, D.L. Michalopoulos, R.E. Smalley, J. Chem. Phys. **78**, 2866–2881 (1983)
44. S.C. Richtsmeier, E.K. Parks, K. Liu, L.G. Pobo, S.J. Riley, J. Chem. Phys. **82**, 3659–3665 (1985)
45. L.S. Zheng, P.J. Brucat, C.L. Pettiette, S. Yang, R.E. Smalley, J. Chem. Phys. **83**, 4273–4274 (1985)
46. P.J. Brucat, L.S. Zheng, C.L. Pettiette, S. Yang, R.E. Smalley, J. Chem. Phys. **84**, 3078–3088 (1986)
47. S.K. Loh, D.A. Hales, P.B. Armentrout, Chem. Phys. Lett. **129**, 527–532 (1986)
48. M.F. Jarrold, J.E. Bower, J.S. Kraus, J. Chem. Phys. **86**, 3876–3885 (1987)
49. M.B. Knickelbein, S. Yang, S.J. Riley, J. Chem. Phys. **93**, 94–104 (1990)
50. S. Maruyama, L.R. Anderson, R.E. Smalley, Rev. Sci. Instrum. **61**, 3686–3693 (1990)
51. P. Milani, W.A. Deheer, Rev. Sci. Instrum. **61**, 1835–1838 (1990)
52. W.A. De Heer, Rev. Mod. Phys. **65**, 611–676 (1993)
53. R.T. Laaksonen, D.A. Goetsch, D.W. Owens, D.M. Poirier, F. Stepniak, J.H. Weaver, Rev. Sci. Instrum. **65**, 2267–2275 (1994)
54. A. Fielicke, A. Kirilyuk, C. Ratsch, J. Behler, M. Scheffler, G. von Helden, G. Meijer, Phys. Rev. Lett. **93**, 023401 (2004)
55. J.M. Weber, Rev. Sci. Instrum. **76**, 043301 (2005)
56. M. Sun, D.T. Halfen, J. Min, B. Harris, D.J. Clouthier, L.M. Ziurys, J. Chem. Phys. **133**, 174301 (2010)
57. G. Scoles, *Atomic and Molecular Beam Methods* (Oxford University Press, New York, 1988)
58. D.H. Russell, *Gas Phase Inorganic Chemistry* (Plenum, New York, 1989)
59. C.-Y. Ng, T. Baer, I. Powis, *Cluster Ions* (Wiley, Chichester, 1993)
60. M.A. Duncan, *Advances in Metal and Semiconductor Clusters* (Elsevier, Amsterdam, 2001)
61. H. Pauly, *Atom, Molecule and Cluster Beams* (Springer, Berlin, 2000)
62. R.L. Johnston, *Atomic and Molecular Clusters* (Taylor & Francis, London, 2002)
63. M.D. Morse, Chem. Rev. **86**, 1049–1109 (1986)
64. M.M. Kappes, Chem. Rev. **88**, 369–389 (1988)
65. D.C. Parent, S.L. Anderson, Chem. Rev. **92**, 1541–1565 (1992)
66. U. Heiz, F. Vanolli, L. Trento, W.D. Schneider, Rev. Sci. Instrum. **68**, 1986–1994 (1997)
67. W.A. Deheer, Rev. Sci. Instrum. **61**, 1835–1838 (1990)
68. R.L. Stegman, J. Schriemp, L.R. Hettche, J. Appl. Phys. **44**, 3675–3681 (1973)
69. K. Laihing, R.G. Wheeler, W.L. Wilson, M.A. Duncan, J. Chem. Phys. **87**, 3401–3409 (1987)
70. M.E. Geusic, R.R. Freeman, M.A. Duncan, J. Chem. Phys. **89**, 223–229 (1988)
71. L.A. Bloomfield, M.E. Geusic, R.R. Freeman, W.L. Brown, Chem. Phys. Lett. **121**, 33–37 (1985)
72. J.R. Heath, Y. Liu, S.C. Obrien, Q.L. Zhang, R.F. Curl, F.K. Tittel, R.E. Smalley, J. Chem. Phys. **83**, 5520–5526 (1985)
73. S.C. Obrien, Y. Liu, Q. Zhang, J.R. Heath, F.K. Tittel, R.F. Curl, R.E. Smalley, J. Chem. Phys. **84**, 4074–4079 (1986)
74. E.A. Rohlfing, D.M. Cox, A. Kaldor, J. Chem. Phys. **81**, 3322–3330 (1984)
75. M.E. Geusic, M.F. Jarrold, T.J. McIllrath, R.R. Freeman, W.L. Brown, J. Chem. Phys. **86**, 3862–3869 (1987)
76. J.R. Heath, Q. Zhang, S.C. Obrien, R.F. Curl, H.W. Kroto, R.E. Smalley, J. Am. Chem. Soc. **109**, 359–363 (1987)
77. P.J. Roach, W.H. Woodward, A.C. Reber, S.N. Khanna, A.W. Castleman Jr., Phys. Rev. B **81**, 195404 (2010)
78. J.J. Melko, S.V. Ong, U. Gupta, J.U. Reveles, J. D'Emidio, S.N. Khanna, A.W. Castleman Jr., Chem. Phys. Lett. **500**, 196–201 (2010)
79. E.A. Rohlfing, D.M. Cox, R. Petkovicluton, A. Kaldor, J. Phys. Chem. **88**, 6227–6231 (1984)
80. R.G. Wheeler, K. Laihing, W.L. Wilson, J.D. Allen, R.B. King, M.A. Duncan, J. Am. Chem. Soc. **108**, 8101–8102 (1986)

81. R.G. Wheeler, K. Laihing, W.L. Wilson, M.A. Duncan, J. Chem. Phys. **88**, 2831–2839 (1988)
82. K. Laihing, P.Y. Cheng, M.A. Duncan, J. Phys. Chem. **91**, 6521–6525 (1987)
83. S. Nonose, Y. Sone, K. Onodera, S. Sudo, K. Kaya, J. Phys. Chem. **94**, 2744–2746 (1990)
84. C.S. Yeh, D.L. Robbins, J.S. Pilgrim, M.A. Duncan, Chem. Phys. Lett. **206**, 509–514 (1993)
85. D. Bellert, T. Buthelezi, T. Hayes, P.J. Brucat, Chem. Phys. Lett. **277**, 27–32 (1997)
86. L.R. Brock, A.M. Knight, J.E. Reddic, J.S. Pilgrim, M.A. Duncan, J. Chem. Phys. **106**, 6268–6278 (1997)
87. B.C. Guo, K.P. Kerns, A.W. Castleman Jr., Science **255**, 1411–1413 (1992)
88. B.C. Guo, S. Wei, J. Purnell, S. Buzza, A.W. Castleman Jr., Science **256**, 515–516 (1992)
89. S. Wei, B.C. Guo, J. Purnell, S. Buzza, A.W. Castleman Jr., J. Phys. Chem. **96**, 4166–4168 (1992)
90. S. Wei, B.C. Guo, J. Purnell, S. Buzza, A.W. Castleman Jr., Science **256**, 818–820 (1992)
91. B.C. Guo, K.P. Kerns, A.W. Castleman Jr., J. Am. Chem. Soc. **115**, 7415–7418 (1993)
92. S. Wei, B.C. Guo, J. Purnell, S.A. Buzza, A.W. Castleman Jr., J. Phys. Chem. **97**, 9559–9561 (1993)
93. H.T. Deng, B.C. Guo, K.P. Kerns, A.W. Castleman Jr., Int. J. Mass Spectrom. Ion Processes **138**, 275–281 (1994)
94. J. Purnell, S. Wei, A.W. Castleman Jr., Chem. Phys. Lett. **229**, 105–110 (1994)
95. K.P. Kerns, B.C. Guo, H.T. Deng, A.W. Castleman Jr., J. Am. Chem. Soc. **117**, 4026–4029 (1995)
96. H.T. Deng, K.P. Kerns, A.W. Castleman Jr., J. Am. Chem. Soc. **118**, 446–450 (1996)
97. H. Sakurai, A.W. Castleman Jr., J. Phys. Chem. A **101**, 7695–7698 (1997)
98. H. Sakurai, A.W. Castleman Jr., J. Chem. Phys. **111**, 1462–1466 (1999)
99. H. Sakurai, S.E. Kooi, A.W. Castleman Jr., J. Cluster Sci. **10**, 493–507 (1999)
100. J.F. Ready, *In Industrial Applications of Lasers* (Academic Press, San Diego, CA, 1997)
101. D.G. Leopold, K.K. Murray, A.E.S. Miller, W.C. Lineberger, J. Chem. Phys. **83**, 4849–4865 (1985)
102. D.G. Leopold, J. Ho, W.C. Lineberger, J. Chem. Phys. **86**, 1715–1726 (1987)
103. J. Ho, K.M. Ervin, W.C. Lineberger, J. Chem. Phys. **93**, 6987–7002 (1990)
104. D.M. Mattox, *The Foundations of Vacuum Coating Technology* (Noyes Publications/William Andrew Publishing, Norwich, NY, 2003)
105. H. Haberland, M. Mall, M. Moseler, Y. Qiang, T. Reiners, Y.J. Thurner, Vac. Sci. Technol. A Vac. Surf. Films **12**, 2925–2930 (1994)
106. S. Pratontep, S.J. Carroll, C. Xirouchaki, M. Streun, R.E. Palmer, Rev. Sci. Instrum. **76**, 045103 (2005)
107. Z. Luo, W.H. Woodward, J.C. Smith, A.W. Castleman Jr., Int. J. Mass Spectrom. **309**, 176–181 (2012)
108. J.S. Chapin, Research/Development **25**, 37–40 (1974)
109. Z. Luo, G.U. Gamboa, J.C. Smith, A.C. Reber, J.U. Reveles, S.N. Khanna, A.W. Castleman Jr., J. Am. Chem. Soc. **134**, 18973–18978 (2012)
110. Z.X. Luo, J.C. Smith, W.H. Woodward, A.W. Castleman Jr., J. Phys. Chem. A **116**, 2012–2017 (2012)
111. M. Guo, B. Yin, B. Huang, H. Wu, Z. Luo, J. Mater. Chem. C (2020). https://doi.org10.1039/D1030TC02296A
112. R. Morel, A. Brenac, P. Bayle-Guillemaud, C. Portemont, F. La Rizza, Eur. Phys. J. D **24**, 287–290 (2003)
113. P.D. Davidse, L.I. Maissel, J. Appl. Phys. **37**, 574–579 (1966)
114. Z. Shi, J.V. Ford, S. Wei, A.W. Castleman Jr., J. Chem. Phys. **99**, 8009–8015 (1993)
115. J.M. Soler, N. García, O. Echt, K. Sattler, E. Recknagel, Phys. Rev. Lett. **49**, 1857–1860 (1982)
116. P. Milani, W.A. Deheer, Phys. Rev. B **44**, 8346–8348 (1991)
117. E.A. Rohlfing, J. Chem. Phys. **89**, 6103–6112 (1988)
118. D.R. Lide, *Handbook of Chemistry and Physics*, 79th edn. (CRC, Boca Raton, FL, 1999)
119. R.B. Huang, Q. Zhang, H. Chen, L.S. Zheng, Eur. Phys. J. D **9**, 253–256 (1999)

120. W.H.H. Woodward, Ph.D., The Pennsylvania State University (2011)
121. J.H. Moore, C.C. Davis, M.A. Coplan, S.C. Greer, *Building Scientific Apparatus* (Cambridge University Press, 2009)
122. L. Capron, W.Y. Feng, C. Lifshitz, B.L. Tjelta, P.B. Armentrout, J. Phys. Chem. **100**, 16571–16576 (1996)
123. A.A. Viggiano, S.T. Arnold, R.A. Morris, Int. Rev. Phys. Chem. **17**, 147–184 (1998)
124. D. Caraiman, D.K. Bohme, J. Phys. Chem. A **106**, 9705–9717 (2002)
125. F. Dhooghe, C. Amelynck, J. Rimetz-Planchon, N. Schoon, F. Vanhaecke, Int. J. Mass Spectrom. **290**, 106–112 (2010)
126. R.D. Adams, B. Captain, Acc. Chem. Res. **42**, 409–418 (2009)
127. M. Hidai, S. Kuwata, Y. Mizobe, Acc. Chem. Res. **33**, 46–52 (2000)
128. C.P.G. Butcher, A. Dinca, P.J. Dyson, B.F.G. Johnson, P.R.R. Langridge-Smith, J.S. McIndoe, Angew. Chem. Int. Ed. **42**, 5752–5755 (2003)
129. N. Dietl, R.F. Höckendorf, M. Schlangen, M. Lerch, M.K. Beyer, H. Schwarz, Angew. Chem. Int. Ed. **50**, 1430–1434 (2011)
130. S. Kabashima, S. Kuwata, K. Ueno, M. Shiro, M. Hidai, Angew. Chem. Int. Ed. **39**, 1128–1131 (2000)
131. B.-Z. Zhan, E. Iglesia, Angew. Chem. Int. Ed. **46**, 3697–3700 (2007)
132. C. Zimmermann, M. Melullis, S. Dehnen, Angew. Chem. Int. Ed. **41**, 4269–4272 (2002)
133. J.F. Garvey, W.J. Herron, G. Vaidyanathan, Chem. Rev. **94**, 1999–2014 (1994)
134. M.M. Rohmer, M. Benard, J.M. Poblet, Chem. Rev. **100**, 495–542 (2000)
135. J. Mwakapumba, K.N. Ervin, Int. J. Mass Spectrom. **161**, 161–174 (1997)
136. J.I. Lee, D.C. Sperry, J.M. Farrar, J. Chem. Phys. **121**, 8375–8384 (2004)
137. V.D. Reddy, J. Organomet. Chem. **691**, 27–34 (2006)
138. H. Wadepohl, H. Pritzkow, J. Organomet. Chem. **450**, 9–19 (1993)
139. M. Andersson, J.L. Persson, A. Rosen, J. Phys. Chem. **100**, 12222–12234 (1996)
140. R.D. Adams, B. Captain, M.B. Hall, J.L. Smith, C.E. Webster, J. Am. Chem. Soc. **127**, 1007–1014 (2005)
141. A.M. Argo, J.F. Goellner, B.L. Phillips, G.A. Panjabi, B.C. Gates, J. Am. Chem. Soc. **123**, 2275–2283 (2001)
142. S. Ghosh, S.I. Gorelsky, S.D. George, J.M. Chan, I. Cabrito, D.M. Dooley, J.J.G. Moura, I. Moura, E.I. Solomon, J. Am. Chem. Soc. **129**, 3955–3965 (2007)
143. G. Zhen, L. Peng, Rev. Sci. Instrum. **69**, 1837–1839 (1998)
144. W. Uhl, Rev. Inorg. Chem. **18**, 239–282 (1998)
145. L.A. Posey, M.J. DeLuca, P.J. Campagnola, M.A. Johnson, J. Phys. Chem. **93**, 1178–1181 (1989)
146. Z.A.H., J. Phys. Chem. **97**, 12427 (1993)
147. A.W. Castleman Jr., S. Wei, Annu. Rev. Phys. Chem. **45**, 685–719 (1994)
148. D. Sulzle, K. Seemayer, B. Witulski, H. Hopf, H. Schwarz, Chem. Ber. **124**, 1481–1483 (1991)
149. N. Dietl, M. Schlangen, H. Schwarz, Angew. Chem. Int. Ed. **51**, 5544–5555 (2012)
150. P.J. Roach, W.H. Woodward, A.W. Castleman Jr., A.C. Reber, S.N. Khanna, Science **323**, 492–495 (2009)
151. A.W. Castleman Jr., S.N. Khanna, J. Phys. Chem. C **113**, 2664–2675 (2009)
152. D. Smith, P. Španěl, Mass Spectrom. Rev. **24**, 661–700 (2005)
153. J. De Haeck, N. Veldeman, P. Claes, E. Janssens, M. Andersson, P. Lievens, J. Phys. Chem. A **115**, 2103–2109 (2011)
154. E. Janssens, H.T. Le, P. Lievens, Chem. Eur. J. **21**, 15256–15262 (2015)
155. H.T. Le, S.M. Lang, J. De Haeck, P. Lievens, E. Janssens, Phys. Chem. Chem. Phys. **14**, 9350–9358 (2012)
156. N. Veldeman, P. Lievens, M. Andersson, J. Phys. Chem. A **109**, 11793–11801 (2005)
157. X. Li, A. Grubisic, S. Eustis, D. Wang, T. Lectka, G.F. Ganteför, K.H. Bowen, R. Burgert, H. Schnöckel, Chem. Phys. Lett. **481**, 190–193 (2009)
158. C. Kerpal, D.J. Harding, D.M. Rayner, A. Fielicke, J. Phys. Chem. A **117**, 8230–8237 (2013)
159. R.A.J. O'Hair, Chem. Commun. 1469–1481 (2006)

160. Q.Z. Hu, R.J. Noll, H.Y. Li, A. Makarov, M. Hardman, R.G. Cooks, J. Mass Spectrom. **40**, 430–443 (2005)
161. N.G. Adams, D. Smith, Int. J. Mass Spectrom. Ion Phys. **21**, 349–359 (1976)
162. H.-F. Li, Z.-Y. Li, Q.-Y. Liu, X.-N. Li, Y.-X. Zhao, S.-G. He, J. Phys. Chem. Lett. **6**, 2287–2291 (2015)
163. T.M. Bernhardt, Int. J. Mass Spectrom. **243**, 1–29 (2005)
164. W.D. Price, P.D. Schnier, R.A. Jockusch, E.F. Strittmatter, E.R. Williams, J. Am. Chem. Soc. **118**, 10640–10644 (1996)
165. R. Keller, F. Nöhmayer, P. Spädtke, M.H. Schönenberg, Vacuum **34**, 31–35 (1984)
166. X.-L. Ding, X.-N. Wu, Y.-X. Zhao, S.-G. He, Acc. Chem. Res. **45**, 382–390 (2012)
167. J. Oomens, B.G. Sartakov, G. Meijer, G. Von Helden, Int. J. Mass Spectrom. **254**, 1–19 (2006)
168. O. Hernandez, S. Isenberg, V. Steinmetz, G.L. Glish, P. Maitre, J. Phys. Chem. A **119**, 6057–6064 (2015)
169. I. Savic, D. Gerlich, O. Asvany, P. Jusko, S. Schlemmer, Mol. Phys. **113**, 2320–2332 (2015)
170. S. Daly, M. Krstic, A. Giuliani, R. Antoine, L. Nahon, A. Zavras, G.N. Khairallah, V. Bonacic-Koutecky, P. Dugourd, R.A.J. O'Hair, Phys. Chem. Chem. Phys. **17**, 25772–25777 (2015)
171. S.M. Lang, A. Frank, T.M. Bernhardt, J. Phys. Chem. C **117**, 9791–9800 (2013)
172. E.E. Ferguson, F.C. Fehsenfeld, A.L. Schmel-tekopf, Adv. At. Mol. Phys. **5**, 1–56 (1969)
173. D. Smith, N.G. Adams, in *Gas Phase Ion Chemistry*, ed. by M.T. Bowers (Academic Press, 1979)
174. S.T. Graul, R.R. Squires, Mass Spectrom. Rev. **7**, 263–358 (1988)
175. D.L. Albritton, At. Data Nucl. Tables **22**, 1–99 (1978)
176. Y. Ikezoe, S. Matsuoka, M. Takebe, A. Viggiano, Rapid Commun. Mass Spectrom. (1987)
177. M. Yang, H. Wu, B. Huang, Z. Luo, J. Phys. Chem. A **123**, 6921–6926 (2019)
178. R.B. Bird, W.E. Stewart, E.N. Lightfoot (Wiley, 2007)
179. H. Zhang, H. Wu, Y. Jia, L. Geng, Z. Luo, H. Fu, J. Yao, Rev. Sci. Instrum. **90**(7), 073101 (2019)
180. A.G. Marshall, C.L. Hendrickson, G.S. Jackson, Mass Spectrom. Rev. **17**, 1–35 (1998)
181. D.J. Douglas, A.J. Frank, D.M. Mao, Mass Spectrom. Rev. **24**, 1–29 (2005)
182. K.M. Ervin, P.B. Armentrout, J. Chem. Phys. **83**, 166–189 (1985)
183. P.H. Dawson, *Quadrupole Mass Spectrometry and Its Applications*. American Institute of Physics (1976)
184. S.H. Guan, J. Chem. Phys. **96**, 7959–7964 (1992)
185. S.H. Guan, H.S. Kim, A.G. Marshall, M.C. Wahl, T.D. Wood, X.Z. Xiang, Chem. Rev. **94**, 2161–2182 (1994)
186. O.P. Balaj, C.B. Berg, S.J. Reitmeier, V.E. Bondybey, M.K. Beyer, Int. J. Mass Spectrom. **279**, 5–9 (2009)
187. S. Yang, S.M. Brereton, M.D. Wheeler, A.M. Ellis, Phys. Chem. Chem. Phys. **7**, 4082–4088 (2005)
188. L. Hanley, R. Zimmermann, Anal. Chem. **81**, 4174–4182 (2009)
189. T.J. Kauppila, J.A. Syage, T. Benter, Mass Spectrom. Rev. (2015)
190. J.A. Zimmerman, R.M. O'Malley, Int. J. Mass Spectrom. Ion Processes **99**, 169–190 (1990)
191. D. Schröder, J. Loos, R. Thissen, O. Dutuit, P. Mourgues, H.E. Audier, C. Lifshitz, H. Schwarz, Angew. Chem. Int. Ed. **41**, 2748–2751 (2002)
192. A. Hamachi, T. Okuno, T. Imasaka, Y. Kida, T. Imasaka, Anal. Chem. **87**, 3027–3031 (2015)
193. P. Liu, Y. Hu, G. Zhu, Q. Yang, Y. Tao, Anal. Methods **7**, 6890–6895 (2015)
194. J.Y. Salpin, D. Scuderi, Rapid Commun. Mass Spectrom. **29**, 1898–1904 (2015)
195. M.S.B. Munson, F.H. Field, J. Am. Chem. Soc. **88**, 2621–2630 (1966)
196. W.J. Richter, H. Schwarz, Angew. Chem. Int. Ed. **17**, 424–439 (1978)
197. J.B. Fenn, M. Mann, C.K. Meng, S.F. Wong, C.M. Whitehouse, Science **246**, 64–71 (1989)
198. M.A. Claydon, S.N. Davey, V. Edwards-Jones, D.B. Gordon, Nat. Biotechnol. **14**, 1584–1586 (1996)
199. C. Chen, Z. Lin, Z. Wang, Appl. Phys. B **80**, 1–25 (2005)
200. Z. Zhu, J. Wang, K. Qiu, C. Liu, F. Qi, Y. Pan, Rev. Sci. Instrum. **85**, 046110 (2014)

201. L. Hua, Q. Wu, K. Hou, H. Cui, P. Chen, W. Wang, J. Li, H. Li, Anal. Chem. **83**, 5309–5316 (2011)
202. F. Mühlberger, J. Wieser, A. Morozov, A. Ulrich, R. Zimmermann, Anal. Chem. **77**, 2218–2226 (2005)
203. Y. Li, F. Qi, Acc. Chem. Res. **43**, 68–78 (2009)
204. H. Guo, L. Zhang, L. Deng, L. Jia, Y. Pan, F. Qi, J. Phys. Chem. A **114**, 3411–3417 (2010)
205. G. Liu, G. Wang, Y. Zhu, H. Zhang, G. Zhang, X. Wang, Y. Zhou, W. Zhang, H. Liu, L. Zhao, Rev. Sci. Instrum. **79**, 023105 (2008)
206. H. Wu, M. Guo, M. Yang, Z. Luo, K. Hansen, Chem. Commun. (Camb.) **55**, 4015–4018 (2019)
207. H. Wu, C. Yuan, H. Zhang, G. Yang, C. Cui, M. Yang, W. Bian, H. Fu, Z. Luo, J. Yao, Anal. Chem. **90**, 10635–10640 (2018)
208. H. Zhang, A.C. Reber, L. Geng, D. Rabayda, H. Wu, Z. Luo, J. Yao, S.N. Khanna, CCS Chemistry **1**, 571–581 (2019)
209. D. Normile, Science **355**, 235 (2017)
210. A. Wolf, in *Eighth International Conference on Atomic and Molecular Data and Their Applications*, vol. 1545, ed. by J.D. Gillaspy, W.L. Wiese, Y.A. Podpaly (2013), pp. 110–114
211. A. Wolf, R. von Hahn, A. Grieser, D.A. Orlov, H. Fadil, C.P. Welsch, V. Andrianarijaona, A. Diehl, C.D. Schroter, J.R.C. Lopez-Urrutia, M. Rappaport, X. Urbain, T. Weber, V. Mallinger, C. Haberstroh, H. Quack, D. Schwalm, J. Ullrich, D. Zajfman, in *Beam Cooling and Related Topics*, vol. 821, ed. by S. Nagaitsev, R.J. Pasquinelli (2006), p. 473
212. S. Menk, P. Bertier, Y. Enomoto, T. Masunaga, T. Majima, Y. Nakano, T. Azuma, Rev. Sci. Instrum. **89** (2018)
213. Y. Nakano, Y. Enomoto, T. Masunaga, S. Menk, P. Bertier, T. Azuma, Rev. Sci. Instrum. **88** (2017)
214. G.E. Johnson, T. Priest, J. Laskin, ACS Nano **6**, 573–582 (2012)
215. G.E. Johnson, C. Wang, T. Priest, J. Laskin, Anal. Chem. **83**, 8069–8072 (2011)
216. G.E. Johnson, Q. Hu, J. Laskin, in *Annu. Rev. Anal. Chem.*, vol. 4, ed. by R.G. Cooks, E.S. Yeung (2011), pp. 83–104
217. K. Judai, K. Sera, S. Amatsutsumi, K. Yagi, T. Yasuike, S. Yabushita, A. Nakajima, K. Kaya, Chem. Phys. Lett. **334**, 277–284 (2001)
218. T. Matsumoto, S. Nagaoka, K. Ikemoto, M. Mitsui, M. Ara, H. Tada, A. Nakajima, Eur. Phys. J. D **52**, 99–102 (2009)
219. M. Mitsui, S. Nagaoka, T. Matsumoto, A. Nakajima, J. Phys. Chem. B **110**, 2968–2971 (2006)
220. S. Nagaoka, T. Matsumoto, E. Okada, M. Mitsui, A. Nakajima, J. Phys. Chem. B **110**, 16008–16017 (2006)
221. K.A. Wepasnick, X. Li, T. Mangler, S. Noessner, C. Wolke, M. Grossmann, G. Gantefoer, D.H. Fairbrother, K.H. Bowen, J. Phys. Chem. C **115**, 12299–12307 (2011)
222. S.C. Nanita, Z. Takats, R.G. Cooks, S. Myung, D.E. Clemmer, J. Am. Soc. Mass Spectrom. **15**, 1360–1365 (2004)
223. J.T. Lau, W. Wurth, H.U. Ehrke, A. Achleitner, Low Temp Phys **29**, 223–227 (2003)
224. W. Harbich, Philos. Mag. B Phys. Condensed Matter Stat. Mech. Electron. Opt. Magn. Prop. **79**, 1307–1320 (1999)
225. R. Alayan, L. Arnaud, M. Broyer, E. Cottancin, J. Lerme, S. Marhaba, J.L. Vialle, M. Pellarin, Phys. Rev. B **76**, 075424 (2007)
226. C. Brechignac, P. Cahuzac, F. Carlier, M. de Frutos, A. Masson, C. Mory, C. Colliex, B. Yoon, Phys. Rev. B **57**, R2084–R2087 (1998)
227. S. Fedrigo, W. Harbich, J. Buttet, Phys. Rev. B **58**, 7428–7433 (1998)
228. M. Moseler, H. Hakkinen, U. Landman, Phys. Rev. Lett. **89**, 176103 (2002)
229. S.J. Davila, D.O. Birdwell, G.F. Verbeck, Rev. Sci. Instrum. **81**, 034104 (2010)
230. G. Li, J. Cyriac, L. Gao, R.G. Cooks, Surf. Interface Anal. **43**, 498–501 (2011)
231. K. Bromann, H. Brune, C. Felix, W. Harbich, R. Monot, J. Buttet, K. Kern, Surf. Sci. **377**, 1051–1055 (1997)

232. B. Yoon, V.M. Akulin, P. Cahuzac, F. Carlier, M. de Frutos, A. Masson, C. Mory, C. Colliex, C. Brechignac, Surf. Sci. **443**, 76–88 (1999)
233. V.N. Popok, I. Barke, E.E.B. Campbell, K.-H. Meiwes-Broer, Surf. Sci. Rep. **66**, 347–377 (2011)
234. K. Ohtomi, Thin Solid Films **464**, 150–154 (2004)
235. C. Brechignac, P. Cahuzac, F. Carlier, M. deFrutos, A. Masson, C. Colliex, C. Mory, B. Yoon, Z. Phys. D At. Mol. Clusters **40**, 516–519 (1997)
236. J.N. Louris, R.G. Cooks, J.E.P. Syka, P.E. Kelley, G.C. Stafford, J.F.J. Todd, Anal. Chem. **59**, 1677–1685 (1987)
237. N. Cruz Hernández, J.F. Sanz, Appl. Surf. Sci. **238**, 228–232 (2004)
238. H. Esrom, Appl. Surf. Sci. **168**, 1–4 (2000)
239. S. Shironita, M. Goto, T. Kamegawa, K. Mori, H. Yamashita, Catal. Today **153**, 189–192 (2010)
240. Z. Luo, J.C. Smith, W.H. Woodward, A.W. Castleman Jr., Presented in part at the *Gordon Research Conferences: Clusters, Nanocrystals & Nanostructures* (Mount Holyoke College, South Hadley, MA, July 24–29, 2011)
241. L. Gao, M.E. Lyn, D.E. Bergeron, A.W. Castleman, Int. J. Mass Spectrom. **229**, 11–17 (2003)
242. P. Kemper, A. Kolmakov, X. Tong, Y. Lilach, L. Benz, M. Manard, H. Metiu, S.K. Buratto, M.T. Bowers, Int. J. Mass Spectrom. **254**, 202–209 (2006)
243. A. Sandoval, A. Aguilar, C. Louis, A. Traverse, R. Zanella, J. Catal. **281**, 40–49 (2011)
244. P. Miquel, P. Granger, N. Jagtap, S. Umbarkar, M. Dongare, C. Dujardin, J. Mol. Catal. A Chem. **322**, 90–97 (2010)
245. J.M. Horn, Z. Song, D.V. Potapenko, J. Hrbek, M.G. White, J. Phys. Chem. B **109**, 44–47 (2004)
246. W.H. Woodward, M.M. Blake, Z. Luo, P.S. Weiss, A.W. Castleman Jr., J. Phys. Chem. C **115**, 5373–5377 (2011)
247. H.-F. Ji, R. Majithia, X. Yang, X. Xu, K. More, J. Am. Chem. Soc. **130**, 10056–10057 (2008)
248. E.J. Lee, M.G. Ko, B.Y. Kim, S.K. Park, H.D. Kim, J.W. Park, J. Korean Phys. Soc. **49**, 1243–1246 (2006)
249. M.A. Rottgen, K. Judai, J.-M. Antonietti, U. Heiz, S. Rauschenbach, K. Kern, Rev. Sci. Instrum. **77**, 013302–013306 (2006)
250. A. Schaefer, A. Sandell, L.E. Walle, V. Zielasek, M. Schowalter, A. Rosenauer, M. Baumer, Surf. Sci. **604**, 1287–1293 (2010)
251. M. Han, Z.Y. Wang, P.P. Chen, S.W. Yu, G.H. Wang, Nucl. Instrum. Methods Phys. Res. Sect. B Beam Interact. Mater. Atoms **135**, 564–569 (1998)
252. L. Liu, A. Corma, Chem. Rev. **118**, 4981–5079 (2018)
253. Y. Dai, T.J. Gorey, S.L. Anderson, S. Lee, S. Lee, S. Seifert, R.E. Winans, J. Phys. Chem. C **121**, 361–374 (2016)
254. A. von Weber, S.L. Anderson, Acc. Chem. Res. **49**, 2632–2639 (2016)
255. J. Timoshenko, A. Halder, B. Yang, S. Seifert, M.J. Pellin, S. Vajda, A.I. Frenkel, J. Phys. Chem. C **122**, 21686–21693 (2018)
256. F.R. Negreiros, A. Halder, C.R. Yin, A. Singh, G. Barcaro, L. Sementa, E.C. Tyo, M.J. Pellin, S. Bartling, K.H. Meiwes-Broer, S. Seifert, P. Sen, S. Nigam, C. Majumder, N. Fukui, H. Yasumatsu, S. Vajda, A. Fortunelli, Angew. Chem. Int. Ed. **57**, 1209–1213 (2018)
257. C.L. Bentley, M. Kang, P.R. Unwin, J. Am. Chem. Soc. 2179–2193 (2018)
258. H. Wang, X.-K. Gu, X. Zheng, H. Pan, J. Zhu, S. Chen, L. Cao, W.-X. Li, J. Lu, Sci. Adv. **5**, eaat6413 (2019)
259. M. Zhou, S. Bao, A.J. Bard, J. Am. Chem. Soc. **141**, 7327–7332 (2019)
260. C. Yin, F.R. Negreiros, G. Barcaro, A. Beniya, L. Sementa, E.C. Tyo, S. Bartling, K.-H. Meiwes-Broer, S. Seifert, H. Hirata, N. Isomura, S. Nigam, C. Majumder, Y. Watanabe, A. Fortunelli, S. Vajda, J. Mater. Chem. A **5**, 4923–4931 (2017)
261. B. Yang, C. Liu, A. Halder, E.C. Tyo, A.B.F. Martinson, S. Seifer, P. Zapol, L.A. Curtiss, S. Vajda, J. Phys. Chem. C **121**, 10406–10412 (2017)

262. A. Halder, C. Liu, Z. Liu, J.D. Emery, M.J. Pellin, L.A. Curtiss, P. Zapol, S. Vajda, A.B.F. Martinson, J. Phys. Chem. C **122**, 9965–9972 (2018)
263. A. Halder, L.A. Curtiss, A. Fortunelli, S. Vajda, J. Chem. Phys. **148**, 110901 (2018)
264. J. Laskin, G.E. Johnson, J. Warneke, V. Prabhakaran, Angew. Chem. Int. Ed. **57**, 16270–16284 (2018)
265. S.A. Miller, H. Luo, S.J. Pachuta, R.G. Cooks, Science **275**, 1447–1450 (1997)
266. M. Nesselberger, M. Roefzaad, R.F. Hamou, P.U. Biedermann, F.F. Schweinberger, S. Kunz, K. Schloegl, G.K. Wiberg, S. Ashton, U. Heiz, K.J. Mayrhofer, M. Arenz, Nat. Mater. **12**, 919–924 (2013)
267. K. Wettergren, F.F. Schweinberger, D. Deiana, C.J. Ridge, A.S. Crampton, M.D. Rotzer, T.W. Hansen, V.P. Zhdanov, U. Heiz, C. Langhammer, Nano Lett. **14**, 5803–5809 (2014)
268. M. Rondelli, G. Zwaschka, M. Krause, M.D. Rötzer, M.N. Hedhili, M.P. Högerl, V. D'Elia, F.F. Schweinberger, J.-M. Basset, U. Heiz, Acs Catal. **7**, 4152–4162 (2017)
269. Y. Lei, F. Mehmood, S. Lee, J. Greeley, B. Lee, S. Seifert, R.E. Winans, J.W. Elam, R.J. Meyer, P.C. Redfern, D. Teschner, R. Schlögl, M.J. Pellin, L.A. Curtiss, S. Vajda, Science **328**, 224–228 (2010)
270. C. Cui, Z. Luo, J. Yao, CCS Chem. **1**, 215–225 (2019)
271. H. Zhang, C. Cui, Z. Luo, J. Phys. Chem. C **124**, 6260–6266 (2020)
272. G. Kwon, G.A. Ferguson, C.J. Heard, E.C. Tyo, C. Yin, J. DeBartolo, S. Seifert, R.E. Winans, A.J. Kropf, J. Greeley, R.L. Johnston, L.A. Curtiss, M.J. Pellin, S. Vajda, ACS Nano **7**, 5808–5817 (2013)
273. J.M. Lisy, in *Progress in Experimental and Theoretical Studies of Clusters* (2015), pp. 79–119
274. H. Häkkinen, B. Yoon, U. Landman, X. Li, H.-J. Zhai, L.-S. Wang, J. Phys. Chem. A **107**, 6168–6175 (2003)
275. F.A. Akin, C.C. Jarrold, J. Chem. Phys. **118**, 5841 (2003)
276. F.A. Akin, C.C. Jarrold, J. Chem. Phys. **118**, 5841–5851 (2003)
277. H.F. Bettinger, Angew. Chem. Int. Ed. **49**, 670–671 (2010)
278. W.J. Chen, H.J. Zhai, Y.F. Zhang, X. Huang, L.S. Wang, J. Phys. Chem. A **114**, 5958–5966 (2010)
279. U. Das, K. Raghavachari, C.C. Jarrold, J. Chem. Phys. **122**, 014313 (2005)
280. S.R. Desai, H.B. Wu, C.M. Rohlfing, L.S. Wang, J. Chem. Phys. **106**, 1309–1317 (1997)
281. K. Dumbuya, G. Cabailh, R. Lazzari, J. Jupille, L. Ringel, M. Pistor, O. Lytken, H.P. Steinrück, J.M. Gottfried, Catal. Today **181**, 20–25 (2012)
282. M.A. Duncan, A.M. Knight, Y. Negishi, S. Nagao, K. Judai, A. Nakajima, K. Kaya, J. Phys. Chem. A **105**, 10093–10097 (2001)
283. S.M.E. Green, S. Alex, N.L. Fleischer, E.L. Millam, T.P. Marcy, D.G. Leopold, J. Chem. Phys. **114**, 2653–2668 (2001)
284. A. Grubisic, H. Wang, Y.J. Ko, K.H. Bowen, J. Chem. Phys. **129**, 054302 (2008)
285. R.F. Gunion, S.J. DixonWarren, W.C. Lineberger, M.D. Morse, J. Chem. Phys. **104**, 1765–1773 (1996)
286. G.L. Gutsev, C.W. Bauschlicher, H.-J. Zhai, L.-S. Wang, J. Chem. Phys. **119**, 11135 (2003)
287. T. Herranz, X.Y. Deng, A. Cabot, P. Alivisatos, Z. Liu, G. Soler-Illia, M. Salmeron, Catal. Today **143**, 158–166 (2009)
288. J.K.L. Knappenberger, P.A. Clayborne, J.U. Reveles, M.A. Sobhy, J.C.E. Jones, U.U. Gupta, S.N. Khanna, I. Iordanov, J. Sofo, J.A.W. Castleman, ACS Nano **1**, 319–326 (2007)
289. K.L. Knappenberger Jr., C.E. Jones, M.A. Sobhy, I. Iordanov, J. Sofo, A.W. Castleman Jr., J. Phys. Chem. A **110**, 12814–12821 (2006)
290. S. Lee, G.W. Zajac, D.W. Goodman, Top. Catal. **38**, 127–132 (2006)
291. R.Z. Li, H.G. Xu, G.J. Cao, Y.C. Zhao, W.J. Zheng, J. Chem. Phys. **135**, 134307 (2011)
292. S.G. Li, H.J. Zhai, L.S. Wang, D.A. Dixon, J. Phys. Chem. A **113**, 11273–11288 (2009)
293. X. Li, S.S. Liu, W. Chen, L.-S. Wang, J. Chem. Phys. **111**, 2464 (1999)
294. X. Li, L.S. Wang, J. Chem. Phys. **111**, 8389–8395 (1999)
295. X. Li, L.S. Wang, A.I. Boldyrev, J. Simons, J. Am. Chem. Soc. **121**, 6033–6038 (1999)
296. X. Li, L.S. Wang, N.A. Cannon, A.I. Boldyrev, J. Chem. Phys. **116**, 1330–1338 (2002)

297. N.J. Mayhall, D.W. Rothgeb, E. Hossain, K. Raghavachari, C.C. Jarrold, J. Chem. Phys. **130**, 124313 (2009)
298. A. Pramann, K. Koyasu, A. Nakajima, K. Kaya, J. Phys. Chem. A **106**, 4891–4896 (2002)
299. A. Pramann, K. Koyasu, A. Nakajima, K. Kaya, J. Chem. Phys. **116**, 6521–6528 (2002)
300. A. Pramann, Y. Nakamura, A. Nakajima, K. Kaya, J. Phys. Chem. A **105**, 7534–7540 (2001)
301. D.W. Rothgeb, E. Hossain, A.T. Kuo, J.L. Troyer, C.C. Jarrold, J. Chem. Phys. **131**, 044310 (2009)
302. J.C. Schultz, F.A. Houle, J.L. Beauchamp, J. Am. Chem. Soc. **106**, 7336–7347 (1984)
303. D.H.A. ter Steege, C. Lagrost, W.J. Buma, D.A. Leigh, F. Zerbetto, J. Chem. Phys. **117**, 8270–8280 (2002)
304. M. Vonderach, O.T. Ehrler, P. Weis, M.M. Kappes, Anal. Chem. **83**, 1108–1115 (2011)
305. H.Q. Wang, H.F. Li, J. Chem. Phys. **137**, 164304 (2012)
306. L.S. Wang, X. Li, J. Chem. Phys. **112**, 3602–3608 (2000)
307. L.-S. Wang, H.-S. Cheng, J. Fan, J. Chem. Phys. **102**, 9480 (1995)
308. L.-S. Wang, C.-F. Ding, X.-B. Wang, S.E. Barlow, Rev. Sci. Instrum. **1999**, 70 (1957)
309. X.B. Wang, L.S. Wang, Annu. Rev. Phys. Chem. **60**, 105–126 (2009)
310. Y.L. Wang, H.J. Zhai, L. Xu, J. Li, L.S. Wang, J. Phys. Chem. A **114**, 1247–1254 (2010)
311. T. Watanabe, T. Tsukuda, J. Phys. Chem. C **117**, 6664–6668 (2013)
312. T. Waters, X.-B. Wang, L.-S. Wang, Coord. Chem. Rev. **251**, 474–491 (2007)
313. H. Wu, L.-S. Wang, J. Phys. Chem. A **102**, 9129–9135 (1998)
314. H.B. Wu, S.R. Desai, L.S. Wang, J. Phys. Chem. A **101**, 2103–2111 (1997)
315. H.G. Xu, Z.G. Zhang, Y. Feng, J.Y. Yuan, Y.C. Zhao, W.J. Zheng, Chem. Phys. Lett. **487**, 204–208 (2010)
316. J. Yang, X.B. Wang, X.P. Xing, L.S. Wang, J. Chem. Phys. **128**, 201102 (2008)
317. X. Yang, T. Waters, X.B. Wang, R.A.J. O'Hair, A.G. Wedd, J. Li, D.A. Dixon, L.S. Wang, J. Phys. Chem. A **108**, 10089–10093 (2004)
318. J.Y. Yuan, H.G. Xu, Z.G. Zhang, Y. Feng, W.J. Zheng, J. Phys. Chem. A **115**, 182–186 (2011)
319. H.J. Zhai, S.R. Liu, X. Li, L.S. Wang, J. Chem. Phys. **115**, 5170–5178 (2001)
320. H.J. Zhai, L.S. Wang, J. Chem. Phys. **125**, 164315 (2006)
321. H.J. Zhai, L.S. Wang, P. Jena, G.L. Gutsev, C.W. Bauschlicher, J. Chem. Phys. **120**, 8996–9008 (2004)
322. Z.G. Zhang, H.G. Xu, X.Y. Kong, W.J. Zheng, J. Phys. Chem. A **115**, 13–18 (2011)
323. Z.G. Zhang, H.G. Xu, Y.C. Zhao, W.J. Zheng, J. Chem. Phys. **133**, 154314 (2010)
324. Y. Matsumoto, K. Honma, Phys. Chem. Chem. Phys. **13**, 13962–13971 (2011)
325. S. Gilb, K. Hartl, A. Kartouzian, J. Peter, U. Heiz, H.G. Boyen, P. Ziemann, Eur. Phys. J. D **45**, 501–506 (2007)
326. A. Kartouzian, M. Thaemer, T. Soini, J. Peter, P. Pitschi, S. Gilb, U. Heiz, J. Appl. Phys. **104** (2008)
327. A. Del Vitto, G. Pacchioni, K.H. Lim, N. Rosch, J.M. Antonietti, M. Michalski, U. Heiz, H. Jones, J. Phys. Chem. B **109**, 19876–19884 (2005)
328. A. Kartouzian, M. Thaemer, U. Heiz, Phys. Status Solidi B Basic Solid State Phys. **247**, 1147–1151 (2010)
329. J.M. Antonietti, M. Michalski, U. Heiz, H. Jones, K.H. Lim, N. Rosch, A. Del Vitto, G. Pacchioni, Phys. Rev. Lett. **94**, 213402 (2005)
330. M.A. Duncan, Int. Rev. Phys. Chem. **22**, 407–435 (2003)
331. Y. Hu, J. Guan, E.R. Bernstein, Mass Spectrom. Rev. **32**, 484–501 (2013)
332. E.R. Bernstein, Int. J. Mass Spectrom. **377**, 248–262 (2015)
333. D. Oepts, A.F.G. van der Meer, P.W. van Amersfoort, Infrared Phys. Technol. **36**, 297–308 (1995)
334. P. Gruene, D.M. Rayner, B. Redlich, A.F.G. van der Meer, J.T. Lyon, G. Meijer, A. Fielicke, Science **321**, 674–676 (2008)
335. N.X. Truong, M. Haertelt, B.K.A. Jaeger, S. Gewinner, W. Schöllkopf, A. Fielicke, O. Dopfer, Int. J. Mass Spectrom. **395**, 1–6 (2016)

336. N.X. Truong, B.K.A. Jaeger, S. Gewinner, W. Schollkopf, A. Fielicke, O. Dopfer, J. Phys. Chem. C **121**, 9560–9571 (2017)
337. H.B. Fu, Y.J. Hu, E.R. Bernstein, J. Chem. Phys. **124**, 024302 (2006)
338. H.B. Fu, Y.J. Hu, E.R. Bernstein, J. Chem. Phys. **125**, 014310 (2006)
339. S.-G. He, Y. Xie, Y. Guo, E.R. Bernstein, J. Chem. Phys. **126**, 194315 (2007)
340. Y.J. Hu, H.B. Fu, E.R. Bernstein, J. Chem. Phys. **125**, 154306 (2006)
341. Y.J. Hu, H.B. Fu, E.R. Bernstein, J. Chem. Phys. **125**, 184308 (2006)
342. F. Dong, S. Heinbuch, Y. Xie, J.J. Rocca, E.R. Bernstein, J. Phys. Chem. A **113**, 3029–3040 (2009)

Chapter 3
Metal Cluster Reacting with Oxygen

Cluster-oxygen reactions on metal and metal-alloy systems have been widely investigated [1–6] Through reactions with oxygen (a strong etchant), several interesting clusters of metals (e.g., magnesium, aluminium and their alloys) with resistence to oxygen etching have been identified as magic species, even though bulk metals of these clusters are highly reactive [7–10] These magic species (at closed electron shells) are also marked by large HOMO–LUMO gaps, high atom-removal energies (or, fragmentation energies), spin excitation energies and/or ionization energies [11, 12]. Advances in this field have been intensified not only because of the importance of cluster reactions with oxygen providing insights into the reactivity and property of metals at reduced sizes, but also enabling using the knowledge gained through gas-phase reactivity of metal clusters to facilitate the understanding of interdisciplinary frontiers including condensed phase physics and surface chemistry. This chapter[1] will highlight the oxygen etching effect and oxygen-addition reactions, cluster odd–even alternation phenomenon, as well as superoxo and peroxo states involved in metal cluster reactivity with oxygen.

3.1 Oxygen Etching Effect

The observation of oxygen etching effect dates back to very early cluster investigations. Figure 3.1 shows one of the earliest cluster reaction studies between aluminum and oxygen by *Jarrold* and *Bower* in 1986 [13]. In this study, the cationic clusters Al_n^+ (n = 4–25) were generated by a LaVa source, ionized by a high-energy electron beam and expanded into the vacuum system where specific cluster sizes were mass-selected by a quadrupole mass filter; focused into a low-energy ion beam and then

[1] This chapter is partly reproduced from Chem. Rev. 2016, 116, 14456–14492.

Z. Luo and S. N. Khanna, *Metal Clusters and Their Reactivity*,
https://doi.org/10.1007/978-981-15-9704-6_3

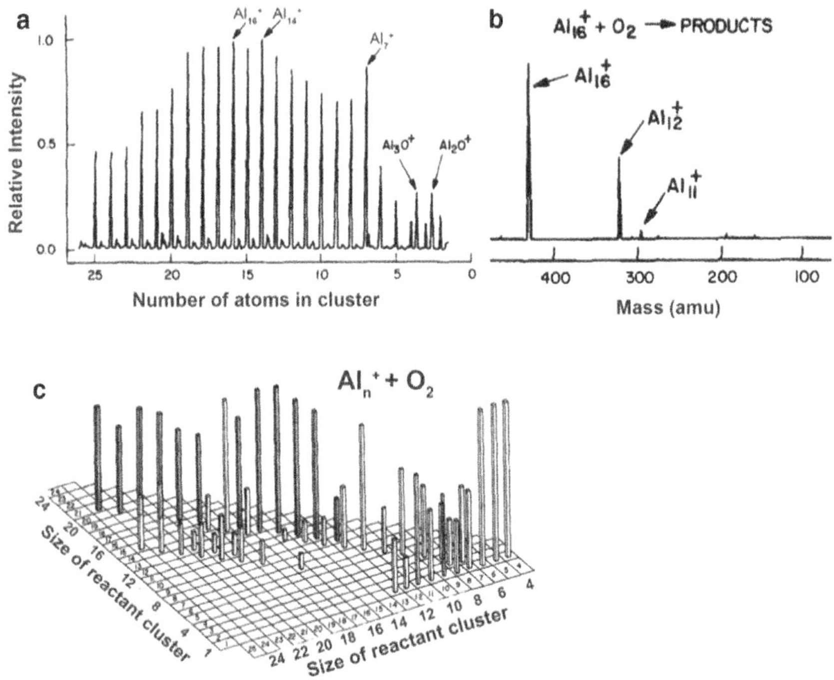

Fig. 3.1 **a** Typical mass spectrum of cationic aluminum cluster ions from the LaVa source. **b**The result of mass selecting Al_{16}^+ and introducing oxygen into the gas cell. The center of mass collision energy was 96.49 kJ/mol. **c** A histogram showing the product distributions from the reactions between Al_n^+ (n = 4–25) and oxygen. No oxygen-containing product ions were observed. The peaks in the histogram due to Al_{n-4}^+ are shaded. Reproduced with permission from Ref. [13]. Copyright 1986 American Institute of Physics

passed through a cell where the reactant gas was introduced [13]. Products and unreacted cluster ions were guided into a quadrupole mass analyzer and finally detected using a collision dynode and electron multiplier. A typical mass spectrum of nascent aluminum cluster ions showed that Al_7^+ appeared with relatively more abundance than its neighbors (Fig. 3.1a) indicating its prominent stability. Later investigations have further demonstrated the magic behavior of the cationic cluster Al_7^+ with 20 valence electrons [14].

The gas-phase reaction of Al_n^+ clusters with oxygen demonstrated that, there were no oxygen-containing product ions being observed for all the clusters studied; instead, mass-selected Al_n^+ clusters were found to dissociate into smaller clusters. For example, the reaction of Al_{16}^+ with oxygen led to products Al_{12}^+ (90%) and Al_{11}^+ (10%), as shown in Fig. 3.1b. Figure 3.1c displays a histogram of the products from reactions between aluminum cluster ions Al_n^+ (n = 4–25) and oxygen. Al^+ was found as a major product in such mass-selected reactions for the clusters up to n = 13; while at Al_{13}^+ a transition occurs; and for n > 13 a loss of Al_4 was found to be

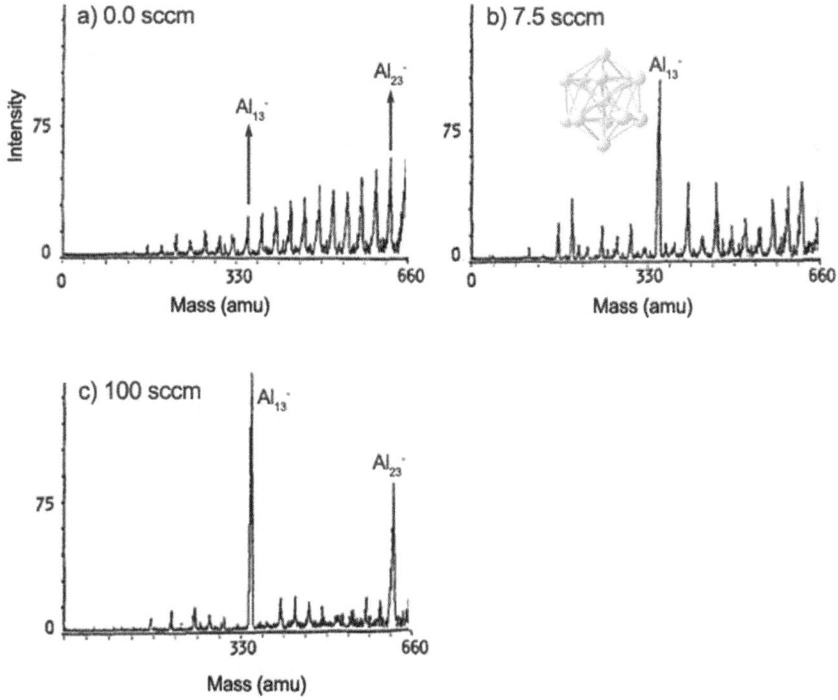

Fig. 3.2 Oxygen etching reactions of anionic Al clusters. Reproduced with permission from Ref. [15]. Copyright 1989 American Institute of Physics

predominant products with rare exception. In brief, the mass loss (i.e., etching effect) for cationic aluminum clusters to react with oxygen follows a pathway as "$Al_n^+ + O_2 \rightarrow Al_m^+ + Al_{n-m}O_2$".

For a similar system, Castleman et al. reported a study of the reactivity of anionic aluminum clusters with oxygen, [15] as shown in Fig. 3.2. Interestingly it was found that the small aluminum cluster anions containing up to twelve Al atoms were rather reactive toward oxygen, but a few selected aluminum clusters including Al_{13}^-, Al_{23}^- and Al_{37}^- were resistant to oxygen etching. Considering that aluminum generally has three valence electrons, the number of free electrons in an anionic cluster is 3n + 1 and hence the observations of Al_{13}^-, Al_{23}^- and Al_{37}^- could be accounted for by shell closings at 40, 70 and 112 electrons predicted by the jellium model [16, 17]. This finding unambiguously revealed that the electronic shell filling of Al clusters could account for mass abundances seen in experiments, that is, electronic properties directly affect the overall stability and chemical reactivity of metal clusters [8]. This relationship stems from the fact that the ground state of an oxygen molecule is spin-triplet. Any activation of the oxygen molecule requires the filling of the minority spin states that results in a change in the spin multiplicity from triplet to singlet. Since the overall spin is conserved in free systems, such a transition requires a spin excitation from singlet to triplet multiplicity in clusters with filled shells. This spin excitation is linked to the HOMO–LUMO gap that creates an energy barrier when reacting with oxygen. In practice, clusters with HOMO–LUMO gaps exceeding 1.2 eV are found to be generally non-reactive to oxygen [18]. Subsequent investigations of Al clusters reacting with ground-state molecular oxygen and singlet atomic oxygen have further explained the reason why the mass abundance of selective clusters could be enhanced, [19] simply being caused by successive fragmentation of larger clusters along with the formation of a very stable molecule Al_2O, written as,

$$Al_{n+4}^{-} + O_2 \rightarrow Al_n^{-} + 2Al_2O \tag{3.1}$$

This reactivity rationalizes that individual clusters could resist to oxygen etching as dominant peaks and even exhibit increased abundance (such as Al_{13}^{-}) in mass spectrum due to fragments of larger unstable species. The inertness of magic Al clusters could be used as models to study the chemisorption of O_2 on Al(111) surface, which is recognized as a model reaction for surface oxidation and catalysis [20].

Among others, Roach et al. [21] reported a study of oxygen etching on alloy metal clusters Al_nCu^{-}, and found that a few Al–Cu clusters with fully-filled subshells and large HOMO–LUMO gaps were also resistant to oxygen etching. For example, $Al_{22}Cu^{-}$ was observed as the largest peak in the products (Fig. 3.3), which was interpreted by its relatively large vertical spin excitation (VSE) energy and HOMO–LUMO gap, resulting from the geometry distortion which leads to a spliting of the shells in a spherical jellium, named as a crystal-field-like splitting of the electronic shells.

Further insights of Al-based cluster reactivity have been attained when examining aluminum-magnesium alloys. Since magnesium is divalent while aluminum is trivalent, the Al–Mg alloy clusters offer larger variations over the electron counts than pure aluminum clusters and hence provide more rigorous grounds to investigate the underlying principles of oxygen etching effect. As expected, $Al_5Mg_2^{-}$ and $Al_{11}Mg_3^{-}$ which correspond to magic numbers of 20 and 40 electrons are found to exhibit reasonable stability (Fig. 3.4); but $Al_7Mg_3^{-}$, $Al_{11}Mg^{-}$ and $Al_{11}Mg_2^{-}$ with electron counts of 28, 36, and 38 respectively displayed unexpected stability. The stabilities of these non-magic numbers of $Al_nMg_m^{-}$ clusters can be understood via a crystal-field-like splitting of degenerated shells due to the geometrical distortions of the clusters. These studies reinforced the importance of near-free electron gas (NFEG) model (with closed shell n = 2, 8, 20, 40...) in rationalizing electronic structure and stability of metal clusters. At the same time, also raised is a pending question regarding the development of a comprehensive model for metal clusters, that could account for "magic numbers" according to the shell model (like noble gas), but also can successfully predict stability of all metal clusters where shell model fails. The study of magic numbers in Al–Mg clusters and their inertness towards oxygen is also important in industrial anticorrosion and aerospace manufacturing [22].

Similar to the drastic etching effect observed for aluminum and Al–Mg clusters, the chemical reactivity of cobalt cluster anions Co_n^{-} (n = 2–8) toward O_2 by flow tube reactor was also found to have rapid rate coefficients, leading to fragmentation

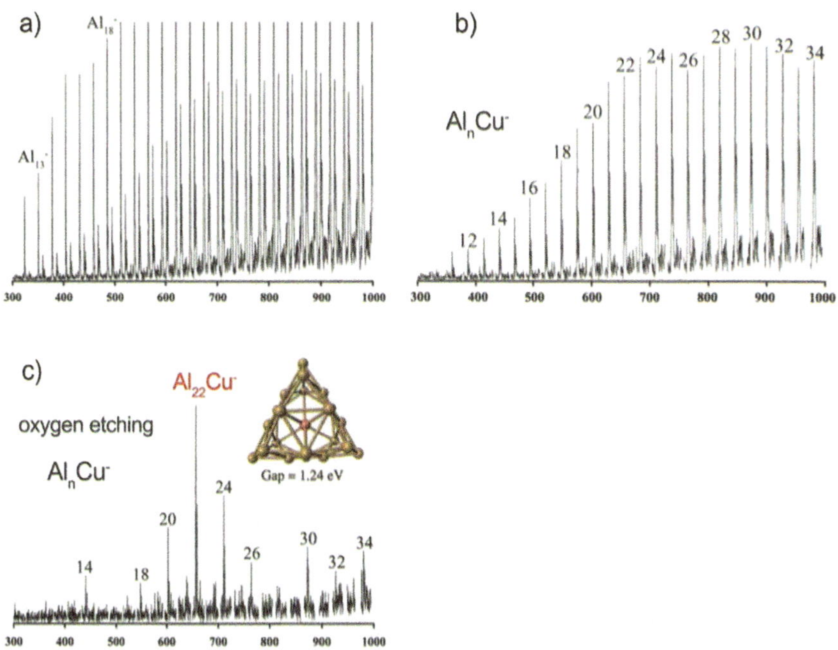

Fig. 3.3 **a** Nascent distribution of $Al_n^-$and Al_nCu^- clusters. **b** Nascent Al_nCu^-distribution. **c** Oxygen-etched Al_nCu^-distribution (1.25% partial pressures of oxygen). Peaks corresponding to pure aluminum clusters were superficially removed from (**b**) and (**c**) to discriminate Al_nCu^-

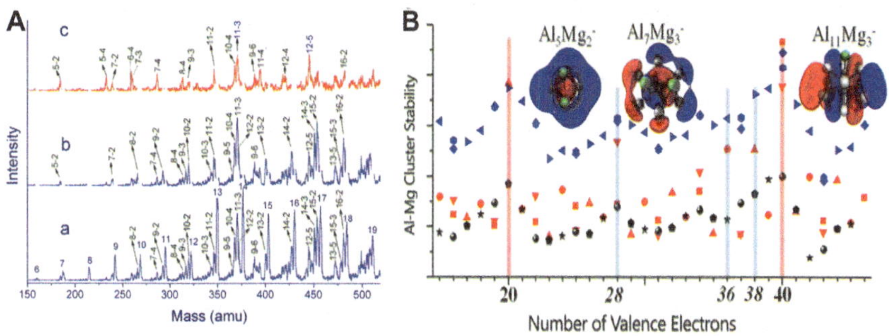

Fig. 3.4 **A** Nascent distribution of Al_n^- and $Al_nMg_m^-$ (a), nascent $Al_nMg_m^-$ distribution (b), and oxygen-etched $Al_nMg_m^-$ distribution (c), where the peaks corresponding to pure aluminum clusters were superficially removed to discriminate $Al_nMg_m^-$. **B** A diagram showing the stability of Al–Mg alloy clusters correlative with the number of valence electrons

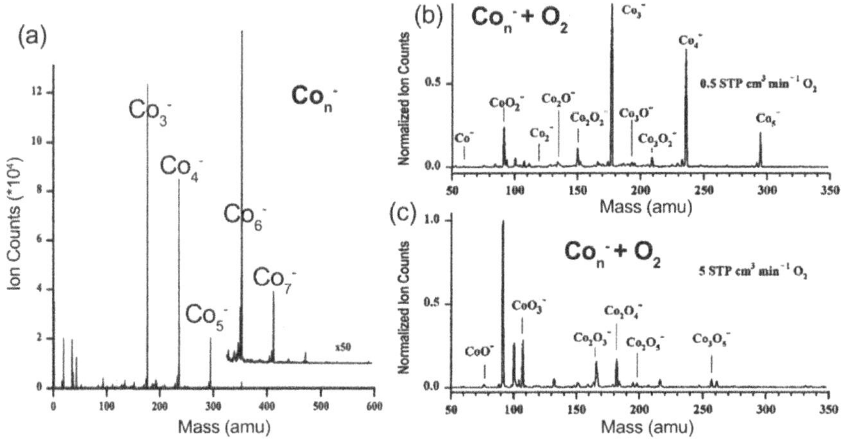

Fig. 3.5 Mass spectra of Co_n^- clusters at the absence **a** and presence of different flow rates of oxygen: **b** 0.5 STP cm^3min^{-1}, **c** 5.0 STP cm^3min^{-1}. Reproduced from Ref. [25]. Copyright 1997 American Chemical Society

of parent clusters independent of the cluster sizes, as shown in Fig. 3.5. Especially in the case of large gas flow rate (Fig. 3.5c), the parent clusters disappeared and the peaks shifted to low mass region. Among the products of $Co_nO_x^-$ species, an intense peak was found at CoO_2^- which has been ascertained as a stable ionic molecule. Besides, other cobalt oxide anions containing one or two cobalt atoms were also observed. It was demonstrated that the primary reaction processes were those in which oxidation occurred by removing one (or two) cobalt atom from the cluster to form a small metal oxide anion, leaving the rest of the cobalt cluster as neutral products, written as:

$$Co_n^- + O_2 \rightarrow Co_{n-1} + CoO_2^- \tag{3.2}$$

The reason why the reactivity of cobalt clusters with oxygen differs from that of aluminum and Al–Mg clusters is partly due to the intrinsic activity of different metals, and those with relatively low boiling point (e.g., aluminum and magnesium) readily combust. What's more, Al clusters react and readily form very stable molecules Al_2O, while Co clusters react and produce anionic CoO_2^-. The valence electrons in Al–Mg clusters form a nearly free electron gas while the reactivity of Co_n clusters is driven by 3d-states that are localized. Oxygen-etching reactions are an important probe to identify magic cluster species with shell closing, such as the aforementioned typical example of Al_{13}^- for which the electronic orbitals mimic the atomic orbitals of Cl^-. That is, 40 valence electrons form a completely filled shell and hence make the Al_{13}^- cluster behave as a superatomic noble gas. Several other superatomic cluster species with outstanding stability have also been ascertained through the examination of whether or not surviving in oxygen-etching reactions, including a few aluminum iodides such as $Al_{13}I^-$, Al_7I^-, [23] $Al_{13}I_{2n}^-$ and $Al_{14}I_{2n+1}^-$ [24].

Among others, the neutral vanadium clusters have also been studied and were found to undergo similar etching reactions with oxyge. Figure 3.6 presents a typical mass abundance of the neutral vanadium clusters at the presence of different reactant oxygen. As is shown, V_nO and V_nO_2 series are formed gradually with increased amount of O_2 introduced into the flow tube reactor. It is notable that the dissociation channels dominate the reaction of neutral vanadium clusters with oxygen, especially at the presence of sufficient reactant oxygen. Such dissociation can also be observed in the collision of V_n with He. The major difference between O_2 reaction and He collision is the intensity change of V_1 after and before the reaction, indicating that the loss of V atom from V_n clusters is preferred in the reaction of $V_n + O_2$. So, the

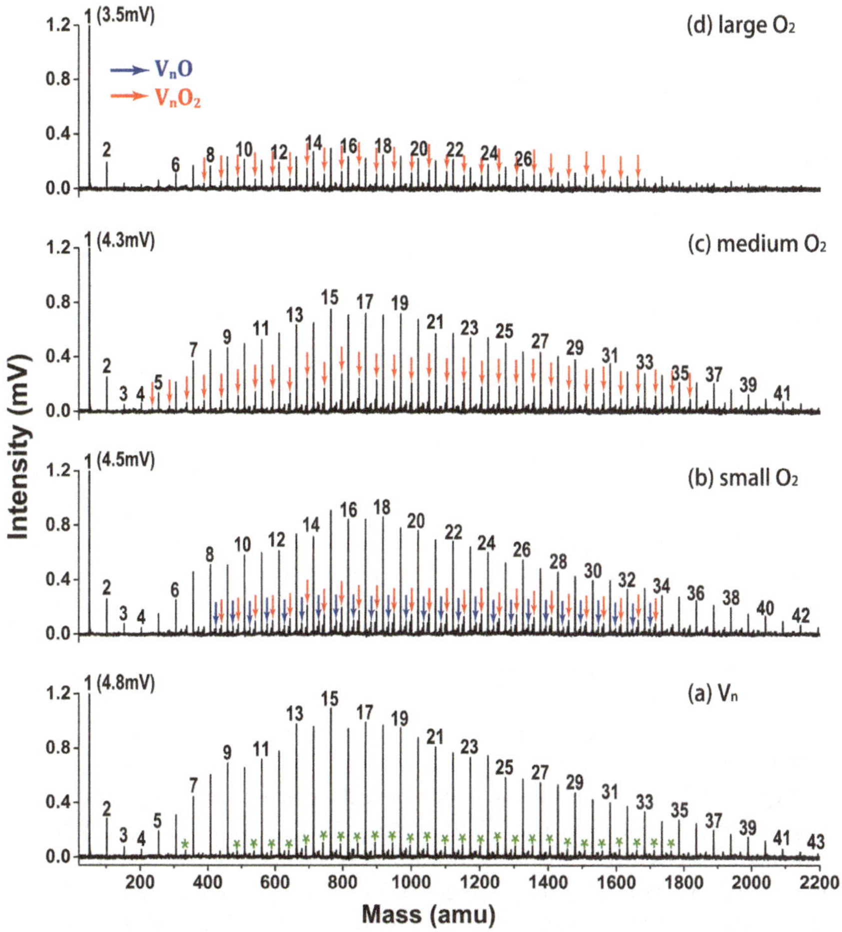

Fig. 3.6 **a** A typical distribution of neutral V_n (n = 1–40) clusters photoionized by 177.3 nm deep ultraviolet (DUV) laser. **b–d** TOF mass spectra of V_n (n = 1–40) reacting with different amount of 5% O_2/He added into the reaction tube

formation of V_nO_2 and the production of V atom is the main reaction pathways for V_n reacting with O_2, that is, $V_n + O_2 \rightarrow V_{n-1}O_2 + V$.

3.2 Oxygen Addition

In addition to oxygen etching effect, [26–28] the reactivities involving oxygen addition onto metal clusters have also been well illustrated in the past 20 years [29–31]. To some extent, the addition of molecular oxygen to metal clusters is actually the initial reaction channel (followed by etching effect, etc.) for most metal clusters, such as the aforementioned vanadium cluster neutrals. The addition of oxygen is often associated with an oxidation process, i.e., loss of electrons, charge distribution rearrangement and orbital hybridization. Also, the oxygen addition/adsorption could follow the principle of increasing valence of metals but bear dramatic selectivity depending on cluster sizes, bonding energies and electronic behaviors. For example, selective anionic Au_n^- with an odd number of electrons (i.e., even number of Au atoms) showed significant O_2 uptake, whereas some other clusters either showed very weak reactivity or did not exhibit any propensity for reactions with oxygen [32–34]. In comparison, neutral and cationic gold clusters were found to be inert toward oxygen with rare exceptions (e.g., Au_{10}^+) [30]. Considering an oxygen double bond, there is much larger electron density within an oxygen molecule than the molecular outer end, allowing for relatively strong electron-withdrawing ability of the oxygen atom. Therefore, noble metal clusters with additional unpaired electrons are favorable for such oxygen-addition reactivity.

In addition to the investigations revealing the oxygen-addition reactivity of Au clusters, there were also a few reports showing similarity of copper clusters in reacting with oxygen [35]. The study by *T. H. Lee* and *K. M. Ervin* demonstrated that the addition of molecular oxygen through "$Cu_n^- + O_2 \rightarrow Cu_nO_2^-$" is a primary reaction channel for most copper cluster anions [29]. Further reactions were also observed for small Cu_n^- clusters (n = 2–5) allowing for the formation of a subsequent product $Cu_nO_4^-$. Further, the effective bimolecular rate coefficients for the reactions of Cu_n^-, Ag_n^- and Au_n^- with oxygen have been measured at a certain buffer gas pressure, as shown in Fig. 3.7. Other than oxygen addition, collision-induced dissociation (or fragmentation) of coinage metal clusters was also included in such gas-phase reactions [36].

Oxygen-addition was also found to dominate the reactivity of cationic Co_n^+ clusters [37]. The cationic Co_n^+ (n = 2–9) clusters displayed a high reactivity toward O_2 by taking successive oxidation pathways. It is notable that the oxygen addition on Co_n^+ clusters does not follow a direct attachment; instead, the primary reaction mainly results in a replacement of a Co atom by an O_2 molecule. Also, the formed oxide clusters allow for successive reactions toward oxygen, which resembles the etching effect observed for anionic Co_n^- as discussed above. Whereas, there is huge difference between cationic and anionic cobalt clusters, probably because the latter

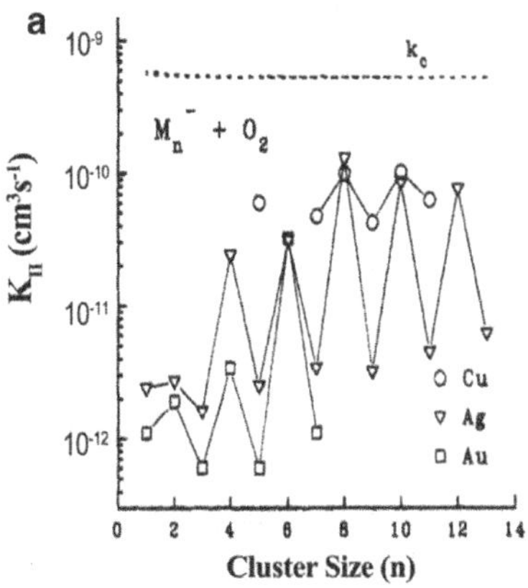

b

Cluster	K_{II}($10^{-11}cm^3s^{-1}$)	K_c($10^{-10}cm^3s^{-1}$)	K_{II}/K_c
Ag^-	0.24 ± 0.3	5.92	0.004 ± 0.005
Ag_2^-	0.27 ± 0.16	5.57	0.005 ± 0.003
Ag_3^-	0.16 ± 0.09	5.45	0.003 ± 0.001
Ag_4^-	2.4 ± 0.4	5.39	0.044 ± 0.007
Ag_5^-	0.25 ± 0.19	5.36	0.005 ± 0.003
Ag_6^-	3.1 ± 0.6	5.33	0.059 ± 0.010
Ag_7^-	0.34 ± 0.20	5.31	0.006 ± 0.004
Ag_8^-	12.9 ± 2.5	5.30	0.24 ± 0.05
Ag_9^-	0.32 ± 0.25	5.29	0.006 ± 0.005
Ag_{10}^-	8.6 ± 1.1	5.28	0.160 ± 0.020
Ag_{11}^-	0.5 ± 0.3	5.27	0.009 ± 0.006
Ag_{12}^-	7.6 ± 1.1	5.27	0.14 ± 0.02
Ag_{13}^-	0.6 ± 0.3	5.26	0.012 ± 0.006
Cu_5^-	6.1 ± 1.7	5.46	0.11 ± 0.03
Cu_7^-	4.7 ± 0.3	5.39	0.088 ± 0.006
Cu_8^-	10 ± 3	5.36	0.19 ± 0.06
Cu_9^-	4.2 ± 1.0	5.34	0.08 ± 0.02
Cu_{10}^-	10 ± 3	5.33	0.19 ± 0.06
Cu_{11}^-	6.3 ± 1.8	5.32	0.12 ± 0.03
Au^-	≤0.11 ± 0.03	5.61	≤0.002 ± 0.0005
Au_2^-	0.19 ± 0.12	5.41	0.004 ± 0.002
Au_3^-	≤0.06 ± 0.03	5.34	≤0.001 ± 0.0005
Au_4^-	0.34 ± 0.09	5.31	0.006 ± 0.002
Au_5^-	≤0.06 ± 0.04	5.29	≤0.001 ± 0.0007
Au_6^-	3.2 ± 0.5	5.27	0.060 ± 0.009
Au_7^-	≤0.11 ± 0.09	5.26	≤0.002 ± 0.001

Fig. 3.7 **a** Measured effective bimolecular rate coefficients for the reactions of copper cluster anions (circles), silver cluster anions (triangles), and gold cluster anions (squares) with oxygen as buffer gas at a pressure of ~60 Pa. **b** A list showing the effective bimolecular rate coefficients for Cu_n^-, Ag_n^- and Au_n^- reacting with oxygen at ~60 Pa buffer gas pressure. K_{II}, reaction rate. K_c, collision rate. Reproduced from Ref. [29]. Copyright 1994 American Chemical Society

readily form a stable molecule CoO_2^- in reacting with oxygen. For cationic counterparts, successive oxidation reactions were found to virtually terminate when the formed $Co_nO_m^+$ clusters displayed stoichiometric structures of $(CoO)_4(CoO_2)_n^+$ (n = 0–3), or $Co_{2,3}O_{4,5}^+$. The loss of a Co atom in the reaction of Co clusters with O_2 (usually termed as a switching reaction) was demonstrated to be consistent with Stevenson's rule [37–39] According to Stevenson's rule, [38] the ionic products in decomposition prefer to bear lower IP values, and the coordination of ligands can reduce the IP of metal clusters [40]. Examination of the reaction rate constants for the mass-selected cobalt clusters illustrated a strong correlation between the cluster sizes and their reactivity, suggesting that the geometric structures and the release of stable molecules could be a major factor in determining the cobalt cluster reactivity [37].

3.3 Superoxo and Peroxo States

Along with oxygen intake reactions for metal clusters, it is important to note that the bonding mechanism between metal clusters and oxygen could involve charge transfer with a concomitant activation of the O–O bond. The excitation of triplet dioxygen ($^3\Sigma_g^-$) to the more reactive singlet state ($a^1\Delta_g$) can be achieved by the interaction with metal clusters leading to the formation of superoxo state ($O_2^{-\bullet}$) and peroxo state (O_2^{2-}) complexes due to chemisorption and charge transfer [41]. For example, Klacar et al. [42] examined the reactivity of oxygen with Ag_n clusters containing up to 9 atoms and found that the molecular oxygen preferred a dissociation mode for its adsorption on larger sized clusters (e.g., containing more than 5 Ag atoms), where the activation of O–O bond was initiated at a superoxo state.

There could be coexistence of superoxo and peroxo states, but a transition from the former to the latter determines the O–O bond activation within a cluster-oxygen reaction process [44]. For this, *Wang*, *Zeng* and coworkers [43] reported an in-depth study of O_2 chemisorption on even-sized Au_n^- clusters (Fig. 3.8). They demonstrated spectroscopic and electronic evidences of the transition from superoxo to peroxo chemisorption for Au_8^-. It was noted that both superoxo and peroxo states coexisted in the cluster beam of $O_2Au_8^-$, and the superoxo form involves a single O-Au bond (η^1–O_2) while the peroxo form involves two O–Au bonds (η^2–O_2). Also the superoxo $O_2Au_8^-$exhibited low binding energy within the O_2-induced PES feature, while the peroxo $O_2Au_8^-$displayed sharper and higher binding energy feature. It was also noted that, although both two-dimensional (2D) and 3D isomers of Au_{12}^-coexist in the cluster beam, O_2 prefers the peroxo binding with the 3D isomer of Au_{12}^-[43]. Superoxo and peroxo states have also been found as important chemical processes associated with other metal cluster reactions [34, 45].

3.4 Cluster Odd–Even Alternation

As has been shown above, with sufficient oxygen flowing, a dramatic loss of Al cluster signal is readily discerned for most species especially even-numbered clusters (i.e., odd number of electrons), but odd-numbered clusters (i.e., even number of electrons) could survive in the rich-pressure condition displaying slightly enlarged mass abundance. This odd/even alternation was also seen in other experiments where most of the even-atom clusters reacted away in sharp contrast to the odd, indicative of a paired electron effect [46, 47]. Taking a glance over the studies of metal cluster in reacting with oxygen, there are actually abundant investigations showing the odd–even alternation [25, 28, 32, 48–58]. For example, among the Au_n^-cluster reactions with O_2, molecular oxygen addition was found to be the main pathway for the even-sized clusters, while the odd-sized clusters were inert toward O_2 [29, 32, 34]. Such even–odd alternation correlates with a similar pattern in the electron affinities of Au_n clusters, validating that electron transfer between $Au_n^-$and O_2dominates their reactivity [32, 43]. Experimental evidences have also been obtained via photoelectron spectroscopy, revealing that even-sized gold clusters favour O_2 chemisorption by noting the distinguishable O–O vibrational fingerprints [59, 60].

Bernhardt et al. [53, 61] reported the reactivity of anionic silver clusters Ag_n^- with O_2, as shown in Fig. 3.9. It was demonstrated that, among the Ag_n^- (n = 1–11) clusters they studied, the even-atom anions (i.e., odd number of valence electrons) were more reactive than the odd-atom cluster anions for the reaction of the first O_2.

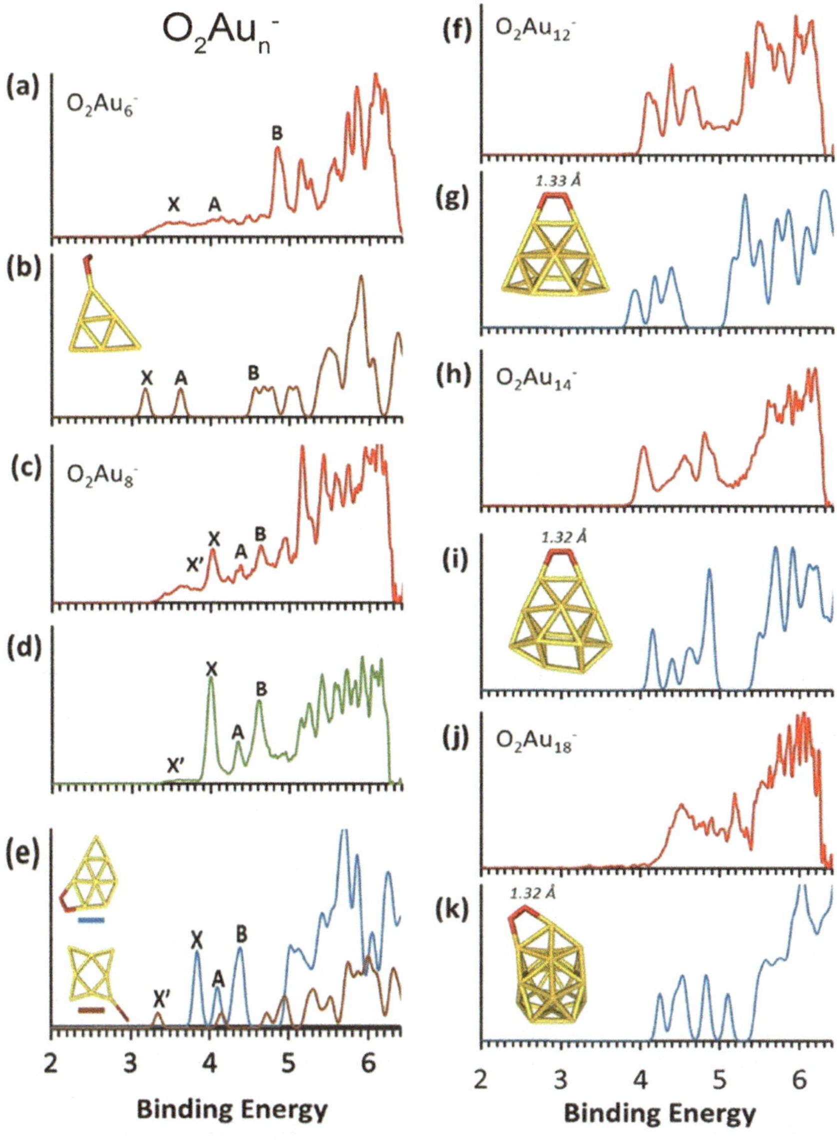

Fig. 3.8 Comparison of the experimental PES spectra of $O_2Au_6^-$ (**a**), $O_2Au_8^-$ (**c** and **d**), $O_2Au_{12}^-$ (**f**), $O_2Au_{14}^-$ (**h**), and $O_2Au_{18}^-$ (**j**) with the computed spectra and the corresponding structures (**b**), (**e**), (**g**), (**i**), (**k**). The experimental spectra of $O_2Au_8^-$ (**c** and **d**) were obtained under two different experimental conditions. The simulated spectra of the superoxo and peroxo isomers are represented in brown and blue colours, respectively. Reproduced from Ref. [43]. Copyright 2012 American Chemical Society

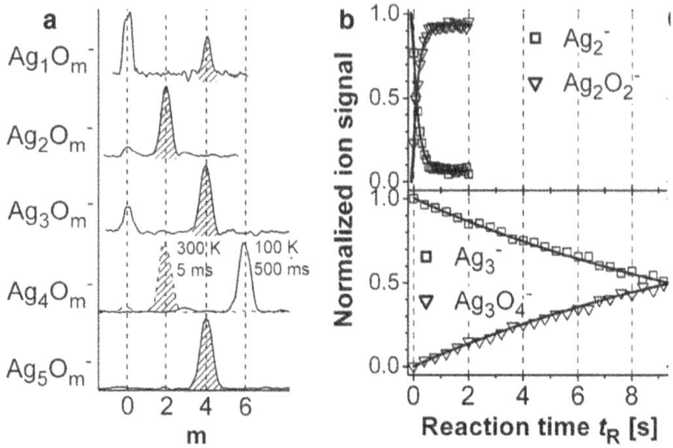

Fig. 3.9 **a** Product ion mass spectra after reaction of Ag_n^- with O_2. Ion intensities are plotted as a function of the number of adsorbed oxygen atoms m. **b** Examples of measured oxidation kinetics for Ag_2^- and Ag_3^- at 300 K. Open symbols: experimental data; solid lines: kinetic fit. Reproduced from Ref. [53]. Copyright 2004 American Chemical Society

Even-atom clusters were found to react readily with molecular oxygen by generating $Ag_nO_2^-$ products except Ag_4^-, while the odd clusters reacted to bind two O_2 molecules except for the single atom. The corresponding kinetic data were examined (Fig. 3.8b) and a sequential reaction channel was proposed as,

$$Ag_n^- + 2O_2 \xrightarrow{k_1} Ag_nO_2^- + O_2 \xrightarrow{k_1} Ag_nO_4^- \tag{3.3}$$

A collaborative effort between Castleman and Khanna groups [12] has given further insights into the odd–even selectivity for silver clusters reacting with oxygen, as shown in Fig. 3.9. It was illustrated that the necessity or not for Ag clusters to become spin excited (and hence to accommodate the triplet spin of oxygen) plays a determining role in their reactivity. This is consistent with the experimental findings that odd-electron silver clusters reacted with oxygen while the even-electron systems were relatively inert (Fig. 3.10A). Furthermore, an anionic 13-atom cluster was found to exhibit unexpected stability against reactivity with oxygen, which was rationalized by comparing the reactivity of Ag_{13}^- with proximate even-electron clusters such as Ag_{15}^-. The inertness of Ag_{13}^- is associated with its large spin excitation energy, a crystal-field-like splitting of the orbitals caused by the unique triangular bilayer structure, as well as a relatively large gap despite not having a magic number of valence electrons, as shown in Fig. 3.10B.

These investigations revealed that the reactivity of metal clusters with oxygen is correlated with the excitation needed to activate an O–O bond. Silver and oxygen have negligible spin–orbit effects, and hence their reactions follow the Wigner-Witmer rules of spin conservation. For the odd-electron systems, the spin of the extra electron could align opposite to the majority spin electrons of the $^3O_2$ molecule and the spin

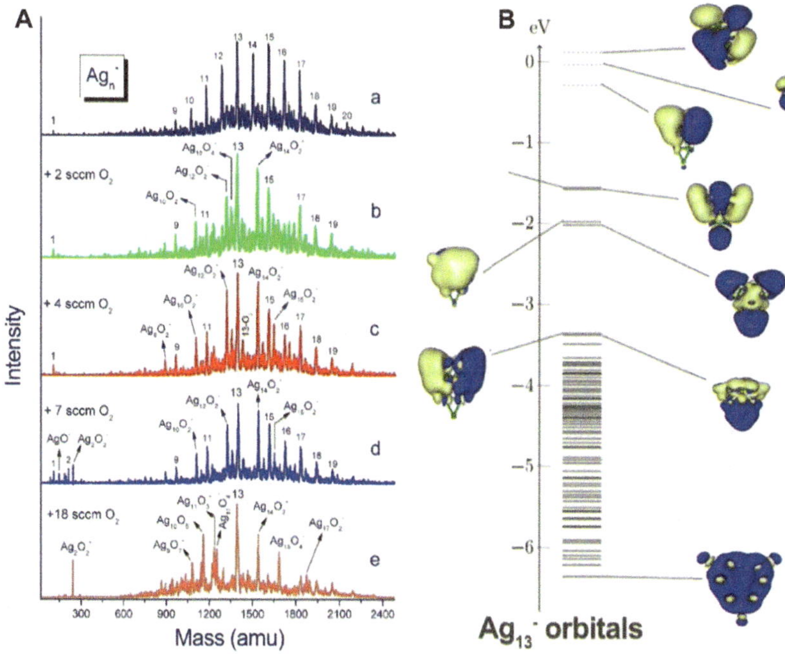

Fig. 3.10 **A** The mass spectrum of silver cluster anions produced via a MagS-source (a) and the spectra after exposure to different quantities of oxygen (b–e). **B** Calculated molecular orbitals of Ag_{13}^-

conservation does not require any spin excitation of the metal counterpart; whereas for even-electron cluster systems, the spin multiplicity of oxygen is decreased, the reacting cluster has to follow the spin excitation of the remaining portion to conserve the total spin. The spin effect will be further demonstrated (see Sect. 5.1).

3.5 Competition of Oxygen Etching Versus Oxygen Addition

Incorporating ideas from previously published studies, it has been ascertained that the dominant reaction pathway for anionic aluminium clusters [19, 62] and anionic cobalt clusters, [25, 62] with oxygen are often subjecting to oxygen-etching effect [12, 22, 63]. Vanadium is an interesting polyvalent element with several oxidation states ranging from $-1(d^6)$ to $+5(d^0)$, [64] allowing for varying forms of vanadium oxide compounds. From collision-induced dissociation (CID) of vanadium oxide clusters, [65] it has been inferred that the $V_nO_m^{+,0}$ species typically consist of stable dioxide (VO_2), trioxide (VO_3), and pentoxide (V_2O_5) [66–68].

Utilizing a customized reflection time-of-flight mass spectrometer (Re-TOFMS) [69], we have prepared well-resolved vanadium clusters $V_n^{+/0}$ ($n = 1$–30), on which we conducted a comprehensive study on the reactivity with oxygen. It is illustrated that, cationic V_n^+ clusters readily react with oxygen leading to the production of both etched building blocks and oxygen-rich $V_nO_m^+$($n < m$) species profiting from the ion–molecule attraction [70]. As for the neutral V_n clusters, the etching effect of $V_n + O_2 \rightarrow V_{n-1}O_2 + V$ dominates the reaction pathway [69].

Figure 3.11 presents the mass spectra of the cationic V_n^+ clusters at the absence and presence of different amount of reactant oxygen. As a whole, the mass abundances of all V_n^+ clusters are reduced, due to the existence of collision-induced dissociation and reaction to form oxides [71, 72]. Considering that $V_2O_5^+$ and VO^+ finds local maximum mass abundance ascribing to its prominent stability and regularity of the small vanadium oxides, we propose a further insight into the reactivity of vanadium clusters with O_2, by generating $VO^{+,0}$ or $V_2O_5^{+,0}$, so that the varied $V_nO_m^+$ formation mechanism can be written as,

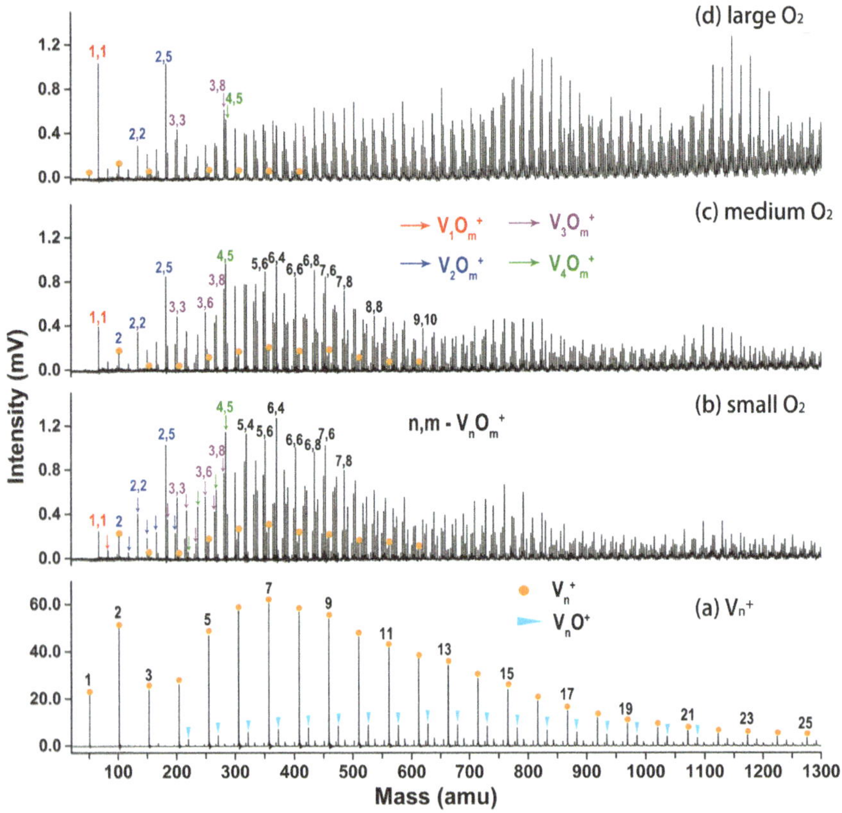

Fig. 3.11 **a** A typical distribution of cationic V_n^+ ($n = 1$–25) clusters. **b–d** TOF mass spectra of V_n^+($n = 1$–25) reacting with different amount of 20% O_2/He added into the reaction tube

$$V_n^+ + yO_2 \rightarrow [V_{n-1}O_{2y-1}]^{0,+} + VO^{+,0} \tag{3.4}$$

$$V_n^+ + yO_2 \rightarrow [V_{n-2}O_{2y-5}]^{0,+} + V_2O_5^{+,0} \tag{3.5}$$

These reaction pathways actually are well consistent with the previously established reactivity for aluminium clusters ($Al_n^- + O_2 \rightarrow Al_{n-4}^- + 2Al_2O$) [19, 62]. Nevertheless, the multi-pathways for the vanadium clusters render the reactivity with oxygen to be more complex. The mass spectrometric observation in Fig. 3.11d suggests that the formation of VO^+ (i.e., Eq. 3.4) and $V_2O_5^+$ (Eq. 3.5) could be the most dominant reaction pathways. Besides, there could be also similar reaction channels based on similar etching-effect to produce VO_2 and VO_3 (Eqs. 3.6 and 3.7), which allows for successive reactions along with direct oxygen addition.

$$V_n^+ + yO_2 \rightarrow [V_{n-1}O_{2y-2}]^{0,+} + VO_2^{+,0} \tag{3.6}$$

$$V_n^+ + yO_2 \rightarrow [V_{n-1}O_{2y-3}]^{0,+} + VO_3^{+,0} \tag{3.7}$$

On the other hand, it is notable that varied $V_nO_m^+$ ($m \geq n$) are produced, with dependence on the quantity of oxygen being introduced. The observation of fruitful $V_nO_m^+$ species on the large mass range manifest the existence of growth channel (i.e., successive oxygen addition process) for V_n^+ in reacting with O_2. The mass abundance of the typical distribution of $V_nO_m^+$ clusters (Fig. 3.12a) displays different series with odd–even alternation [73] regarding the numbers of vanadium atoms. These series can be classified into five groups on a basis of equidifferent V_2O_5 building blocks, i.e. the least abundance series $[V_{2\eta}O_{5\eta} \cdot VO]^+$ (labelled with blue circles), the first dominant $[V_{2\eta}O_{5\eta} \cdot VO_2]^+$ (blue dots), the second least $[V_{2\eta}O_{5\eta} \cdot VO_3]^+$ (half-filled blue dots), the second abundance of $[V_{2\eta}O_{5\eta} \cdot V_2O_4]^+$ (red dots), and the complete building blocks $(V_2O_5)_\eta^+$ which takes on the third mass abundance among these five series as marked in half-filled red dots. Besides, there are plenty of weak peaks, indicating coexistence of other vanadium oxides such as V_2O, and V_2O_2, etc. The growth mechanism of the $V_nO_m^+$ clusters could be simply summarized as,

$$[\eta VO_3]^{+/0} + [\eta VO_2]^{0/+} \rightarrow [V_{2\eta}O_{5\eta}]^+(A_\eta B_\eta) \tag{3.8}$$

$$V_{2\eta}O_{5\eta} + [VO_2]^+ \overset{\text{fast}\rightarrow}{\rightarrow} [V_{2\eta+1}O_{5\eta+2}]^+(A_{\eta+1}B_\eta) \tag{3.9}$$

$$V_{2\eta}O_{5\eta} + [VO_3]^+ \rightarrow [V_{2\eta+1}O_{5\eta+3}]^+(A_\eta B_{\eta+1}) \tag{3.10}$$

$$V_{2\eta}O_{5\eta}{}^{+/0} + 2[VO_2]^{0/+} \rightarrow [V_{2\eta+2}O_{5n+4}]^+(A_{\eta+2}B_\eta) \tag{3.11}$$

$$V_{2\eta}O_{5\eta} + x[VO]^+ \overset{\text{slow}\rightarrow}{\rightarrow} [V_{2\eta+x}O_{5n+x}]^+(C_xA_\eta B_\eta) \tag{3.12}$$

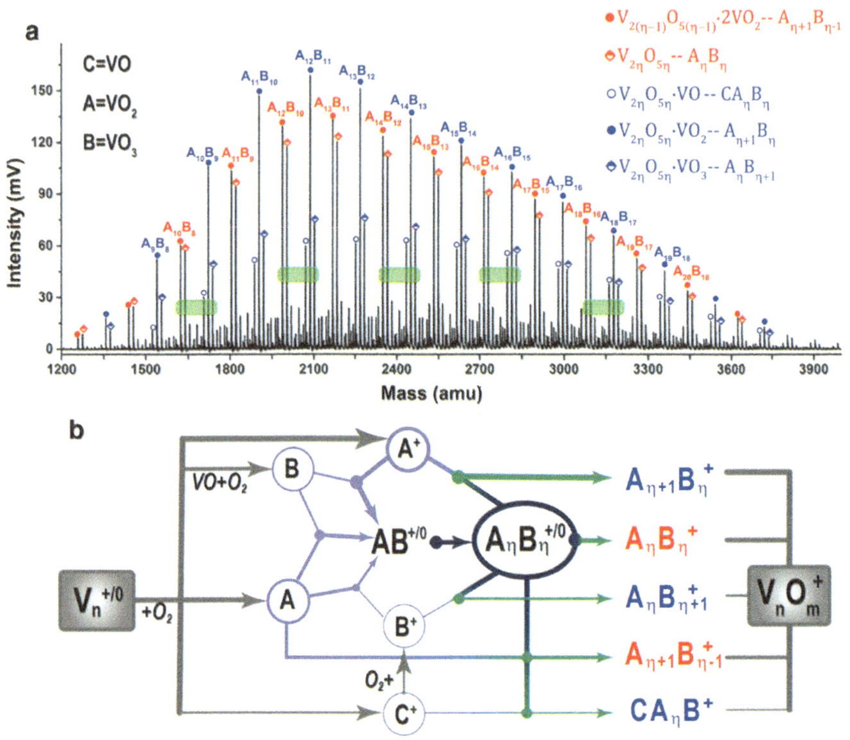

Fig. 3.12 **a** A typical Re-TOF mass spectrum of cationic $V_nO_m^+$ clusters obtained by using 10% O_2/He as buffer gas. **b** Proposed ternary-building-block mechanism for the formation of various $[(VO_2)_X(VO_3)_Y(VO)_Z]^+$

Considering that the $(V_2O_5)_\eta^+$ series can also be expressed with further detailed by VO_2 and VO_3 moieties, the five series can be actually summarized into $(VO)_x(VO_2)_y(VO_3)_z^+$ series. So, we can take them as basic elements or bricks, and from each of these individual parts. At this point, a ternary-code-like mechanism with general formula $C_xA_yB_z$ (A = VO_2, B = VO_3, C = VO) is proposed to elucidate the growth of $V_nO_m^+$ clusters (Fig. 3.12b). The availability to use VO^+, VO_2^+ and VO_3^+ as basic units to chemically synthesize controllable $V_nO_m^+$ clusters brings forth feasibility of ternary code-like cluster building of highly oxygen-rich vanadium oxides for oxygen-exchange related catalysts and functional materials [74].

References

1. A. Shekhar, W.Q. Wang, R. Clark, R.K. Kalia, A. Nakano, P. Vashishta, Appl. Phys. Lett. **102**, 221904 (2013)
2. C.H. Kjaergaard, M.F. Qayyum, A.J. Augustine, L. Ziegler, D.J. Kosman, K.O. Hodgson, B. Hedman, E.I. Solomon, Biochemistry **52**, 3702–3711 (2013)
3. A. Baksi, T. Pradeep, B. Yoon, C. Yannouleas, U. Landman, ChemPhysChem **14**, 1272–1282 (2013)
4. C. van der Linde, S. Hemmann, R.F. Hoeckendorf, O.P. Balaj, M.K. Beyer, J. Phys. Chem. A **117**, 1011–1020 (2013)
5. A. Desireddy, B.E. Conn, J. Guo, B. Yoon, R.N. Barnett, B.M. Monahan, K. Kirschbaum, W.P. Griffith, R.L. Whetten, U. Landman, T.P. Bigioni, Nature **501**, 399–402 (2013)
6. B.C. Guo, K.P. Kerns, A.W. Castleman Jr., Int. J. Mass Spectrom. Ion Processes **117**, 129–144 (1992)
7. R.E. Leuchtner, A.C. Harms, A.W. Castleman Jr., J. Chem. Phys. **94**, 1093–1101 (1991)
8. A.W. Castleman Jr., S.N. Khanna, J. Phys. Chem. C **113**, 2664–2675 (2009)
9. R. Burgert, H. Schnockel, A. Grubisic, X. Li, S.T. Stokes, K.H. Bowen, G.F. Gantefor, B. Kiran, P. Jena, Science **319**, 438–442 (2008)
10. A.C. Reber, S.N. Khanna, P.J. Roach, W.H. Woodward, A.W. Castleman Jr., J. Am. Chem. Soc. **129**, 16098–16101 (2007)
11. A. Halder, V.V. Kresin, J. Chem. Phys. **143**, 164313 (2015)
12. Z. Luo, G.U. Gamboa, J.C. Smith, A.C. Reber, J.U. Reveles, S.N. Khanna, A.W. Castleman Jr., J. Am. Chem. Soc. **134**, 18973–18978 (2012)
13. J.E. Bower, J. Chem. Phys. **85**, 5373–5375 (1986)
14. D.M. Cox, D.J. Trevor, R.L. Whetten, E.A. Rohlfing, A. Kaldor, J. Chem. Phys. **84**, 4651–4656 (1986)
15. R.E. Leuchtner, A.C. Harms, A.W. Castleman Jr., J. Chem. Phys. **91**, 2753–2754 (1989)
16. W. Ekardt, Phys. Rev. B **29**, 1558–1564 (1984)
17. M. Brack, Rev. Mod. Phys. **65**, 677–732 (1993)
18. D.E. Bergeron, P.J. Roach, A.W. Castleman Jr., N. Jones, S.N. Khanna, Science **307**, 231–235 (2005)
19. W.H. Woodward, N. Eyet, N.S. Shuman, J.C. Smith, A.A. Viggiano, A.W. Castleman Jr., J. Phys. Chem. C **115**, 9903–9908 (2011)
20. Q.H. Yuan, J.B. Li, X.L. Fan, W.M. Lau, Z.-F. Liu, Chem. Phys. Lett. **489**, 16–19 (2010)
21. P.J. Roach, W.H. Woodward, A.C. Reber, S.N. Khanna, A.W. Castleman Jr., Phys. Rev. B **81**, 195404 (2010)
22. Z. Luo, C.J. Grover, A.C. Reber, S.N. Khanna, A.W. Castleman, J. Am. Chem. Soc. **135**, 4307–4313 (2013)
23. D.E. Bergeron, P.J. Roach, A.W. Castleman Jr., N.O. Jones, J.U. Reveles, S.N. Khanna, J. Am. Chem. Soc. **127**, 16048–16053 (2005)
24. D.E. Bergeron, A.W. Castleman Jr., T. Morisato, S.N. Khanna, J. Chem. Phys. **121**, 10456–10466 (2004)
25. E. Kapiloff, K.M. Ervin, J. Phys. Chem. A **101**, 8460–8469 (1997)
26. M.B.A. Torres, E.M. Fernández, L.C. Balbás, J. Phys. Chem. A **112**, 6678–6689 (2008)
27. M. Stratakis, H. Garcia, Chem. Rev. **112**, 4469–4506 (2012)
28. M.L. Kimble, A.W. Castleman Jr., R. Mitrić, C. Bürgel, V. Bonačić-Koutecký, J. Am. Chem. Soc. **126**, 2526–2535 (2004)
29. T.H. Lee, K.M. Ervin, J. Phys. Chem. **98**, 10023–10031 (1994)
30. D.M. Cox, R. Brickman, K. Creegan, A. Kaldor, Z. Phys, D-Atoms. Molecules Clusters **19**, 353–355 (1991)
31. M. Neumaier, M. Olzmann, B. Kiran, K.H. Bowen, B. Eichhorn, S.T. Stokes, A. Buonaugurio, R. Burgert, H. Schnöckel, J. Am. Chem. Soc. **136**, 3607–3616 (2014)
32. B.E. Salisbury, W.T. Wallace, R.L. Whetten, Chem. Phys. **262**, 131–141 (2000)

33. J. Hagen, L.D. Socaciu, M. Elijazyfer, U. Heiz, T.M. Bernhardt, L. Woste, Phys. Chem. Chem. Phys. **4**, 1707–1709 (2002)
34. B. Yoon, H. Häkkinen, U. Landman, J. Phys. Chem. A **107**, 4066–4071 (2003)
35. E. Florez, W. Tiznado, F. Mondragón, P. Fuentealba, J. Phys. Chem. A **109**, 7815–7821 (2005)
36. V.A. Spasov, T.H. Lee, K.M. Ervin, J. Chem. Phys. **112**, 1713–1720 (2000)
37. B.C. Guo, K.P. Kerns, A.W. Castleman Jr., J. Phys. Chem. **96**, 6931–6937 (1992)
38. D.P. Stevenson, Discuss. Faraday Soc. **10**, 35–45 (1951)
39. B.C. Guo, K.P. Kerns, A.W. Castleman Jr., J. Phys. Chem. **96**, 4879–4883 (1992)
40. P.D. Dao, K.I. Peterson, A.W. Castleman Jr., J. Chem. Phys. **80**, 563–564 (1984)
41. G.L. Gutsev, B.K. Rao, P. Jena, J. Phys. Chem. A **104**, 11961–11971 (2000)
42. S. Klacar, A. Hellman, I. Panas, H. Grönbeck, J. Phys. Chem. C **114**, 12610–12617 (2010)
43. R. Pal, L.-M. Wang, Y. Pei, L.-S. Wang, X.C. Zeng, J. Am. Chem. Soc. **134**, 9438–9445 (2012)
44. M. Hayyan, M.A. Hashim, I.M. AlNashef, Chem. Rev. **116**, 3029–3085 (2016)
45. M.-Y. Jia, Z. Luo, S.-G. He, M.-F. Ge, J. Phys. Chem. A **118**, 8163–8169 (2014)
46. M.F. Jarrold, J.E. Bower, Chem. Phys. Lett. **144**, 311–316 (1988)
47. R.L. Hettich, J. Am. Chem. Soc. **111**, 8582–8588 (1989)
48. S.E. Deutsch, J.T. Miller, K. Tomishige, Y. Iwasawa, W.A. Weber, B.C. Gates, J. Phys. Chem. **100**, 13408–13415 (1996)
49. C. Ashman, S.N. Khanna, M.R. Pederson, Chem. Phys. Lett. **324**, 137–142 (2000)
50. M. Bienati, V. Bonacic-Koutecky, P. Fantucci, J. Phys. Chem. A **104**, 6983–6992 (2000)
51. A. Pramann, Y. Nakamura, A. Nakajima, K. Kaya, J. Phys. Chem. A **105**, 7534–7540 (2001)
52. A.W. Castleman Jr., Abstr. Pap. Am. Chem. Soc. **226**, U284–U285 (2003)
53. J. Hagen, L.D. Socaciu, J. Le Roux, D. Popolan, T.M. Bernhardt, L. Wöste, R. Mitrić, H. Noack, V. Bonačić-Koutecký, J. Am. Chem. Soc. **126**, 3442–3443 (2004)
54. A.W. Castleman Jr., M.L. Kimble, N.A. Moore, Abstr. Pap. Am. Chem. Soc. **230**, U2845–U2845 (2005)
55. W. Wang, R. Clark, A. Nakano, R.K. Kalia, P. Vashishta, Appl. Phys. Lett. **95**, 261901 (2009)
56. L.-H. Tian, Y.-X. Zhao, X.-N. Wu, X.-L. Ding, S.-G. He, T.-M. Ma, ChemPhysChem **13**, 1282–1288 (2012)
57. O. Gehret, M.P. Irion, Chem. Eur. J. **2**, 598–603 (1996)
58. E.R. Fisher, J.L. Elkind, D.E. Clemmer, R. Georgiadis, S.K. Loh, N. Aristov, L.S. Sunderlin, P.B. Armentrout, J. Chem. Phys. **93**, 2676–2691 (1990)
59. D. Stolcic, M. Fischer, G. Ganteför, Y.D. Kim, Q. Sun, P. Jena, J. Am. Chem. Soc. **125**, 2848–2849 (2003)
60. Y.D. Kim, M. Fischer, G. Ganteför, Chem. Phys. Lett. **377**, 170–176 (2003)
61. L.D. Socaciu, J. Hagen, J. Le Roux, D. Popolan, T.M. Bernhardt, L. Woste, S. Vajda, J. Chem. Phys. **120**, 2078–2081 (2004)
62. Z. Luo, A.W. Castleman Jr., S.N. Khanna, Chem. Rev. **116**, 14456–14492 (2016)
63. Z. Luo, A.C. Reber, M. Jia, W.H. Blades, S.N. Khanna, A.W. Castleman, Chem. Sci. **7**, 3067–3074 (2016)
64. F.A. Cotton, G. Wilkinson, in *Advanced Inorganic Chemistry*, 4th edn. (Wiley, New York, 1980), p. 709.
65. R.C. Bell, K.A. Zemski, K.P. Kerns, H.T. Deng, A.W. Castleman Jr., J. Phys. Chem. A **102**, 1733–1742 (1998)
66. M. Foltin, G.J. Stueber, E.R. Bernstein, J. Chem. Phys. **111**, 9577–9586 (1999)
67. R.C. Bell, K.A. Zemski, D.R. Justes, A.W. Castleman Jr., J. Chem. Phys. **114**, 798–811 (2001)
68. Y. Matsuda, E.R. Bernstein, J. Phys. Chem. A **109**, 3803–3811 (2005)
69. H. Zhang, H. Wu, Y. Jia, L. Geng, Z. Luo, H. Fu, J. Yao, Rev. Sci. Instrum. **90**, 073101 (2019)
70. H. Zhang, H. Wu, L. Geng, Y. Jia, M. Yang, Z. Luo, Phys. Chem. Chem. Phys. **21**, 11234–11241 (2019)
71. J. Xu, M.T. Rodgers, J.B. Griffin, P.B. Armentrout, J. Chem. Phys. **108**, 9339–9350 (1998)
72. M. Engeser, T. Weiske, D. Schroder, H. Schwarz, J. Phys. Chem. A **107**, 2855–2859 (2003)
73. X. Wu, A.K. Ray, J. Chem. Phys. **110**, 2437–2445 (1999)
74. P. Jena, Q. Sun, *Chem. Rev. (Washington, DC, USA)*, **118**, 5755–5870 (2018)

Chapter 4
Halogenation of Metal Clusters

It has been proposed a unified view of principles could be avaliable to determine the stability of ligand-protected metal clusters via wet chemistry and halogen-passivated metal cluster in gas phase [1–3]. Typicl examples not only include $Au_{102}(SR)_{44}$, $Au_{39}(PR_3)_{14}X_6$, $Au_{11}(PR_3)_7X_3$ and $Au_{13}(PR_3)_{10}X_2$ where X is either a halogen or a thiolate [1], but also shed light on the notable $Al_{13}I_{2n}$ and $Al_{14}I_y$ series where the presence of iodine at balanced sites give rise to enhanced stability of the aluminum iodide clusters [3–5]. These clusters generally have a filled spherical electronic shell, a compact/symmetric core, balanced charge distribution, and complete steric protection, along with a relatively large HOMO–LUMO energy gap, pertaining to superatomic metal cluster characteristic [2].

Similar to halogenation of organic compounds, the halogenation in cluster reactions also embodies several pathways rendering the addition of one or more halogen atoms to a cluster. The stoichiometry of halogenation in cluster reactivity depends on both electronic and geometric structural features of the cluster, as well as of the specific halogen-donor reactant. Halogenation reactions are important in chemical synthesis processes, and halide products are useful intermediates serving as branching points in the synthesis of numerous functionalized materials [6–13].

In this chapter, we illustrate how metal clusters and even metal cluster complexes undergo halogenation with inorganic halogen reactants. It is worth noting that, a novel 13-atoms aluminum cluster has been recognized as superhalogen. In this regard, halogenation of such superatom clusters will fuel more research interest in this area. It has been recognized that the utilization of superstorms in materials synthesis contributes to a new generation of nanostructured materials [14–30].

Z. Luo and S. N. Khanna, *Metal Clusters and Their Reactivity*,
https://doi.org/10.1007/978-981-15-9704-6_4

4.1 Reaction with Alkyl Halide

Several halogenation pathways have been addressed in metal cluster reactivity, analogous to those demonstrated in organic synthesis, including free radical halogenation [31–40], electrophilic halogenation [41–43], and halogen addition reaction [44–47], etc. An early investigation regarding clusters by Bergeron and Castleman [48] found that the aluminum cluster anions readily react with methyl iodide vapor in a fast-flow tube apparatus, yielding primarily I^- and clusters series of Al_nI^-, as shown in Fig. 4.1. With thermodynamic considerations, it was found that abundant I^- ions were produced associated with a neutralization reaction to form Al_nCH_3 clusters, read as,

$$Al_n^- + CH_3I \leftrightarrow Al_nCH_3I^- \quad (4.1)$$

$$Al_nCH_3I^- \rightarrow Al_nCH_3 + I^- \quad (4.2)$$

$$Al_nCH_3I^- \rightarrow Al_nI^- + CH_3 \quad (4.3)$$

$$Al_nCH_3I^- \rightarrow Al_nCH_3^- + I \quad (4.4)$$

It is noteworthy that, unavailable dissociation reactions have been ascertained, in particular, the reaction "$Al_{13}^- + CH_3I \rightarrow Al_{13} + CH_3 + I^-$" requires approximately ~140 kJ/mol energy of interaction between the cluster and the methyl group; even the cluster of smallest electron-affinity, Al_2^-, requires ~45 kJ/mol of interaction energy to make such a reaction favourable as "$Al_2^- + CH_3I \rightarrow Al_2 + CH_3 + I^-$". This is also consistent with the conclusion in subsequent investigations by Kim et al. [49].

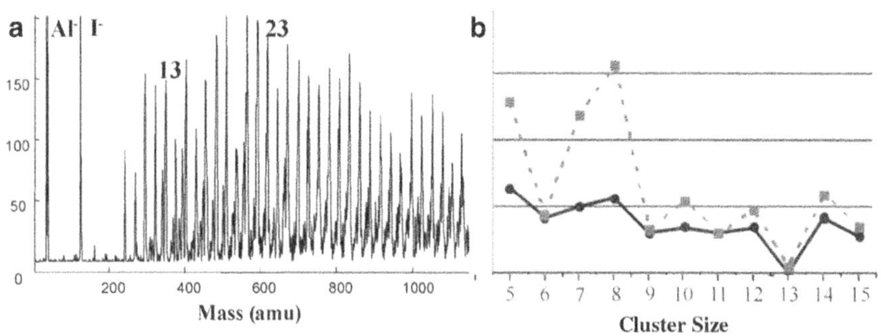

Fig. 4.1 **a** Mass spectra of Al_n^- clusters reacting with 100 sccm helium-diluted MeI. The I^- peak grows rapidly upon introduction of MeI, and Al_nI^- peaks can be seen between Al_n^- series; **b** Comparison of relative rates of disappearance for Al_n^- (dash line linked dots) to relative inverse vertical detachment energies (solid line linked dots) as measured by Cha et al. [50]. Identical trends are interpreted as evidence of the formation of covalently bonded Al_nCH_3 species

4.2 Metal Clusters Reacting with HX

Utilizing flow-tube reaction apparatus, the halogenation of aluminum clusters have been also studied using HX as reactants [4]. Figure 4.2 displays an interesting result for aluminum clusters reacting with HI, where the reaction products demonstrated a key mechanism involving acid etching and I^- addition. Similar reactions with HCl and HBr have also been investigated, and consequently similar acid-etching pathways were found, where Al_nX^- generation was found to be energetically favorable [51]. Although HCl and HBr are less reactive than HI towards Al_n^-, similar trends in reactivity were addressed. Also found was that, the lowest energy structure for $Al_{13}I^-$ (also $Al_{13}Br^-$ and $Al_{13}Cl^-$) was found to feature icosahedral Al_{13} units with the halogen atom located at an on-top site, indicating that the halogen incorporation into Al_{13} could leave the original icosahedral Al_{13} unit and electronic property unperturbed. The charge density of the highest occupied molecular orbital in these $Al_{13}X^-$ clusters is dependent on the identity of X.

Further, tandem reaction experiments allowed to explore the stability of halogenation products by checking their resistance to oxygen etching. As results, enhanced stability of $Al_{13}I^-$ and $Al_{13}I_2^-$ has been unambiguously confirmed, arguing that super-halogen behavior of Al_{13} in these clusters. A simplest explanation was proposed for the occurrence of the magic $Al_{13}I^-$ from HI through an acid-etching reaction pathway,

$$Al_n^- + HX \rightarrow Al_{n-2}^- + AlH + AlX \qquad (4.5)$$

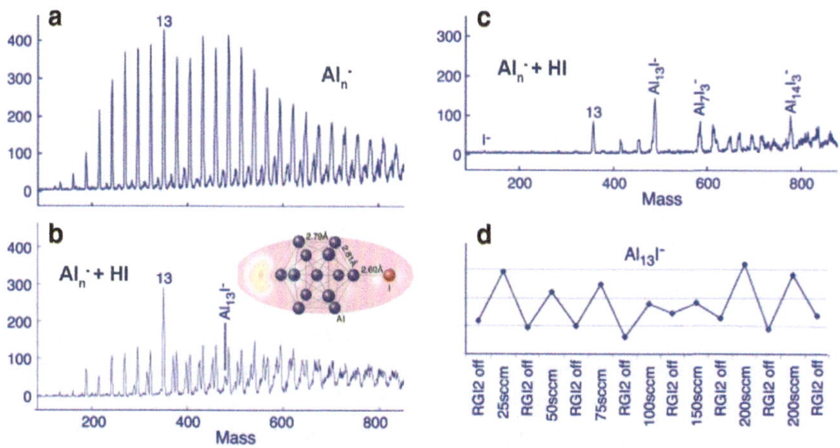

Fig. 4.2 **a–c** Mass spectra showing the reaction of aluminium clusters with HI: 0 sccm (**a**), 25 sccm (**b**), and 200 sccm (**c**) of 10% HI seeded in He. Inset showing the lowest energy structure for $Al_{13}I^-$. (**d**) Growth of $Al_{13}I^-$ peak in the presence of oxygen demonstrates the cluster's stability. In all panels, the y axis is peak intensity (in arbitrary units)

$$Al_n^- + HX \rightarrow Al_{n-1}X^- + AlH \tag{4.6}$$

These two equations explained how the etching effect proceeds within the reactions of "$HX + Al_n^- \rightarrow$ " in forming Al_nI^- species. For a certain Al cluster, these two reaction pathways could refer to successive and circulatory reactions. For closed-shell clusters such as Al_{13}^-, there could be significantly populated mass abundance due to both initial population and the subsequent adsorption and dissociation of HI. The eenergy diagram of a complete set of reaction steps for the Al_{13}^- has led to a better understanding on the reactive mechanism, demonstrated as,

$$Al_{13}^- + HX \leftrightarrow Al_{13}HX^- \rightarrow \tag{4.7}$$

$$Al_{13}HX^- + HX \rightarrow Al_{13}X_2^- + H_2 \tag{4.8}$$

$$Al_{13}HX^- + HX \rightarrow Al_{13}X^- + H_2 + X \tag{4.9}$$

Here the reactions are favoured due to the weakly bonded $Al_{13}HX^-$ system which will be unstable and allow for further reactions with HI. This is the essential difference for the Al_n^- reacting with HI compared with CH_3I as mentioned in the above section. It is notable that Eq. (4.8) (forming $Al_{13}I_2^-$) is more energetically favourable due to the extreme exothermicity of H_2 generation [4, 51].

4.3 Reactivity and Stability of $Al_nI_x^-$ Clusters

4.3.1 Selective $Al_nI_x^-$ Surviving Oxygen Etching

As having demonstrated in the above investigation of Al_n^- reacting with CH_3I and HX, several $Al_nI_m^-$ clusters have been found to exhibit reasonable stability. In order to examine selectively stable and unstable species in $Al_nI_m^-$ clusters, researchers have given an further insight by conducting gas-phase reactions of $Al_nI_m^-$ clusters with oxygen [51], which is known as an effective method to explore magic cluster species. Two classes of gas-phase $Al_nI_m^-$ clusters (i.e., $Al_{13}I_x^-$ and $Al_{14}I_y^-$) were identified with magic stability selectively showing resistance to oxygen-etching. In specific, the $Al_{13}I_x^-$ clusters exhibit pronounced stability for those of even-numbered iodine atoms; while interestingly, $Al_{14}I_x^-$ series exhibit enhanced stability for those with odd numbers of I atoms, as shown in Fig. 4.3.

First-principle calculations suggested that the reaction "$Al_{13}^- + I_2 \rightarrow Al_{13}I_2^-$" is energetically favorable by 3.63 eV; in comparison, the reaction "$Al_{14}^- + I_2 \rightarrow Al_{13}I^- + AlI$" was also described to be energetically favorable by 2.25 eV, which indicates an example of I_2 etching towards Al_n^- clusters. Also the reaction "$Al_{13}^- + I_2 \rightarrow Al_{13}I^- + I$" is energetically allowed by 0.31 eV [51], and there are other

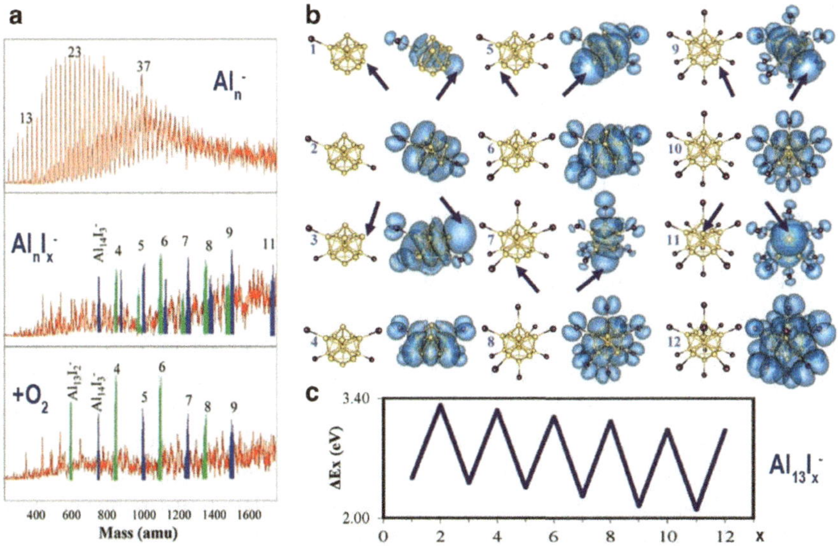

Fig. 4.3 **a** Mass spectra of Al cluster anions reacted with I_2 vapor and then etched by O_2. Peaks shaded green fall into the $Al_{13}I_x^-$ family, whereas peaks shaded blue fall into the $Al_{14}I_x^-$ family. In all panels, the y axis is peak intensity (in arbitrary units). **b** Lowest energy structures and charge maps for $Al_{13}I_x^-$ (x values from 1 to 12). The areas of high charge density, or active sites, are indicated by arrows. **c** ΔE_x, the energy to remove one I atom from $Al_{13}I_x^-$ for x values from 1 to 12

etching pathways for the $Al_nI_x^-$ to react with I_2, such as,

$$Al_nI_x^- + I_2 \rightarrow Al_{n-1}I_{x+1}^- + AlI \tag{4.10}$$

$$Al_nI_x^- + I_2 \rightarrow Al_{n-2}I_x^- + 2AlI \tag{4.11}$$

$$Al_nI_x^- + I_2 \rightarrow Al_{n-1}I_{x-1}^- + AlI_3 \tag{4.12}$$

Theoretical investigations revealed that such reactivity can be understood in terms of the spherical shell jellium model, and the enhanced stability of $Al_{13}I_x^-$ is largely associated with complementary pairs of I atoms occupying the on-top sites on the opposing Al atoms of the Al_{13}^- core. Analysis of the HOMO charge density reveals that mostly the Millikan charge distribution of the additional electron is localized at the Al_{13} moiety, suggesting superhalogen and poly-halide character [4, 5].

4.3.2 $Al_nI_x^-$ Reacting with Methyl Iodide

Al_n^- clusters readily react with iodine to generate a cluster distribution in which $Al_{13}I_x^-$ and $Al_{14}I_x^-$ constitute the major peaks (Fig. 4.4A-a); however, at the presence of methyl iodide, a drastic change in the mass spectrum was observed, as shown in Fig. 4.4A-b, where kinetically-mediated etching reactions are prominent. With reasonable relevance to the reactions between bare aluminum clusters and methyl iodide [48], it is interesting and important to find that, among the $Al_{13}I_x^-$ series, clusters with odd values of x were found to be reactive while those with even x were rather stable [5, 52]. However, there are also sharp differences, for example, the emergence of Al_7I^- as the dominant product in this reaction.

There is a similar case. Considering that the laser for a LaVa source is enough to create proper plasma around the target material and can be used to dissociate iodine gas which is concomitantly introduced through the source [53], it is possible to create a gaseous medium-containing clusters of varying amounts of different elements [22, 54]. Attempts have been made to increase the population of I-rich species by employing I_2 sublimation vessel subjected to proper heating and He buffer-gas flow. As a result, the mass distribution, as shown in Fig. 4.4B, finds coexistence of $Al_{13}I_x^-$ clusters with I-rich $Al_nI_x^-$ species, such as AlI_4^-, $Al_2I_5^-$, $Al_3I_7^-$, $Al_3I_8^-$ and $Al_4I_9^-$. In addition to the observation of I^- and I_3^-, prominent products were found to survive the etching reaction with methyl iodide including polyhalide-like $Al_{13}I_{2x}^-$ cluster series [52].

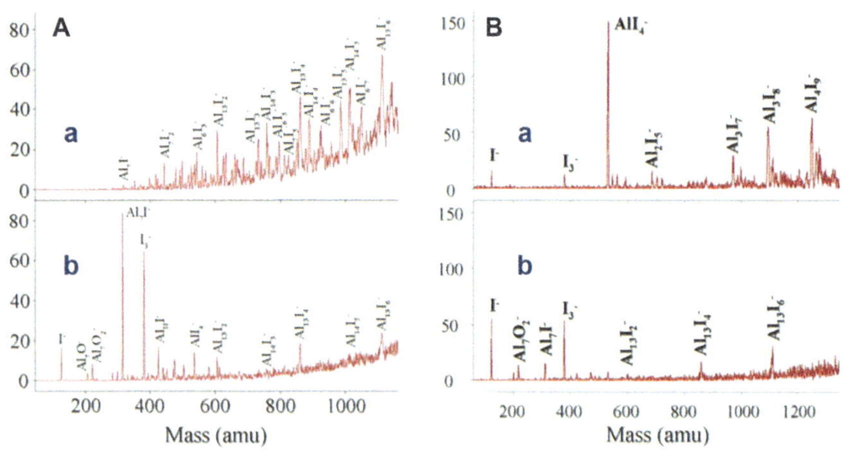

Fig. 4.4 **A** $Al_nI_x^-$ clusters generated via reaction of Al_n^- with I_2 (a) and reacted with methyl iodide (b). Both spectra are shown on identical scales without normalization. Unlabeled peaks correspond to clusters with atomic oxygen or CH_3 present. **B** Mass spectrum of I-rich $Al_nI_x^-$ clusters generated by increased heating of the I_2 vessel (a) and reacted with methyl iodide (b)

4.3.3 $Al_nI_x^-$ Reacting with Methanol

Further insights into the reactivity of $Al_nI_x^-$ clusters have been performed by comprehensive theoretical and experimental investigations [3]. Two series of stable clusters $Al_{13}I_n^-$ and $Al_{14}I_m^-$ have been produced and allowed to react with methanol (Fig. 4.5). As results, clean etching spectra were noted, and it was interestingly noted that $Al_{14}I_3^-$ is reactive but all the $Al_{13}I_m^-$ clusters ($m = 0–3$) and $Al_7I_2^-$ were found to be unreactive with methanol. On the other hand, it was found that, the lowest energy reaction pathways with methanol for the ground states of Al_{13}^-, $Al_{13}I^-$, and $Al_{13}I_2^-$ are slow with necessary activation energy at 0.25, 0.23 and 0.2 eV respectively; whereas, Al_{14}^-, $Al_{14}I^-$, and $Al_{14}I_3^-$ are reactive without necessary activation energy, as shown in Fig. 4.6.

Calculation results also demonstrated that an $Al_{13}I_2^-$ isomer with two adjacent ligands on the closed geometric shell of Al_{13}^- could be highly reactive in the gas phase, with a binding energy of 0.67 eV (which is more than twice that of the binding energy of $Al_{13}I_2^-$ in the ground state) and a transition energy of 0.42 eV indicating that this higher energy isomer is a strong Lewis acid site [55]. It is notable that $Al_{14}I_3^-$ reacts with methanol at the adatom site despite it has a closed electronic shell and is resistant to oxygen etching. This is because the oxygen atom of methanol donates its electron pair to the adatom and then the hydrogen bonds to an adjacent aluminum atom which is marked by an occupied orbital density. Note that molecular oxygen does not donate its electron pair and hence a survival of $Al_{14}I_3^-$ to oxygen etching due to the closed geometric shell [5]. These findings not only explain the cluster stability and reactivity of $Al_nI_m^-$ cluster upon a joint electronic and geometric consideration, but also provide insights into the origin of stability and mechanism that prevents the ligand-stabilized metal clusters from dissociation or etching-like reactions.

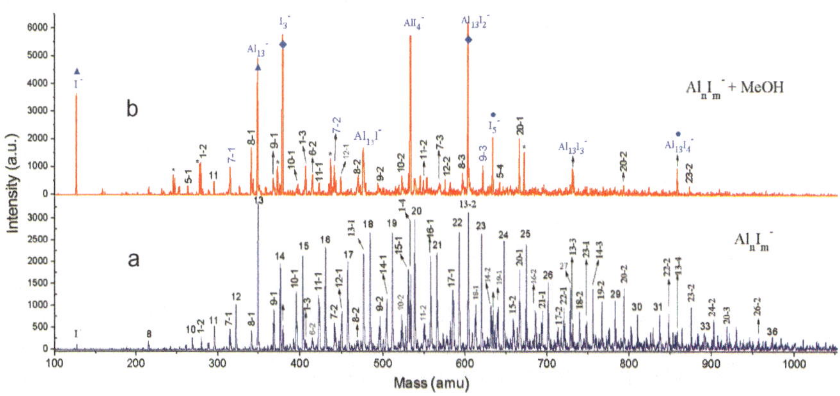

Fig. 4.5 The reactivity of $Al_nI_m^-$ clusters with MeOH. **a** The nascent mass spectrum of $Al_nI_m^-$ clusters. **b** The spectrum of $Al_nI_m^-$ clusters after methanol etching. A few peaks marked with * refer to $Al_n(CH_3OH)_m^-$; while the peaks marked with ▲,◆,• display the I^-, Al_{13}^-; I_3^-, $Al_{13}I_2^-$; I_5^-, and $Al_{13}I_4^-$ species respectively. a.u. = arbitrary units

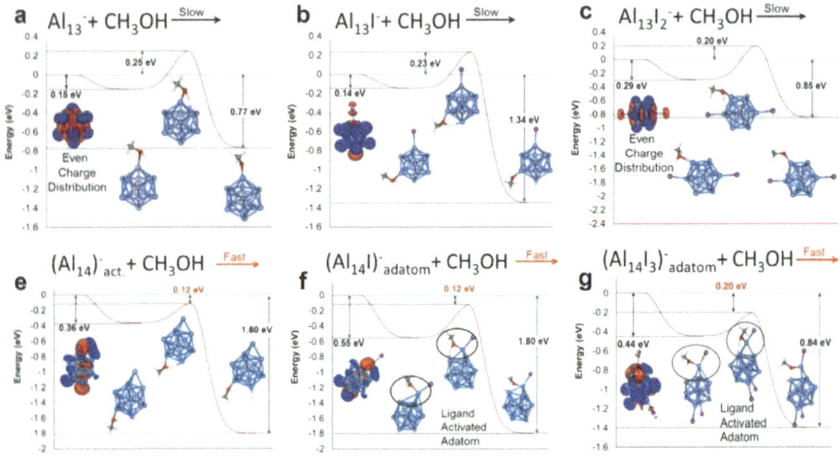

Fig. 4.6 (Upper) Reaction pathways of Al_{13}^-, $Al_{13}I^-$ and $Al_{13}I_2^-$ with methanol. The lowest energy reaction pathways with methanol for the ground states of (**a**) Al_{13}^-, (**b**) $Al_{13}I^-$, and (**c**) $Al_{13}I_2^-$. (Below) Reaction pathways of Al_{14}^-, $Al_{14}I^-$ and $Al_{14}I_3^-$ clusters with methanol. The lowest energy reaction pathways with methanol for the ground states of (**d**) Al_{14}^-, (**e**) $Al_{14}I^-$, and (**f**) $Al_{14}I_3^-$. Al atoms are shown in light blue, I atoms in purple. HOMOs are red and LUMOs blue

4.4 Ionic Crystal Growth of $Cu_nCl_{n+1}^-$

Within the middle school chemistry textbook, there is an interesting experiment that, copper scraps burn in chlorine and result in copper chloride by noting the final products in water to display blue solution. But in lab experiments, it is known that chlorine is slightly more selective in reacting with most metals and heavier nonmetals. In 1986, the surface chlorination behavior of copper has been investigated using X-ray photoemission technique [56]. It was found that, a surface layer of $CuCl_x$ could be formed at Cl_2 exposure, with a likely varying value of x (0–2) as a function of the gas pressure and the exposure time. Regarding this topic, a study towards the reactivity of copper clusters with chlorine has unveiled the intrinsic nature [57]. Figure 4.7 shows the mass spectrum of Cu_n^- clusters, produced via the MagS-source, in the absence and presence of chlorine. The inset images show the different isotope combinations of copper, as well as a sketch showing the building blocks $Cu_nCl_{n+1}^-$ in growing ionic crystals. It is notable that when chlorine was introduced into the fast-flow reactor, all the original Cu_n^- clusters disappeared due to the violent gas-phase reactivity between chlorine and copper cluster anions except a weak intensity of Cu_7^- which is known to be a stable species with a closed shell of eight electrons according to the jellium model [15, 24, 58]. Among the observed $Cu_nCl_{n+1}^-$ species, $CuCl_2^-$ dominates the products, and the following $Cu_2Cl_3^-$, $Cu_3Cl_4^-$, $Cu_4Cl_5^-$, $Cu_5Cl_6^-$, and $Cu_6Cl_7^-$ in the mass spectra exhibit an exponential decay with increasing values of n with independence of the quantities of chlorine being introduced. With an examination on the integral intensity for the $Cu_nCl_{n+1}^-$ species, a fitting curve based on exponential function was

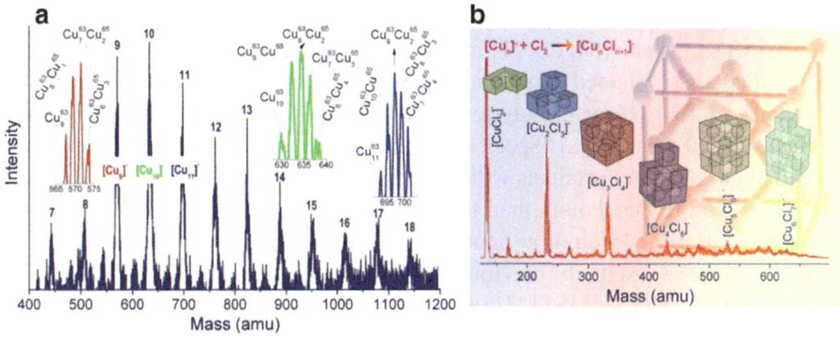

Fig. 4.7 A mass spectrum of Cu cluster anions obtained by MagS source (**a**), and its reaction with chlorine, leading to the $[Cu_nCl_{n+1}]^-$ products (**b**), where the insets showed the structural sketch in forming the crystal

concluded, which coincides with the reaction kinetics through the proposed reaction channels, as shown in Fig. 4.8.

In addition to the mass spectrometric observation, it is noteworthy that alkali halide clusters have also been extensively studied and provided a better understanding on the physical basis of ionic crystal growth [59–62]. As the group IB coinage metals (i.e., Cu, Ag, Au) have electronic configurations characterized by a closed d-shell and a single s-valance electron, it is believed that the coinage metal clusters may react in a similar fashion to alkali metals [63–68]. Extensive investigations both

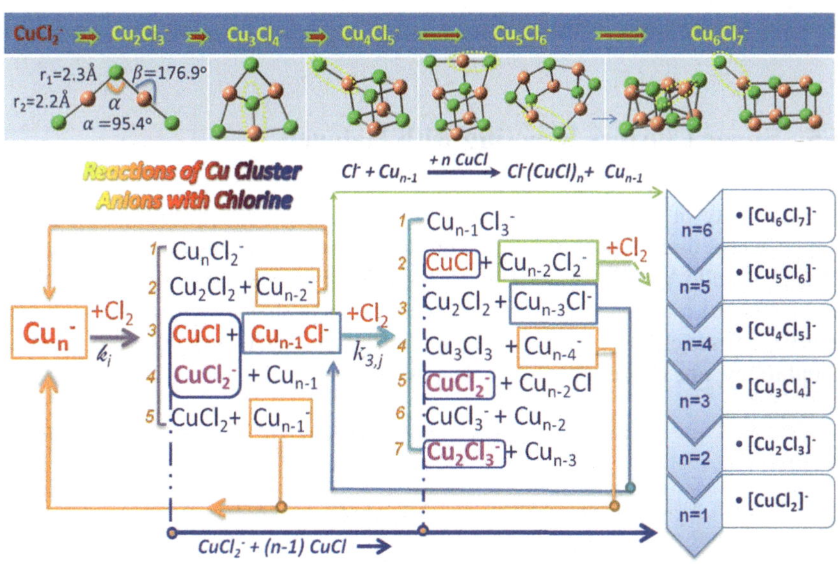

Fig. 4.8 The proposed reaction channels of $[Cu_n]^-$ clusters with chlorine

by experimental and theoretical have been done on the structures and properties of coinage metal clusters; [64, 68–71] in particular, their mass abundance spectra have been studied and explained by the one-electron shell model [72, 73].

The $Cu_nCl_{n+1}^-$ species closely resemble the alkali halide cluster cations where the intensities of the $M_nX_{n-1}^+$ species (such as $Na_nCl_{n-1}^+$ and $Cs_nI_{n-1}^+$) were found to be greater than other products with different stoichiometries [59, 60, 62, 74–78]. There is one more metal atom than the halogen in the alkali halide cluster cations, whereas the number of chlorine atoms is one more than the copper atoms in the $Cu_nCl_{n+1}^-$ species. Actually previous investigations using electrospray ionization mass spectrometry (ESI–MS) [43] on alkali halide clusters have observed both the $M_nX_{n-1}^+$ and $M_nX_{n+1}^-$ showing unusually high intensities [60, 61]. The existence of $Cu_nCl_{n+1}^-$ species, similar to $M^+(MX)_n$, reveals the 'F-center' localization in bulk alkali halide crystals [79–81], but there is reasonable difference because the interaction of excess electrons plays an important role in the outcome of the observed halide clusters [81–85]. There is no observation of large values of n (>6) in the $Cu_nCl_{n+1}^-$ species due to vigorous reactions and multiple collisions in the fast-flow reaction apparatus [86] The optimized chemical structures of the $Cu_nCl_{n+1}^-$ species (Fig. 4.8) demonstrate reasonable stability with a cubic structure arrangement, especially $Cu_4Cl_5^-$, which mimics previously reported result on structures and stabilities of $Na_nCl_{n-1}^+$ and $Cs_nI_{n-1}^+$ [59]. Abundant studies of mass spectrometric results and ultraviolet absorption analysis on charged alkali halide clusters have shown that the alkali metal halide clusters have a strong tendency to assume cubic nanocrystal arrangements that resemble portions of bulk simple cubic, rock salt lattices [59, 87–92]. Nevertheless, it is worth mentioning that an alkali halide cluster does not necessarily have the NaCl crystal structure during the initial stages of growth, for instance, it can be ring growth [93].

4.5 Silver Clusters Reacting with Halogen

The reactivity of silver cluster anions with chlorine has also been clearly demonstrated, where three classes of reaction products are observed, including $[Ag_nCl_{n+1}]^-$, $[Ag_nCl_2]^-$ and $[Ag_nCl]^-$. Among them, $[Ag_nCl_{n+1}]^-$ species were observed only in the small mass range ($n \leq 4$), likely due to reactions in analogy with aforementioned $[Cu_nCl_{n+1}]^-$ series. When a $[Ag_n]^-$ cluster reacts with a Cl_2 molecule, the first-step could follow one of the following channels:

$$[Ag_n]^- + Cl_2 \rightarrow Ag_{n-1} + [AgCl_2]^- \rightarrow \quad (4.13)$$

$$[Ag_n]^- + Cl_2 \rightarrow [Ag_nCl_2]^- \rightarrow \quad (4.14)$$

$$[Ag_n]^- + Cl_2 \rightarrow AgCl + [Ag_{n-1}Cl]^- \rightarrow \quad (4.15)$$

where Eq. (4.13) is responsible for the formation of $[AgCl_2]^-$, while Eqs. (4.14) and (4.15) indicate likely pathways in forming $[Ag_nCl_2]^-$ and $[Ag_nCl]^-$; also an additional arrow in each equation indicates possible successive reactions. Note that the $[Ag_nCl]^-$ and $[Ag_nCl_2]^-$ species appearing in the larger mass range ($8 \leq n \leq 14$) display an odd–even alternation. This agrees with previous theoretical findings that the calculated incremental binding energies, spin excitation energies, and HOMO–LUMO gaps of $[Ag_n]^-$ clusters all display an even/odd oscillation, which corresponds to their even/odd selective reactivity [94]. As found, the intensity ratio of $[Ag_8Cl]^-$ to $[Ag_8]^-$ is larger than that of the other observed $[Ag_nCl]^-$ clusters and their correlated $[Ag_n]^-$ product clusters when $8 \leq n \leq 14$. In view of the atomic electron configurations, Ag:[Kr]$4d^{10}5s^1$, the Ag_8^- cluster exhibits 9 valence electrons and the delocalized nearly free electron gas (NFEG) orbitals are described as $|1S^2 \mid 1P^2 \mid 1P^4 \mid 2S^1|$. Therefore, it is expected that Ag_8^- will behave similar to an alkali-metal atom and hence their reactivity towards chlorine is expected to follow the "harpoon model" [95, 96] with products of "$Ag_8Cl^- + Cl$". The well-known example of harpoon mechanism was proposed to explain the reaction of K atoms with Br_2 where the K atom plucks a Br atom out of the Br_2 molecule. An electron leaps from the metal atom (i.e., a harpoon) to the halogen, resulting in a coulomb attraction between the metal and halogen, and hence the cross section is extended for their reactive encounter [96–101]. This will be discussed in Chap. 11 [102].

References

1. M. Walter, J. Akola, O. Lopez-Acevedo, P.D. Jadzinsky, G. Calero, C.J. Ackerson, R.L. Whetten, H. Grönbeck, H. Häkkinen, Proc. Natl. Acad. Sci. USA **105**, 9157–9162 (2008)
2. Z. Luo, A.W. Castleman, Acc. Chem. Res. **47**, 2931–2940 (2014)
3. Z. Luo, A.C. Reber, M. Jia, W.H. Blades, S.N. Khanna, A.W. Castleman, Chem. Sci. **7**, 3067–3074 (2016)
4. D.E. Bergeron, A.W. Castleman Jr., T. Morisato, S.N. Khanna, Science **304**, 84–87 (2004)
5. D.E. Bergeron, P.J. Roach, A.W. Castleman Jr., N. Jones, S.N. Khanna, Science **307**, 231–235 (2005)
6. S. Kamiguchi, S. Nagashima, K.I. Komori, M. Kodomari, T. Chihara, J. Cluster Sci. **18**, 414–430 (2007)
7. Y.J. Ko, H. Wang, K. Pradhan, P. Koirala, A.K. Kandalam, K.H. Bowen, P. Jena, J. Chem. Phys. **135**, 244312 (2011)
8. F. Poineau, C.D. Malliakas, P.F. Weck, B.L. Scott, E.V. Johnstone, P.M. Forster, E. Kim, M.G. Kanatzidis, K.R. Czerwinski, A.P. Sattelberger, J. Am. Chem. Soc. **133**, 8814–8817 (2011)
9. C. Paduani, P. Jena, J. Nanopart. Res. **14**, 1035 (2012)
10. B. Xu, R. Wang, X. Wang, Nanoscale **4**, 2713–2719 (2012)
11. J.A.B. Abdalla, I.M. Riddlestone, R. Tirfoin, N. Phillips, J.I. Bates, S. Aldridge, Chem. Commun. **49**, 5547–5549 (2013)
12. A. Desireddy, B.E. Conn, J. Guo, B. Yoon, R.N. Barnett, B.M. Monahan, K. Kirschbaum, W.P. Griffith, R.L. Whetten, U. Landman, T.P. Bigioni, Nature **501**, 399–402 (2013)
13. Y.W. Li, S.H. Zhang, Q. Wang, P. Jena, J. Chem. Phys. **138**, 054309 (2013)
14. W.D. Knight, K. Clemenger, W.A. Deheer, W.A. Saunders, M.Y. Chou, M.L. Cohen, Phys. Rev. Lett. **52**, 2141–2143 (1984)
15. R.E. Leuchtner, A.C. Harms, A.W. Castleman Jr., J. Chem. Phys. **91**, 2753–2754 (1989)

16. R.W. Farley, A.W. Castleman Jr., J. Chem. Phys. **92**, 1790–1795 (1990)
17. R.E. Leuchtner, A.C. Harms, A.W. Castleman Jr., J. Chem. Phys. **94**, 1093–1101 (1991)
18. S.N. Khanna, P. Jena, Phys. Rev. B **51**, 13705–13716 (1995)
19. C. Ashman, S.N. Khanna, F. Liu, P. Jena, T. Kaplan, M. Mostoller, Phys. Rev. B **55**, 15868–15873 (1997)
20. S. Burkart, N. Blessing, B. Klipp, J. Muller, G. Gantefor, G. Seifert, Chem. Phys. Lett. **301**, 546–550 (1999)
21. B.K. Rao, S.N. Khanna, P. Jena, J. Cluster Sci. **10**, 477–491 (1999)
22. B.D. Leskiw, A.W. Castleman Jr., Chem. Phys. Lett. **316**, 31–36 (2000)
23. H. Kawamata, Y. Negishi, A. Nakajima, K. Kaya, Chem. Phys. Lett. **337**, 255–262 (2001)
24. B.D. Leskiw, A.W. Castleman Jr., C. Ashman, S.N. Khanna, J. Chem. Phys. **114**, 1165–1169 (2001)
25. A. Pramann, A. Nakajima, K. Kaya, J. Chem. Phys. **115**, 5404–5410 (2001)
26. O.C. Thomas, W.J. Zheng, K.H. Bowen, J. Chem. Phys. **114**, 5514–5519 (2001)
27. O.C. Thomas, W.J. Zheng, T.P. Lippa, S.J. Xu, S.A. Lyapustina, K.H. Bowen, J. Chem. Phys. **114**, 9895–9900 (2001)
28. J.A. Alonso, M.J. Lopez, L.M. Molina, F. Duque, A. Mananes, Nanotechnology **13**, 253–257 (2002)
29. O.P. Charkin, D.O. Charkin, N.M. Klimenko, A.M. Mebel, Chem. Phys. Lett. **365**, 494–504 (2002)
30. J.F. Jia, J.Z. Wang, X. Liu, Q.K. Xue, Z.Q. Li, Y. Kawazoe, S.B. Zhang, Appl. Phys. Lett. **80**, 3186–3188 (2002)
31. F. Minisci, F. Fontana, S. Araneo, F. Recupero and L. H. Zhao, *Synlett*, 1996, 119–125.
32. S.M. Hubig, W. Jung, J.K. Kochi, J. Org. Chem. **59**, 6233–6244 (1994)
33. I. Tabushi, Z. Yoshida, Y. Aoyama, J. Hamuro, S. Kojo, J. Am. Chem. Soc. **94**, 1177–1183 (1972)
34. P. Fredricks, J.M. Tedder, J. Chem. Soc. 3520–3525 (1961)
35. C. Walling, B.B. Jacknow, J. Am. Chem. Soc. **82**, 6108–6112 (1960)
36. P.S. Fredricks, J.M. Tedder, J. Chem. Soc. 144–150 (1960)
37. P.C. Anson, P.S. Fredricks, J.M. Tedder, J. Chem. Soc. 918–922 (1959)
38. Y. Wang, G.-X. Li, G. Yang, G. He, G. Chen, Chem. Sci. **7**, 2679–2683 (2016)
39. K. Moriyama, M. Kuramochi, K. Fujii, T. Morita, H. Togo, Angew. Chem. Int. Ed. **55**, 14546–14551 (2016)
40. Y.-J. Han, C.-Y. Lin, M. Liang, Y.-L. Liu, Macromol. Rapid Commun. **37**, 845–850 (2016)
41. A. Vigalok, Acc. Chem. Res. **48**, 238–247 (2015)
42. H. Hart, J.L. Reilly, J.B.C. Jiang, J. Org. Chem. **42**, 2684–2689 (1977)
43. H. Gershon, M.W. McNeil, S.G. Schulman, J. Org. Chem. **37**, 4078–4082 (1972)
44. Z.Y. Yang, J. Org. Chem. **68**, 5419–5421 (2003)
45. J. Adam, P.A. Gosselain, P. Goldfinger, Nature **171**, 704–705 (1953)
46. P.B.D. Delamare, P.W. Robertson, J. Chem. Soc. 2838–2842 (1950)
47. J.B. Conn, G.B. Kistiakowsky, E.A. Smith, J. Am. Chem. Soc. **60**, 2764–2771 (1938)
48. D.E. Bergeron, A.W. Castleman Jr., Chem. Phys. Lett. **371**, 189–193 (2003)
49. J.C. Kim, K.H. Kim, J. Jung, Y.-K. Han, J. Comput. Chem. **29**, 1626–1631 (2008)
50. C. Cha, G. Gantefoer, W. Eberhardt, J. Chem. Phys. **100**, 995 (1994)
51. D.E. Bergeron, A.W. Castleman Jr., T. Morisato, S.N. Khanna, J. Chem. Phys. **121**, 10456–10466 (2004)
52. D.E. Bergeron, P.J. Roach, A.W. Castleman Jr., N.O. Jones, J.U. Reveles, S.N. Khanna, J. Am. Chem. Soc. **127**, 16048–16053 (2005)
53. S.J. Peppernick, K.D.D. Gunaratne, A.W. Castleman Jr., Int. J. Mass Spectrom. **290**, 65–71 (2010)
54. P.J. Roach, A.C. Reber, W.H. Woodward, S.N. Khanna, A.W. Castleman Jr., Proc. Natl. Acad. Sci. U. S. A. **104**, 14565–14569 (2007)
55. M.B. Abreu, C. Powell, A.C. Reber, S.N. Khanna, J. Am. Chem. Soc. **134**, 20507–20512 (2012)

56. W. Sesselmann, T.J. Chuang, Surf. Sci. **176**, 32–66 (1986)
57. Z.X. Luo, J.C. Smith, W.H. Woodward, A.W. Castleman Jr., J. Phys. Chem. A **116**, 2012–2017 (2012)
58. A.C. Harms, R.E. Leuchtner, S.W. Sigsworth, A.W. Castleman Jr., J. Am. Chem. Soc. **112**, 5673 (1990)
59. A. Aguado, A. Ayuela, J.M. Lopez, J.A. Alonso, Phys. Rev. B **58**, 9972–9979 (1998)
60. C. Hao, R.E. March, T.R. Croley, J.C. Smith, S.P. Rafferty, J. Mass Spectrom. **36**, 79–96 (2001)
61. A.T. Blades, M. Peschke, U.H. Verkerk, P. Kebarle, J. Am. Chem. Soc. **126**, 11995–12003 (2004)
62. A. Wakisaka, Faraday Discuss. **136**, 299–308 (2007)
63. C. Massobrio, A. Pasquarello, R. Car, Chem. Phys. Lett. **238**, 215–221 (1995)
64. P. Calaminici, A.M. Köster, N. Russo, D.R. Salahub, J. Chem. Phys. Lett. **105**, 9546–9556 (1996)
65. W.T. Wallace, R.L. Whetten, J. Am. Chem. Soc. **124**, 7499–7505 (2002)
66. L.D. Socaciu, J. Hagen, T.M. Bernhardt, L. Woste, U. Heiz, H. Hakkinen, U. Landman, J. Am. Chem. Soc. **125**, 10437–10445 (2003)
67. G.H. Guvelioglu, P. Ma, X. He, R.C. Forrey, H. Cheng, Phys. Rev. B **73**, 155436 (2006)
68. W. Liu, Y. Zhao, R. Zhang, Y. Li, E.J. Lavernia, Q. Jiang, ChemPhysChem **10**, 3295–3302 (2009)
69. M.B. Knickelbein, Chem. Phys. Lett. **192**, 129–134 (1992)
70. M. Kabir, A. Mookerjee, R.P. Datta, A. Banerjea, A.K. Bhattacharya, Int. J. Mod. Phys. B **17**, 2061–2075 (2003)
71. M. Kabir, A. Mookerjee, A.K. Bhattacharya, Eur. Phys. J. D **31**, 477–485 (2004)
72. I. Katakuse, T. Ichihara, Y. Fujita, T. Matsuo, T. Sakurai, H. Matsuda, Int. J. Mass Spectrom. Ion Processes **67**, 229–236 (1985)
73. I. Katakuse, T. Ichihara, Y. Fujita, T. Matsuo, T. Sakurai, H. Matsuda, Int. J. Mass Spectrom. Ion Processes **74**, 33 (1986)
74. A.W. Castleman Jr., K.H. Bowen, J. Phys. Chem. **100**, 12911–12944 (1996)
75. R. Pfaum, K. Sattler, E. Recknagel, Phys. Rev. B **33**, 1522–1525 (1986)
76. E.C. Honea, M.L. Homer, R.L. Whetten, Int. J. Mass Spectrom. Ion Processes **102**, 213 (1990)
77. L.A. Bloomfield, C.W.S. Conover, Y.A. Yang, Y.J. Twu, N.G. Phillips, Z. Phys. D **20**, 93–96 (1991)
78. J.P.K. Doye, D.J. Wales, Phys. Rev. B **59**, 2292–2300 (1999)
79. F. Seitz, Rev. Mod. Phys. **18**, 384–408 (1946)
80. W.B. Fowler, *Physics of Color Centers*, Academic Ed. (New York, 1968)
81. U. Landman, D. Scharf, J. Jortner, Phys. Rev. Lett. **54**, 1860–1863 (1985)
82. J.E. Campana, T.M. Barlak, R.J. Colton, J.J. DeCorpo, J.R. Wyatt, B.I. Dunlap, Phys. Rev. Lett. **47**, 1046–1049 (1981)
83. D. Scharf, J. Jortner, U. Landman, J. Chem. Phys. **87**, 2716 (1987)
84. K.K. Sunil, K.D. Jordan, J. Phys. Chem. **91**, 1710–1711 (1987)
85. R.N. Barnett, U. Landman, D. Scharf, J. Jortner, Acc. Chem. Res. **22**, 350–357 (1989)
86. V.A. Spasov, T.H. Lee, K.M. Ervin, J. Chem. Phys. **112**, 1713–1720 (2000)
87. F. Honda, G.M. Lancaster, Y. Fukuda, J.W. Rabalais, J. Chem. Phys. **69**, 4931–4937 (1978)
88. T.M. Barlak, J.R. Wyatt, R.J. Colton, J.J. DeCorpo, J.E. Campana, J. Am. Chem. Soc. **104**, 1212–1215 (1982)
89. W. Ens, R. Beavis, K.G. Standing, Phys. Rev. Lett. **50**, 27–30 (1983)
90. H.J. Hwang, D.K. Sensharma, M.A. El-Sayed, Phys. Rev. Lett. **64**, 808–811 (1990)
91. Y.J. Twu, C.W.S. Conover, Y.A. Yang, L.A. Bloomfield, Phys. Rev. B **42**, 5306–5316 (1990)
92. X. Li, R.L. Whetten, J. Chem. Phys. **98**, 6170–6175 (1993)
93. T.P. Martin, Phys. Rep. **95**, 167 (1983)
94. Z. Luo, G.U. Gamboa, J.C. Smith, A.C. Reber, J.U. Reveles, S.N. Khanna, A.W. Castleman Jr., J. Am. Chem. Soc. **134**, 18973–18978 (2012)
95. J. Maya, P. Davidovits, J. Chem. Phys. **59**, 3143–3152 (1973)

96. E.M. Goldfield, A.M. Kosmas, E.A. Gislason, J. Chem. Phys. **82**, 3191–3197 (1985)
97. V.A. Kartoshkin, Opt. Spektrosk. **79**, 26–31 (1995)
98. C.K. Kwan, W.E. Kauppila, S.P. Parikh, T.S. Stein, S. Zhou, Hyperfine Interact. **89**, 33–45 (1994)
99. T.T. Gien, J. Phys. B: At. Mol. Opt. Phys. **24**, 2871–2876 (1991)
100. P. Davidovits, D.L. McFadden (eds.), *Alkali Halide Vapours: Structure, Spectra and Reaction Dynamics* (Academic Press Inc., New York, 1979)
101. M.J. Pilling, P.W. Seakins (eds.), *Reaction Kinetics* (Oxford University Press Inc., New York, 1995)
102. Z.X. Luo, C. Berkdemir, J.C. Smith, A.W. Castleman Jr., Chem. Phys. Lett. **582**, 24–30 (2013)

Chapter 5
The Reactivity with Hydrogen and Nitrogen

In this chapter,[1] we present the reactivity of metal clusters with hydrogen and nitrogen [1–14]. Hydrogen is known as a highly combustible diatomic gas, and the lightest element on the periodic table. Hydrogen readily forms covalent compounds with many non-metallic elements and it is an important reducing agent of metallic ores. The H_2 chemisorption and hydrogen evolution reactions upon metal clusters have been extensively studied in the past decades in view of the broad interest of hydrogen storage and green energy sources [15–35].

Nitrogen is the lightest pnictogen and it is the most abundant uncombined element having an electronegativity of 3.04. An N atom consists of five electrons in its outer shell allowing a triple bond in molecular nitrogen (N_2). The strong $N \equiv N$ triple bond results in difficulty of converting N_2 into other compounds. Nitrogen is usually unreactive at standard temperature and pressure; however, metal lithium (or magnesium) does burn in an N_2 atmosphere, giving rise to lithium nitride (or magnesium nitride).

5.1 The Reactivity with Hydrogen

Metal-hydrogen cluster reactivity has been extensively studied with a focus on iron [19, 36], cobalt [15, 37–43], vanadium and niobium [44–48]. A typical example in Fig. 5.1 illustrates the reactivity of cationic Fe_n^+ and V_n^+ clusters with hydrogen (D_2 was used in order to avoid mass degeneracy) [47, 48]. It was found that the presence of positive charge had a substantial influence on the reaction rate for the majority of iron and vanadium clusters [19, 49, 50], and the kinetics of D_2 chemisorption on Fe_n^+/V_n^+ clusters exhibited a non-monotonic dependence on *n*. It is interesting to mention that, studies of hydrogen chemisorption onto cationic Fe_n^+ ($n = 4$–22) found a generally enhanced (although size-selective) reactivity compared to that for neutral

[1]This chapter is reproduced from Chem. Rev. 2016.

Z. Luo and S. N. Khanna, *Metal Clusters and Their Reactivity*,
https://doi.org/10.1007/978-981-15-9704-6_5

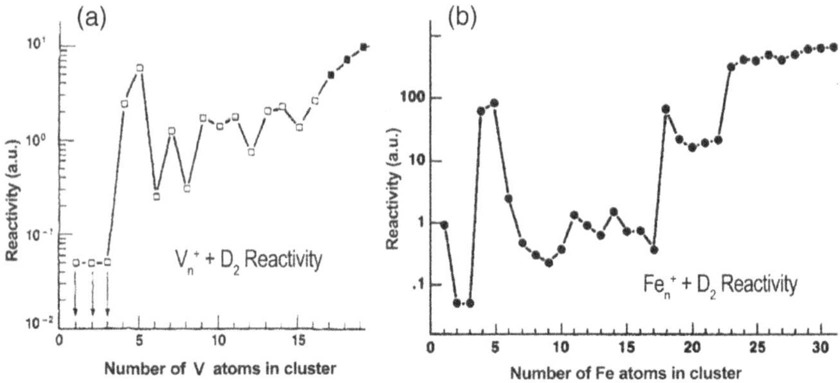

Fig. 5.1 **a** Relative rate constants for reaction of cationic vanadium cluster V_n^+ toward D_2 as a function of cluster size, as determined by bare cluster depletion. The Y-axis represents the reactivity equal to $\ln S_x/[D_2]$, where S_x is the fraction of bare clusters unreacted, i.e., the survival fraction, while $[D_2]$ is a factor directly proportional to the concentrations of D_2 in the reactor. Typical uncertainties are estimated to be ±20%. Reproduced from Ref. [48]. Copyright 1989 American Chemical Society. **b** Reactivity of cationic Fen+ toward D2 under identical conditions. Typical uncertainties are ±10%. Reproduced with permission from Ref. [47]. Copyright 1988 American Institute of Physics

Fe_n. This behavior can be rationalized within a framework of the frontier orbital model of activated chemisorption of hydrogen by invoking an activation barrier and incorporating electrostatic interactions arising from the nonzero charge state of the Fe_n^+ cluster [47].

The rate of D_2 chemisorption on neutral niobium clusters has also been found to exhibit a striking dependence on cluster sizes [15, 36, 39, 51]. For the Nb clusters, Zakin et al. [52] compared the reactivity and chemisorption kinetics of cationic, anionic and neutral species (Nb_n^-, Nb_n and Nb_n^+) via measurements of the relative rates of D_2 activation by these niobium clusters, as shown in Fig. 5.2. It has been concluded that, some of the niobium clusters react with D_2 and exhibit different reaction rates at negative or positive charge state; in particular, the excess charge displays a profound influence for the clusters sized at $7 \leq n \leq 16$ revealing the dependence on electronic structure of niobium clusters in reacting with hydrogen. In addition, the maximum uptake of D_2 by niobium clusters was found to be essentially independent of the charge state but varied with n. This is reasonable, as a high barrier may be present with size selectivity for D-D bond activation of the clusters [18, 39].

Within numerous investigations, it is conclusive that hydrogen chemisorption is the dominant reaction pathway, and the hydrogen co-adsorption could largely increase the photoionization threshold energies of the small metal clusters allowing for dramatic cluster size dependence. The photoionization threshold energies were found to be even larger for multiple H_2-chemisorptive clusters comparing with the related bare metal clusters [1]. Regarding the size dependence of hydrogen chemisorption, a typical study of the reactivity of neutral Al clusters with hydrogen

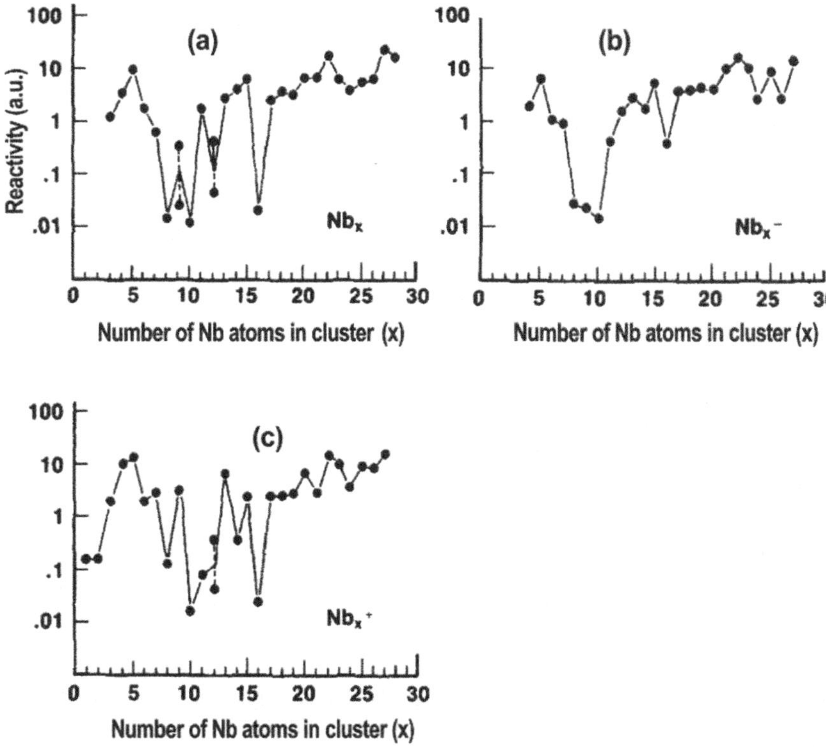

Fig. 5.2 Dependence of D_2 chemisorption reactivity R_X on cluster size, for Nb_X (**a**), Nb_X^- (**b**), and Nb_X^+ (**c**), respectively. Two data points are included for Nb_9, Nb_{12}, and Nb_{12}^+ in order to reflect the rapid and slow components of cluster depletion. The three curves are plotted on the same vertical scale with typical uncertainties ±20%. Reproduced with permission from Ref. [52]. Copyright 1988 American Institute of Physics

demonstrated that clusters with closed electronic shells are less reactive to hydrogen than those with unfilled shells, indicating similar pattern as the metal cluster reactivity with other diatomic gas molecules [1]. Considering the H_2 concentration as a constant throughout the reaction region, the reactivity can be quantified within a simple pseudo-first-order kinetic relationship,

$$-\ln(f_r) = k[H_2]\tau \tag{5.1}$$

where f_r is the fraction of bare cluster remaining, k refers to the rate constant for the addition of the first H_2 to the cluster, $[H_2]$ is the concentration of the hydrogen, and τ is the reaction time approximated as "reaction channel length/flow velocity". Based on Eq. (5.1), the constant k can be determined by measuring the mass spectra for a series of H_2 flows and plotting out the logarithm values of the remaining bare cluster signal versus reagent flow; and then the rate constants can be calculated.

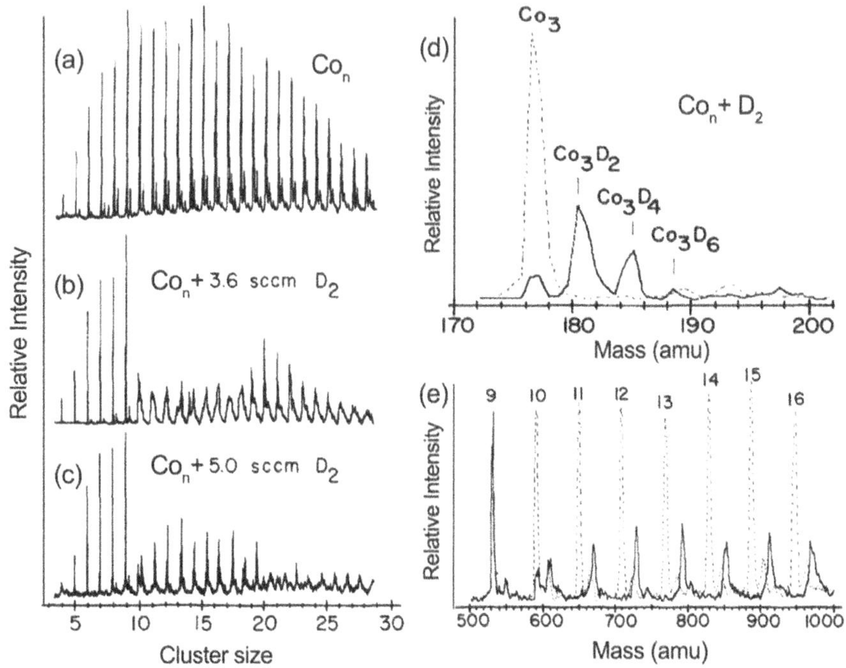

Fig. 5.3 A chemisorption study of D_2 on neutral cobalt clusters. Controlled mass spectrum was performed with only pure helium injected as the reactant gas (**a**); the lower two mass spectra were taken with 3.6 sccm (**b**) and 5.0 sccm (**c**) flow of injected D_2 reactant, respectively. The sharp peaks see in the bottom-most trace for clusters with more than 10 atoms are all due to cobalt clusters with more than one molecule of D_2 chemisorption. An enlarged spectrum shows the detail of D_2 chemisorption experiment on Co trimer (**d**) and the 9–16 atom clusters (**e**), where the dashed mass spectrum is the control experiment (only pure helium involved), while the solid trace is the observed mass spectrum with an average reactant flow of 7.4 and 5.0 sccm D_2 respectively. Reproduced with permission from Ref. [39]. Copyright 1985 American Institute of Physics

In addition, dissociative chemisorption of D_2 was also found to occur in the cases of cobalt and nickel clusters showing vivid sensitivity to the cluster sizes, as shown in Fig. 5.3, where the detailed patterns of reactivity differed for each metal under the same conditions. Among the observed D_2 chemisorption product species, $Co_n(D_2)_m$ formation appears to shut off at five D_2 molecules for Co_{11} and Co_{12}; six for Co_3, Co_{13} and Co_{14}; while up to seven for Co_{15} and Co_{16}.

5.2 The Reactivity with Nitrogen

Nitrogen gas, known as the most stable diatomic molecule, also allows for the addition on certain-sized metal clusters. For example, Lang and Bernhardt studied the reactions of small gold clusters Au_3^+ and Au_5^+ with N_2 in a multi-collision octupole ion trap [53]. Low-temperature reaction behaviors for mass-selected Au_3^+ and Au_5^+ clusters toward N_2 were addressed, as is seen in Fig. 5.4, where the products $Au_nN_m^+$ were marked with (n, m). It is interesting that $Au_3N_6^+$, $Au_5N_8^+$ and $Au_5N_6^+$ were observed as products respectively for mass-selected Au_3^+ and Au_5^+, indicating multiple N_2-molecules adsorption on the Au_n^+ clusters along with size-dependent selectivity. Among others, neutral and cationic cobalt clusters were found to be more favorable to undertake such chemisorption reaction with N_2 and typically form $Co_n(N_2)_m$ species with n and m depending on the environmental temperature [46, 54]. The adsorption of molecular nitrogen on the cobalt cluster surfaces was demonstrated to help determine the geometrical structures of the related small cobalt clusters [54]. What was interesting is that, almost no reaction was observed for nitrogen towards anionic cobalt clusters [55], except for Co_7^- and Co_8^- which adsorbed a single nitrogen molecule. Weak adsorption energies for N_2 on Co_n^- clusters provide smaller amounts of energies (e.g., 20–50 kJ/mol), which could be completely removed by buffer gas collisions before fragmentation occurs.

A few other metals have also been studied showing similar reactivity towards nitrogen [15, 16, 40, 44, 56–59], such as tungsten [60], nickel [61], niobium [10, 12], and molybdenum [62]. Among these, niobium clusters readily react and attach N_2 molecules. Even in the nascent niobium cluster distribution, there could be contamination peaks of $Nb_n(N_2)_m$ in the small mass region [39]. Also well-defined product peaks of $Mo_n(N_2)_{1,2}$ were found to dominate the reaction products within "$Mo_n + N_2$", as shown in Fig. 5.5. The temperature dependence of rate coefficients coincides with the reaction mechanism where initially a weakly-bonded molecular precursor state is formed [62]. Simply, the metal cluster reactivity with nitrogen can be summarized as:

$$M_n + xN_2 \overset{k_a}{\leftrightarrow} M_n(N_2)_x \overset{k_b}{\rightarrow} M_nN_{2x} \tag{5.2}$$

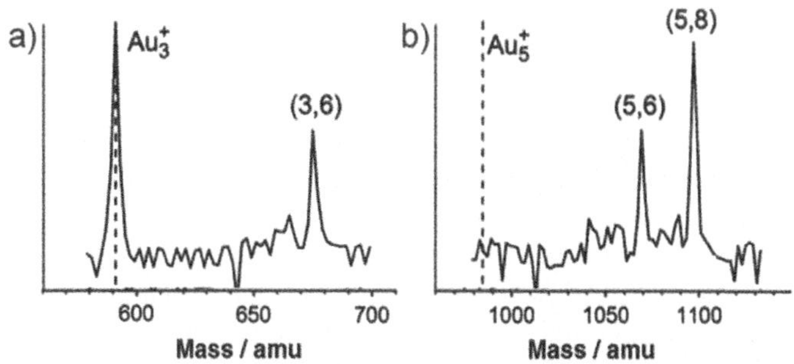

Fig. 5.4 Ion mass distributions of Au_3^+ (**a**) and Au_5^+ (**b**) in the presence of pure N_2at $T_R = 200$ K. The mass peaks are denoted by (x, y) corresponding to complexes of the stoichiometry $Au_xN_y^+$. Reproduced with permission from Ref. [53]. Copyright 1986 American Institute of Physics

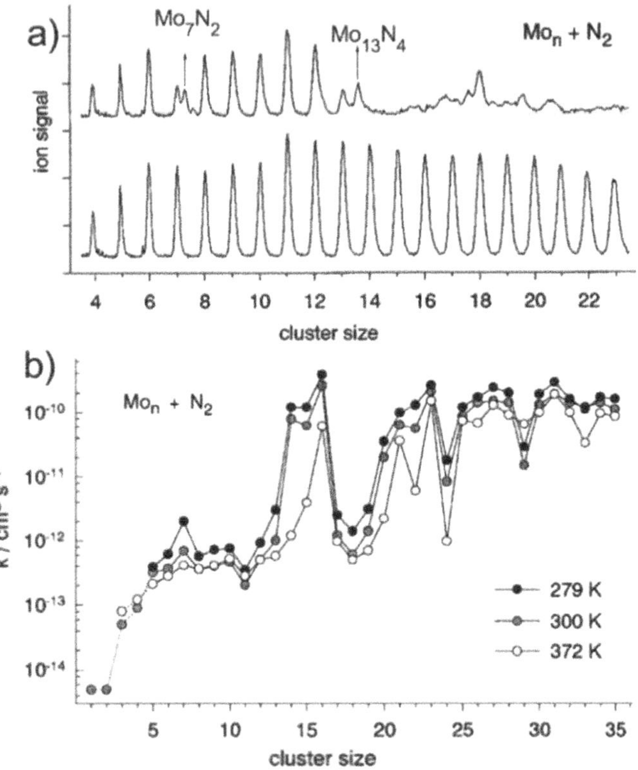

Fig. 5.5 **a** Mass spectra of molybdenum clusters recorded in the presence (top) and absence (bottom) of 0.6% N_2 in a reaction zone at 300 K. Product peaks due to $Mo_7(N_2)_{1,2}$ and $Mo_{13}(N_2)_2$ are prominent in the top trace. **b** Second-order rate coefficients for reaction of molybdenum clusters with N_2 at 279, 300, and 372 K. Points connected by dashed lined are upper limiting values. Reproduced with permission from Ref. [62]. Copyright 1995 American Institute of Physics

where M refers to the symbol for a generic metal; $M_n(N_2)_x$ is the precursor while M_nN_{2x} is the chemisorption product; k_a and k_b are the rate constants for association/dissociation and chemisorption, respectively, depending on the metals and environmental temperature. It is still worth mentioning that, nickel and copper clusters were found to be less reactive or almost nonreactive with nitrogen at room temperature [39]. The tendency of N_2 chemisorption on small metal clusters was due to limited but varied charge transfer between the orbitals of metal clusters and the nitrogen molecules [60]. Different charge transfer rate allows the metal clusters to react with nitrogen slowly or rapidly.

Having discussed the diversity of cluster reactivity of metals toward nitrogen and hydrogen, herewith we also summarize their similarities. Firstly, metals take on similar patterns of reactivity with the H_2 and N_2, i.e., chemisorption or dissociative chemisorption, although with different size dependence. Secondly, the critical factor enabling dissociative chemisorption of H_2 or N_2 on transition metal clusters is applicable to each other. It is worth mentioning that the dissociative chemisorption behavior of H_2 or N_2 on metal surfaces has long been a unique scientific topic in the chemistry of transition metals where the surface d-orbitals play an important role. Through theoretical calculations, researchers have modeled the H_2 chemisorption on surface sites of small metal clusters, and found that the 3d orbital participation is crucial in lowering the activation barrier for the dissociative chemisorption of such diatomic gas molecules [63]. It is still interesting to note that the aforementioned studies of small metal clusters in reacting with hydrogen and nitrogen both display no conspicuous odd–even effect, which largely differs from the metal cluster reactivity with oxygen. As mentioned earlier, oxygen is spin-triplet in its ground state and the two half-filled molecular orbitals are anti-bonding in nature. Therefore, the activation of an oxygen molecule ($^3O_2$) requires to fill its half-filled anti-bonding orbitals and to reduce the multiplicity from triplet to singlet. For clusters with odd number of electrons, the reaction can proceed without changes in the multiplicity of the cluster; while for clusters with even number of electrons, the spin conservation requires a spin excitation of the cluster leading to a barrier resulting in a dominant odd–even effect. Nitrogen and hydrogen tend to chemisorb on metal clusters, where the N–N bond and H–H bond do not have to be broken; and even with dissociative chemisorption, there is no necessity to change the spin multiplicity of the metal clusters.

Nonpolar gas molecules have also been found to be reactive with other metal clusters [45–48, 64] For instance, utilizing a fast flow tube method, Kaya et al. [46] investigated the reactivity of cobalt cluster cations Co_n^+ (n = 2–22) with a few molecules including CH_4, C_2H_4 and C_2H_2 (also N_2, H_2). Their results indicated an interesting cluster size dependence for the reactants CH_4 and C_2H_4 (also N_2, H_2), where $Co_{4,5}^+$ and Co_{10-15}^+ displayed much higher reactivity than their neighboring clusters; however, all these Co_n^+ clusters were found to be highly reactive with C_2H_2 regardless of the cluster sizes. This interesting difference was explained by the activation energies for their chemisorption reactions. Cationic Co_n^+ clusters showed similar size dependence as neutral Co_n clusters, except for Co_4^+ and Co_5^+ that display enhanced reactivity due to active sites induced by the influence of positive charge. On the other hand, the adsorption rate of C_2H_2 to the cobalt cluster surfaces was found to be large enough to attain a stable chemisorption state instead of an activation of the C−H or C−C bond. Therefore, gas-phase reactivity does not necessarily depend on cluster sizes when the chemisorption state is not affected by the different frontier orbital energies.

References

1. S.N. Khanna, P. Jena, Chem. Phys. Lett. **218**, 383–386 (1994)
2. C. Bois, J.A. Cabeza, R.J. Franco, V. Riera, E. Saborit, J. Organomet. Chem. **564**, 201–207 (1998)
3. J.A. Cabeza, I. del Rio, V. Riera, Inorg. Chim. Acta **268**, 131–133 (1998)
4. B. Bergman, E. Rosenberg, Organometallics **21**, 1508–1511 (2002)

5. J.A. Spirko, M.L. Neiman, A.M. Oelker, K. Klier, Surf. Sci. **572**, 191–205 (2004)
6. R.E. Winans, S. Vajda, G.E. Ballentine, J.W. Elam, B. Lee, M.J. Pelline, S. Seifert, G.Y. Tikhonov, N.A. Tomczyk, Top. Catal. **39**, 145–149 (2006)
7. N.K. Jena, K.R.S. Chandrakumar, S.K. Ghosh, J. Phys. Chem. C **113**, 17885–17892 (2009)
8. W. Silva, T.N. Truong, F. Mondragon, J. Alloys Compd. **509**, 8501–8509 (2011)
9. J.A. Cabeza, I. delRio, R.J. Franco, F. Grepioni, V. Riera, Organometallics **16**, 2763–2764 (1997)
10. J. Mwakapumba, K.N. Ervin, Int. J. Mass Spectrom. **161**, 161–174 (1997)
11. B.P. Pozniak, R.C. Dunbar, J. Am. Chem. Soc. **119**, 7343–7349 (1997)
12. A. Berces, P.A. Hackett, L. Lian, S.A. Mitchell, D.M. Rayner, J. Chem. Phys. **108**, 5476–5490 (1998)
13. B.D. Leskiw, A.W. Castleman Jr., C. Ashman, S.N. Khanna, J. Chem. Phys. **114**, 1165–1169 (2001)
14. K. Horn, J. Dinardo, W. Eberhardt, H.J. Freund, E.W. Plummer, Surf. Sci. **118**, 465–495 (1982)
15. M.E. Geusic, M.D. Morse, R.E. Smalley, J. Chem. Phys. **82**, 590–591 (1985)
16. T. Tanabe, H. Adachi, S. Imoto, Jap. J. Appl. Phys. **17**, 49–58 (1978)
17. J.B. Benziger, Appl. Surf. Sci. **6**, 105–121 (1980)
18. J.Y. Saillard, R. Hoffmann, J. Am. Chem. Soc. **106**, 2006–2026 (1984)
19. E.K. Parks, K. Liu, S.C. Richtsmeier, L.G. Pobo, S.J. Riley, J. Chem. Phys. **82**, 5470–5474 (1985)
20. K. Takahashi, S. Isobe, S. Ohnuki, Appl. Phys. Lett. **102**, 113108 (2013)
21. Q. Meng, P.S. May, M.T. Berry, D. Kilin, Int. J. Quantum Chem. **112**, 3896–3903 (2012)
22. N.S. Venkataramanan, A. Suvitha, H. Mizuseki, Y. Kawazoe, Int. J. Quantum Chem. **113**, 1940–1948 (2013)
23. V. Dryza, E.J. Bieske, Int. Rev. Phys. Chem. **32**, 559–587 (2013)
24. Z. Li, Y. Li, J. Li, J. Chem. Phys. **137**, 234704 (2012)
25. T. Miyao, A. Yoshida, H. Yamada, S. Naito, J. Mol. Catal. a-Chem. **378**, 174–178 (2013)
26. P. Gonzalez-Navarrete, M. Calatayud, J. Andres, F. Ruiperez, D. Roca-Sanjuan, J. Phys. Chem. A **117**, 5354–5364 (2013)
27. L. Guo, J. Phys. Chem. A **117**, 3458–3466 (2013)
28. C. Kerpal, D.J. Harding, D.M. Rayner, A. Fielicke, J. Phys. Chem. A **117**, 8230–8237 (2013)
29. H. Xie, X. Li, L. Zhao, Z. Liu, Z. Qin, X. Wu, Z. Tang, X. Xing, J. Phys. Chem. A **117**, 1706–1711 (2013)
30. I. Cabria, M.J. Lopez, S. Fraile, J.A. Alonso, J. Phys. Chem. C **116**, 21179–21189 (2012)
31. C.K. Brozek, M. Dinca, J. Am. Chem. Soc. **135**, 12886–12891 (2013)
32. S. Banerjee, G. Periyasamy, S.K. Pati, Phys. Chem. Chem. Phys. **15**, 8303–8310 (2013)
33. J.A. Santana, N. Roesch, Phys. Chem. Chem. Phys. **14**, 16062–16069 (2012)
34. S.M. Solov'ev, C. Pettenkofer, I.I. Pronin, N.D. Potekhina, V.N. Petrov, Surf. Sci. **608**, 165–172 (2013)
35. J. Moc, Theor. Chem. Acc. **132**, 1378 (2013)
36. J.M. Alford, F.D. Weiss, R.T. Laaksonen, R.E. Smalley, J. Phys. Chem. **90**, 4480–4482 (1986)
37. T.D. Klots, B.J. Winter, E.K. Parks, S.J. Riley, J. Chem. Phys. **95**, 8919–8930 (1991)
38. T.D. Klots, B.J. Winter, E.K. Parks, S.J. Riley, J. Chem. Phys. **92**, 2110–2111 (1990)
39. M.D. Morse, M.E. Geusic, J.R. Heath, R.E. Smalley, J. Chem. Phys. **83**, 2293–2304 (1985)
40. A. Kaldor, D.M. Cox, J. Chem. Soc.-Faraday Trans. **86**, 2459–2463 (1990)
41. J. Ho, L. Zhu, E.K. Parks, S.J. Riley, J. Chem. Phys. **99**, 140–147 (1993)
42. J. Ho, L. Zhu, E.K. Parks, S.J. Riley, Z. Phys. D: At., Mol. Clusters **26**, 331–333 (1993)
43. M. Andersson, J.L. Persson, A. Rosen, J. Phys. Chem. **100**, 12222–12234 (1996)
44. M.R. Zakin, D.M. Cox, R.L. Whetten, D.J. Trevor, A. Kaldor, Chem. Phys. Lett. **135**, 223–228 (1987)
45. M.P. Irion, P. Schnabel, J. Phys. Chem. **95**, 10596–10599 (1991)
46. A. Nakajima, T. Kishi, Y. Sone, S. Nonose, K. Kaya, Z. Phys. D: At., Mol. Clusters **19**, 385–387 (1991)
47. M.R. Zakin, R.O. Brickman, D.M. Cox, A. Kaldor, J. Chem. Phys. **88**, 6605–6610 (1988)

48. M.R. Zakin, D.M. Cox, R.O. Brickman, A. Kaldor, J. Phys. Chem. **93**, 6823–6827 (1989)
49. K. Liu, E.K. Parks, S.C. Richtsmeier, L.G. Pobo, S.J. Riley, J. Chem. Phys. **83**, 2882–2888 (1985)
50. E.K. Parks, B.H. Weiller, P.S. Bechthold, W.F. Hoffman, G.C. Nieman, L.G. Pobo, S.J. Riley, J. Chem. Phys. **88**, 1622–1632 (1988)
51. R.L. Whetten, D.M. Cox, D.J. Trevor, A. Kaldor, Phys. Rev. Lett. **54**, 1494–1497 (1985)
52. M.R. Zakin, R.O. Brickman, D.M. Cox, A. Kaldor, J. Chem. Phys. **88**, 3555–3560 (1988)
53. S.M. Lang, T.M. Bernhardt, J. Chem. Phys. **131**, 024310 (2009)
54. J. Ho, E.K. Parks, L. Zhu, S.J. Riley, Chem. Phys. **201**, 245–261 (1995)
55. E. Kapiloff, K.M. Ervin, J. Phys. Chem. A **101**, 8460–8469 (1997)
56. J.C. Phillips, J. Chem. Phys. **84**, 1951–1952 (1986)
57. W. Guo, D.G. Vlachos, J. Chem. Phys. **138**, 174702 (2013)
58. C. Kerpal, D.J. Harding, J.T. Lyon, G. Meijer, A. Fielicke, J. Phys. Chem. C **117**, 12153–12158 (2013)
59. S.A. Aal, A.S. Shalabi, W.S.A. Halim, Thin Solid Films **545**, 341–352 (2013)
60. Y.D. Kim, D. Stolcic, M. Fischer, G. Gantefor, J. Chem. Phys. **119**, 10307–10312 (2003)
61. E.K. Parks, G.C. Nieman, K.P. Kerns, S.J. Riley, J. Chem. Phys. **108**, 3731–3739 (1998)
62. S.A. Mitchell, L. Lian, D.M. Rayner, P.A. Hackett, J. Chem. Phys. **103**, 5539–5547 (1995)
63. P.E.M. Siegbahn, M.R.A. Blomberg, J.C.W. Bauschlicher, J. Chem. Phys. **81**, 2103–2111 (1984)
64. S. Nonose, Y. Sone, N. Kikuchi, K. Fuke, K. Kaya, Chem. Phys. Lett. **158**, 152–156 (1989)

Chapter 6
Cooperative Active-Sites Mechanism

Akin to the shell structures of atoms, shell-filling concepts from traditional valence bond theory can be applied to the description of cluster stability. In view of this, the result of a chemical interaction could be explained through the energy minimization attained when a cluster closes an incomplete electronic shell, either by direct ionization or through the formation of a covalent/ionic bond. Also it has been widely recognized that the cluster reactivity depends on both geometric and electronic structure, although not all reactions are subject to the same fundamental constraints. However, unexpected stability may also exist for some clusters with neither a spherical geometry nor a closed electron shell according to the NFEG model. Understanding how a specific size and/or shape can affect the affinity of a metal cluster toward a specific reagent will facilitate the efforts to design either stable or reactive materials for specific applications.

In this regard, cooperative active-sites mechanism has found reasonable research interest for chemists to interpret novel cluster reactivitity. Originally, active sites in biology usually refer to the region of an enzyme (a groove or pocket) where certain substrate molecules bind and react [1–3], allowing the residues in the binding site to form hydrogen bonds, hydrophobic interactions, or temporary covalent interactions (van der Waals). In cluster science, complementary Lewis acid/base active sites refer to a location on the cluster surface where one atom acts as a Lewis acid and a second Al atom acts as a Lewis base site [4–8]. This established mechanism well explained the size-selectivity of Al_n^- in reacting with water, and has been recognized to be highly helpful in understanding reactivities between metal clusters and polar molecules [9–22]. In this chapter, we will introduce how the complementary active sites (CAS) mechanism is operative in metal cluster reactivity.

Z. Luo and S. N. Khanna, *Metal Clusters and Their Reactivity*,
https://doi.org/10.1007/978-981-15-9704-6_6

6.1 Reaction of Aluminum Clusters with Water

Abundant theoretical and experimental investigations on the reactivity of metal clusters with water have attracted reasonable research interest partly due to the importance of hydrogen evolution being involved [23–42]. Among the extensive literature reports, there is an interesting study by Castleman, Khanna and their colleagues [7] who reported the size selectivity of aluminum cluster anion reacting with water. They found that identical arrangements of multiple sites in Al_{16}^-, Al_{17}^-, and Al_{18}^- result in the production of H_2 from water attributed to the dissociative chemisorption of water at specific surface sites, as shown in Fig. 6.1. In comparison, Al_{13}^-, Al_{23}^- and Al_{37}^- are considered by the superatom model to have rare gas-like closed electronic shells; however, the observed selective reactivity of Al_n^- with water is inconsistent with the closing of superatom shells by noting their adsorbing water molecules. While Al_{12}^- reacts to form a product $Al_{12}H_2O^-$ of observable intensity, Al_{14}^- and Al_{46}^- have open electronic shells but do not support the product observation with water adsorption, etc. In this regard, they proposed that it's the complementary-active-sites mechanism that causes the size-selective reactivity of aluminium cluster anions with water. The complementary active sites refer to a location on the cluster surface where one Al atom acts as a Lewis acid and a second Al atom acts as a Lewis base.

The first of these, a combination of geometric (sizes, shapes, and adsorption sites) and electronic features (energies, orbitals, and spin effects etc.) was reasonably demonstrated to account for the observed size-selective reactivities [42]. In particular, Al_n^- clusters of certain sizes may harbor distinct active sites in which a

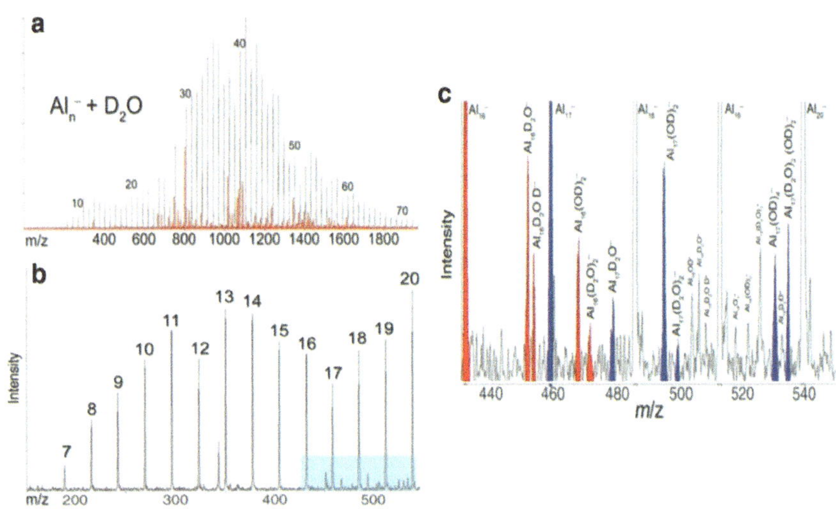

Fig. 6.1 **a** Distribution of Al_n^- (n = 7–73) clusters reacting with D_2O. Nonpure aluminum clusters are shown in red; **b** Reaction of low-mass Al_n^- clusters (n = 7–20) with D_2O. **c** Expansion of the shaded area in (**b**). Red peaks are Al_{16}^- species; blue for Al_{17}^-.[7]

pair of adjacent Al atoms is responsible for the dissociative chemisorption of water molecules. Considering the initial interaction between water and Al_n^- clusters is the nucleophilic attack of a water molecule on the Al surface, this reaction requires the donation of lone-pair electrons from water to the LUMO of the Al cluster (or LUMO + 1 for odd-electron systems as the lone pair interacts most strongly with the levels where both spin states are unoccupied) where the probability density of the vacant orbital protrudes out from the cluster structure into vacuum [7]. Further, the transition state for splitting water in an Eley–Rideal-type mechanism requires the previously dissociated H atom to act as the Lewis acid, rather than an Al atom, which is consistent with the size selectivity for Al_n^- clusters reacting with water whether or not giving rise to H_2 release. A free H atom on the cluster surface with a neighboring Lewis acid site is likely to be attained in many clusters after reacting with a single water molecule, but only those with paired active sites result in the inferred release of H_2. This is the main reason for the reactivity difference of Al_{12}^- (forming $Al_{12}H_2O^-$) compared to Al_{16}^-, Al_{17}^- and Al_{18}^-, as shown in Fig. 6.2 [43].

It is important mentioning that Al_{12}^- bears particularly high binding energy but low LUMO energy level, and it was found to bind water tightly, which is in contrast with the superatomic species Al_{13}^- which has a strikingly high LUMO energy level and hence binds water quite weakly. Moreover, some of their adjacent clusters (such as Al_{11}^- and Al_{13}^-) bear insurmountable energy-transition states and hence also have low reactivity towards water [5].

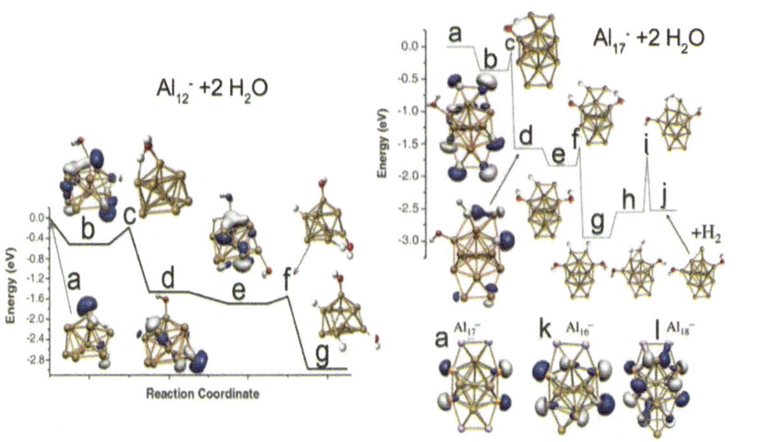

Fig. 6.2 **A** Reaction coordinates for $Al_{12}^- + 2H_2O$: (a) The calculated LUMO + 1 for Al_{12}^-; (b) The HOMO for the chemisorption $[Al_{12}(H_2O)]^-$ complex; (c) A proposed transition state; (d) The dissociatively chemisorbed product and LUMO + 1; (e) The HOMO after a second water is bound to the active site; (f) The transition state for the second water; (g) The final product. (**B**) Reaction coordinate for the formation of H_2 from Al_{17}^-

6.2 Reaction with H_2S and NH_3

Similar experimental investigations have also been undertaken to explore the reactivity of cluster anions with H_2S and NH_3 gases which were introduced into the flow tube reactor, as shown in Fig. 6.3. As predicted above, interestingly, Al_{12}^- always appears very reactive both in the presence of H_2S and NH_3. This is explained by its very prominent complementary active sites, high binding energy of the Al_{12}^- itself but low LUMO energy level. In contrast, Al_{13}^- and Al_{20}^- are less reactive than their neighboring clusters, so it was proposed that the polar S–H and N–H bonds could undertake the same mechanism as O–H bond in water [44].

Among others, the reaction of gold cluster cations Au_n^+ with H_2S was also studied [45], and it was found that initial products were mainly $AuSH^+$ for n = 2, while selective Au_nS^+ and $Au_nSH_2^+$ for other Au_n^+ clusters. In general, the gold clusters cations with even number of atoms were more reactive than adjacent odd n clusters. No reactions of Au^+ and Au_3^+ with H_2S were observed in their study. The low reactivity of Au_n^+ at n = 1, 3, 9, and 11 coincides with the low ionization potential of Au_n and the weak binding energy of Au_n^+–Au. Further sulfuration reactions of Au_nS^+ proceeded to give $Au_nS_m^+$ and finally stopped at those $Au_nS_{m+x}H_2^+$ species when H_2 release did not occur and the maximum number of sulfur atoms m + x increased with the cluster sizes.

Several investigations have also demonstrated the reactivity of metal clusters with ammonia [46–65]. For example, an investigation on the reactivity of aluminum cluster anions with ammonia was examined in *Bowen* group [48]. The results coincided with the aforementioned mechanism on complementary active sites, typicallly with the reactivity on Al_{12}^-. Based on theoretical calculation, it was proposed that the presence of ammonia could lead to geometric distortion of Al_{12}^- together with the dissociation of the NH_3 molecule of which the N atom and H atoms were demonstrated to transfer on the Al cluster resulting the reorganized minimum energy structure. The dissociation of NH_3 molecule resembles the dissociation of H_2O as demonstrated above, supporting the applicability of the complementary-active-sites mechanism in various polar molecular systems [48].

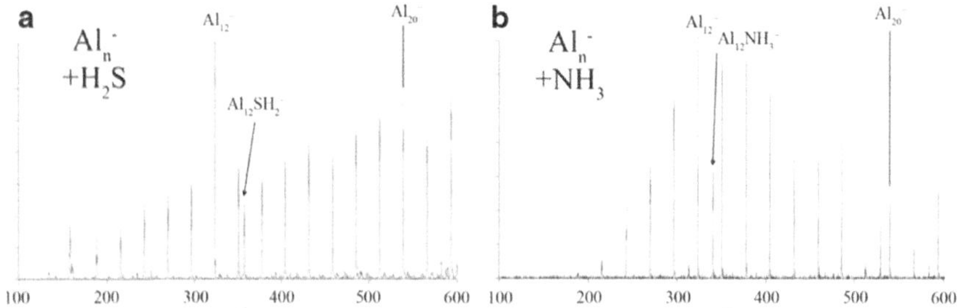

Fig. 6.3 The mass spectra of Al_n^- clusters after reacting with **a** H_2S and **b** NH_3

Recently, utilizing the customized Re-TOF mass spectrometer combined with a 177-nm deep-ultraviolet laser, Luo et al. have been able to observe well-resolved cobalt clusters $Co_n^{\pm/0}$ on which they performed a comprehensive study on cobalt cluster reactions with ammonia (NH_3). As results, the anions Co_n^- were found to be inert, but the neutrals were able to adsorb multiple ammonia molecules. The neutral Co_n clusters, especially those of relatively larger sizes, were observed to can adsorbe multiple NH_3 molecules, i.e., $Co_n + mNH_3 \rightarrow Co_n(NH_3)_m$ [66]. This is in consistence with the previously established coordination chemistry theory on cobalt-ammonia complexes. Besides the adsorptive reactions, it was proposed that an etching-like fragmentation channel (Eq. 6.2) could also exist in the cobalt cluster reaction with NH_3, which is enabled in the presence of multiple NH_3 molecules. What is interesting is that, the cationic Co_n^+ clusters readily reacted with NH_3 resulting in a series of dehydrogenation products; whereas, the dehydrogenation of NH_3 on Co_n^+ clusters was only observed for those of $n \geq 3$, as shown in Fig. 6.4. It is also notable that the dehydrogenation of NH_3 was observed only when more than two NH_3 molecules were present indicative of a co-operative mechanism. In all, these reactions were summarized as [66],

$$Co_n^+ + NH_3 \rightarrow Co_nNH_3{}^+ \tag{6.1}$$

$$Co_n^+ + mNH_3 \rightarrow Co_x(NH_3)_y{}^+ + Co_{n-x}(NH_3)_{m-y}(n > x; m \geq y \geq 0) \tag{6.2}$$

$$Co_n^+ + mNH_3 \rightarrow [Co_n(NH_3)_m]^+ \rightarrow [Co_n(NH_3)_{m-2}(NH_2)_2]^+ + H_2(m \geq 2) \tag{6.3}$$

The diverse reactivities of $Co_n^{\pm/0}$ with ammonia brought forth comprehensive insights into the charge-dependence and size-dependence and cooperativion effect in such metal cluster reactions [66]. The charge-dependence with $Co_n^+ > Co_n > Co_n^-$ was well explained by their altered electrostatic potential for an NH_3 molecule in approaching the $Co_n^{\pm/0}$ clusters, also associated with altered charge distributions, binding energies, and PDOS of dominant orbitals. Among the cationic Co_n^+ clusters, the Co^+ ion and Co_2^+ clusters lack active sites for the second hydrogen transfer, and suffer from insurmountable rate-limiting barrier for dehydrogenation from NH_3, as shown in Fig. 6.5. In contrast, transition states for $Co_{3,6}{}^+$ are surmountable, which concurs with the experimental observation of dehydrogenation products for $Co_{n\geq3}{}^+$ clusters. Besides, the DFT calculations indicate that two co-adsorbed NH_3 molecules benefit to the H_2 evolution, evidencing the importance of cooperative active sites for NH_3 decomposition.

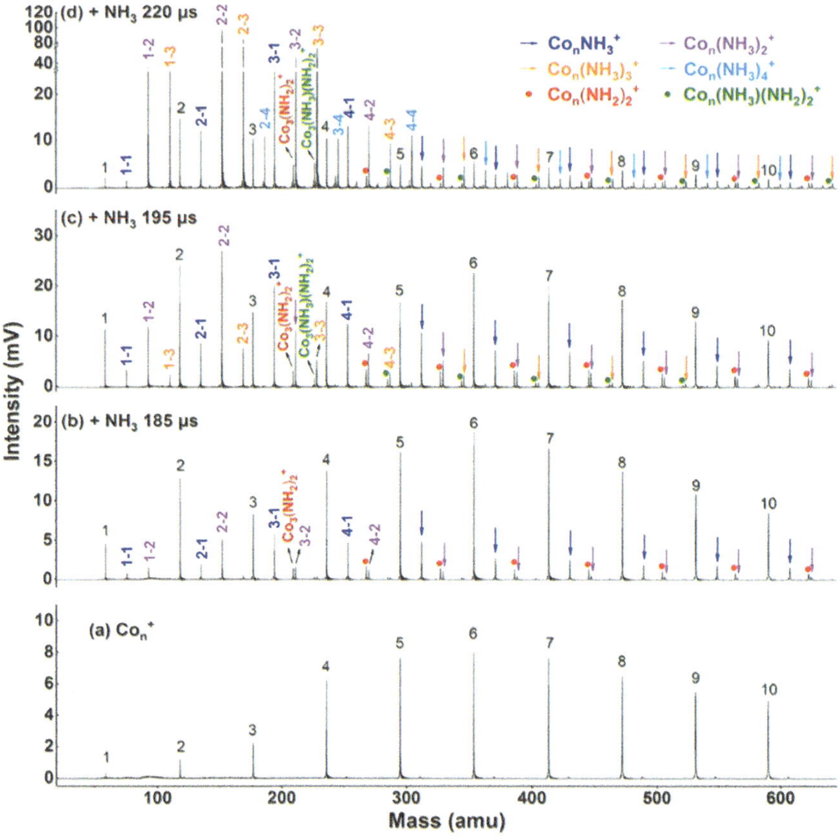

Fig. 6.4 Enlarged area for the Co_n^+ (n = 1–10) clusters (**a**) reacting with different amounts of NH_3 with pulse width of 185, 195 and 220 μs respectively (**b–d**). The blue, purple, orange, and light blue arrows indicate the $Co_n(NH_3)^+$, $Co_n(NH_3)_2^+$, $Co_n(NH_3)_3^+$ and $Co_n(NH_3)_4^+$ clusters; and the red and green dots represent the $Co_n(NH_2)_2^+$ and $Co_n(NH_3)(NH_2)_2^+$ respectively

6.3 Reaction with Alcohols

It is believed that the complementary-active-sites mechanism is not specific to water but rather any molecule with the –OH functional groups [4, 8, 67–79]. The role of complementary active sites in reactions with other hydroxyl containing compounds remains to be an interesting topic. Recently systematic studies on the reactivity of alcohols with aluminum cluster anions were reported. Figure 6.6 shows the reactions of Al cluster anions with methanol and tert-butyl alcohol, respectively. In general, these alcohols (especially methanol) were found to exhibit an etching effect towards the Al clusters with few exceptions (such as Al_{13}^-). The etching of methanol closely resembles the reaction of Al_n^- towards oxygen where selective species such as Al_{13}^- exhibit resistance to the etching effect and the cluster's increasing intensity is due

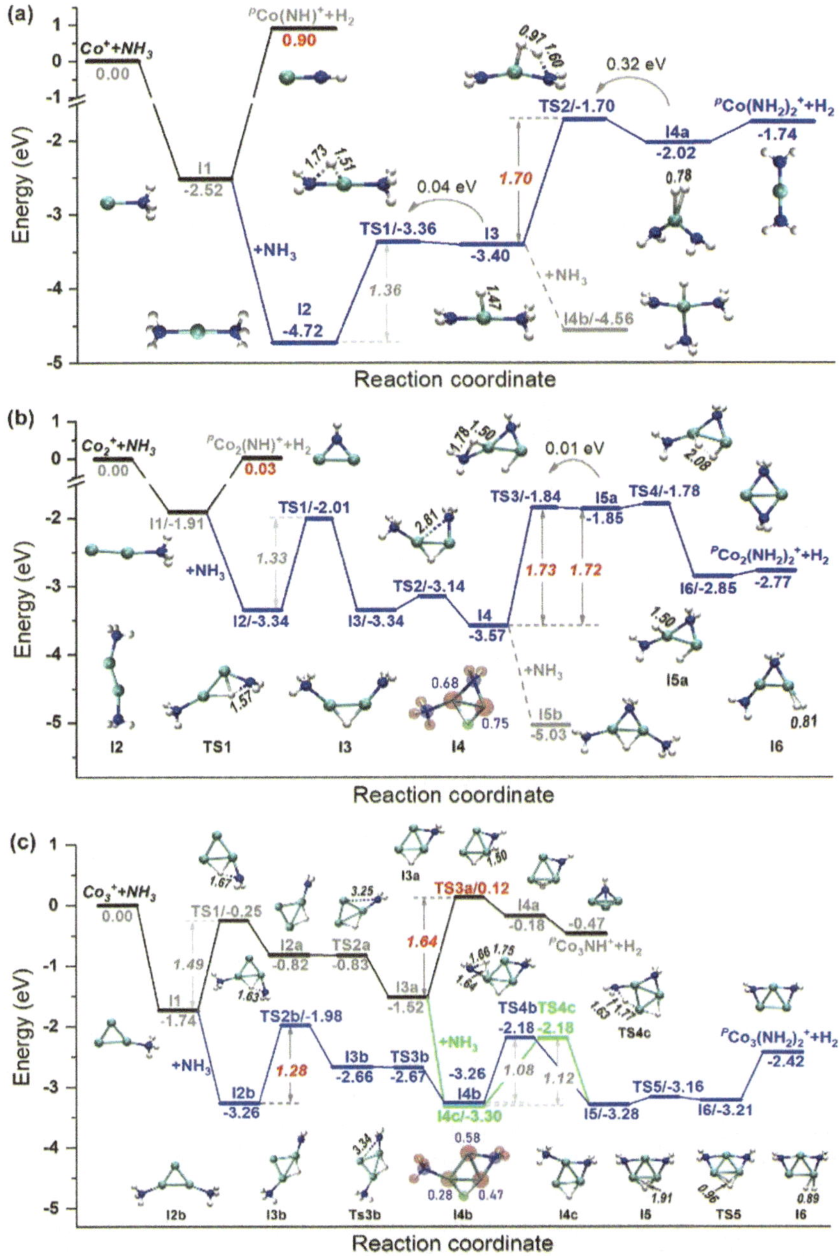

Fig. 6.5 Reaction coordinates of "$Co^+ + NH_3$ (/2NH_3)" (**a**), and "$Co_2^+ + NH_3$(/2NH_3)" (**b**), and "$Co_3^+ + NH_3$ (/2NH_3)" (**c**). Atoms in cyan, blue, and white color represents Co, N, and H, respectively. For the I4 intermediates, the NPA charge distributions of Co atoms in the $Co_{2,3}(NH_3)_2^+$ are displayed. Energies are in eV and bond lengths are in angstrom

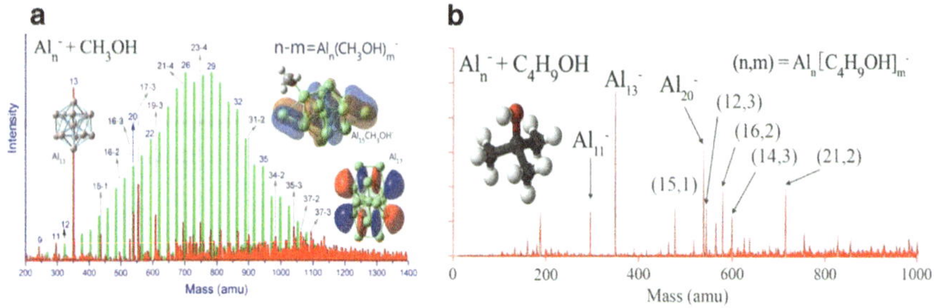

Fig. 6.6 Reaction of Al cluster anions with methanol (**a**), and tert-butyl alcohol (**b**)

to its being a uniquely stable product after the dissociation of larger Al clusters. In addition to the etching effect, the alcohols tended to bind to the Al clusters but were not observed to produce H_2 in the room-temperature fast-flow tube apparatus. Among the Al_n^- species which were also found to undergo the attachment of one or multiple alcohol molecules, Al_{15}^-, Al_{16}^-, Al_{17}^-, Al_{19}^- and Al_{21}^- have been repeatedly proved to be highly reactive species with strong tendencies for the chemisorption of water molecules. Note that Al_{15}^- attaches only one methanol molecule to form $Al_{15}CH_3OH^-$ but Al_{17}^- gives rise to $Al_{17}(CH_3OH)_3^-$. This is because there are more active sites on Al_{17}^- than Al_{15}^- resulting in less steric hindrance for the Al_{17}^- cluster to attach multiple methanol molecules. Such observation coincides with the established theory that complementary active sites support size-selective reactivity of aluminum cluster anions with water. However, it remains to be explored how other reactions are promoted to achieve such active sites and how nanostructures can be tailored with a preponderance of such sites.

6.4 Reaction with Acetone and Formaldehyde

The experimental results in the above section indicate that, although H_2O has a slightly larger O–H bond dissociation energy (118.8 vs. 104.6 kcal/mol respectively) [80] compared to methanol, the –OH group in methanol seems to be not as easy as that in water to bear a cleavage. Experiments (Fig. 6.7a and b) have also been carried out to explore the reactivity of aluminum cluster anions with three carbonyl-containing species of differing bond strength: formaldehyde (743.4 kJ/mol), acetone (771.4 kJ/mol); carbon dioxide (532.2 kJ/mol); and carbon monoxide (1076.4 kJ/mol) [81]. However, C=O bond cleavage was observed for acetone and formaldehyde reacting with a certain sized Al clusters, as shown in Fig. 6.7, while carbon dioxide and carbon monoxide showed no reactivity, even though carbon dioxide has a lower bond dissociation enthalpy than formaldehyde and acetone.

The theoretical calculation results revealed that Al_9^- reacts readily at the complementary active sites and subsequently lose an Al_2O; in contrast, Al_{13}^- does not have active sites while has a barrier to carbonyl cleavage, and also Al_2O release is

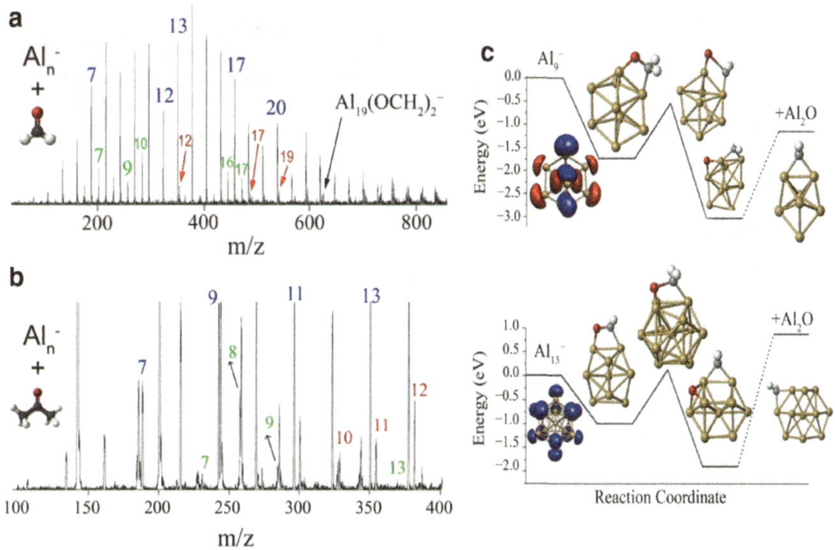

Fig. 6.7 Aluminum cluster anion distribution after reaction with formaldehyde (**a**) and acetone (**b**). Aluminum clusters Al_n^- are labelled with blue numbers, formaldehyde additions $Al_n(OCH_2)^-$ are labelled with red numbers, while oxygen losses $Al_n(CH_2)^-$ are labelled with green numbers. **c** Theoretically determined reaction coordinate diagrams of $Al_n^- + OCH_2$ for $n = 9$ and $n = 13$. For each initial structure, the HOMO and LUMO (LUMO + 1) are shown in red and blue, respectively

endothermic (Fig. 6.7c). This agrees with experimental observations. It was therefore essential that research along this path continued, both in order to determine strong bonds that may be broken in this manner but also to identify what surfaces and clusters may possess the well-patterned Lewis acid/Lewis base sites optimal to promote this type of reactivity [82]. It is notable of particular interest of developing alternative reactants and clusters/metal surfaces that would exhibit this mechanism for species with large bond energies, be it metal oxides, bimetallic interfaces, customized defect sites and step edges, or cluster-assembled materials [83–88].

6.5 Edge Effect

Understanding the emergence of properties from size-selective clusters to nanoparticles is one of the principal goals of both cluster science and nanotechnology. In general, the presence of an active site is a result of irregular charge distribution on the cluster surface, which is most prominent in clusters with geometries that are akin to defects on the cluster surface. Recent investigations of Al_n^- reacting with alcohols have revealed that, at small sizes of cluster, the location of reactive pairs occurs on specific active sites, but at larger sizes the reactive pairs begin to accumulate on the edges between facets, indicating the transition from cluster regime to the nanoscale

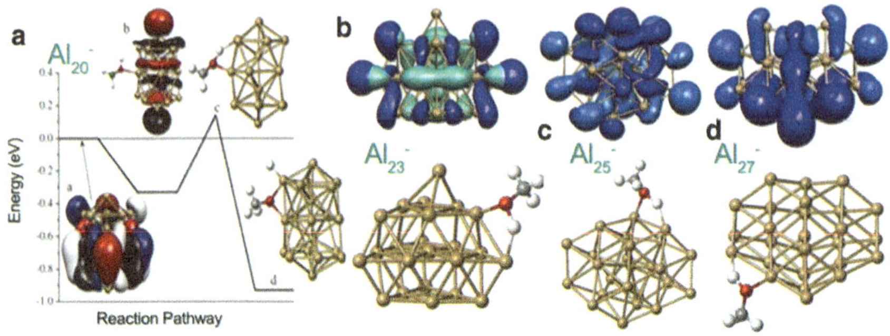

Fig. 6.8 (Left) The reaction pathway for Al_{20}^- with methanol: **a** The Al_{20}^- cluster with LUMO and LUMO + 1 plotted in Red–Black, and Blue-White; **b** Methanol with O–H intact bound to the Al_{20}^- cluster, with HOMO charge density plotted; **c** The transition state for O–H cleavage; **d** the final state where O–H is broken. (Right) The LUMO charge density of Al_{23}^-, Al_{25}^-, and Al_{27}^-, and their energy structures when the O atom is bound to the selected Al atom respectively

where the nanoparticles are universally reactive. The structure of Al_{20}^- (Fig. 6.8) is a defect free double cage structure in which may be thought of as a double icosahedron with an atom embedded. Theoretical calculation results show that the LUMO and LUMO + 1 states are delocalized along the equator of the prolate cluster. As these states are indicative of Lewis acid sites the oxygen prefers to bind here during cleavage. The transition state displays a larger energy than the binding energy of the nondissociated methanol. This high barrier suggests that the HER is inhibited despite the O–H cleavage process is exothermic. In comparison, the structure of Al_{23}^- is a hexagonal packing of the aluminum atoms with threefold longitudinal edges where the Lewis base sites are also located, with some additional density on the more obtuse equatorial surface; Al_{25}^- is triangular and also has well defined edges along different sections of the cluster; similarly, the LUMO of Al_{27}^- are located primarily along the edges of the cluster, indicating the location where the methanol molecule is binding.

6.6 Hydrogen Evolution Mechanism

Hydrogen evolution reaction (HER) has been meticulously studied in view of the potential value of H_2 as a powerful "green" fuel; however, an in-depth and complete understanding of the HER mechanism is still elusive although nearly a century of study and debate. In the ongoing efforts devoted to exploring effective catalysts for water splitting [41, 89–91], Al-based alloys with nanoscaled galvanic microstructure was found to produce hydrogen at the contact with water [92]. As mentioned above, a joint experimental and theoretical study illustrated that gas-phase aluminium cluster anions are able to produce H_2 from water [7], where active sites on the cluster surface, typically Al_{16}^-, Al_{17}^-, and Al_{18}^-, produced hydrogen from water in the fast-flow tube reactor.

In general, a typical HER process on a metal electrode involves the following steps: (i) a discharge reaction, i.e., a Volmer step ($H^+ + e^- \rightarrow H_{ad}$), or the discharge reaction of H_3O^+ ions formed by the dissociation of water ($H_3O^+ + e^- \rightarrow H_{ad} + H_2O$); followed by (ii) either a recombination reaction, i.e., a Tafel step ($H_{ad} + H_{ad} \rightarrow H_2$) and/or an electrochemical desorption reaction, i.e., Heyrovsky reaction ($H^+ + H_{ad} + e^- \rightarrow H_2$; or $H_2O + H_{ad} + e^- \rightarrow H_2 + OH^-$), where H_{ad} refers to an adsorbed H atom [10]. Among others, the HER investigations of individual Al atoms with water has been reported using laser induced fluorescence (LIF) suggesting that the major product appears to be 'AlOH + H' with fragmentation of the HAlOH molecule [89]. Also, the reactivity of Al clusters with multiple water molecules was studied and it was demonstrated that the first step reaction is the generation of a $HAl_nOH(H_2O)_x$ species in which the additional water molecules play a catalytic role [41, 90].

Recently, a further insight was presented on the HER investigation of aluminum clusters with water and methanol/isopropanol mixture reactants. Although aluminum clusters were found to undertake an etching effect and an addition reaction in the presence of methanol-only and isopropanol-only respectively, products with hydrogen released were observed interestingly in the reactions with both "water + methanol" and "isoproanol + water" systems. The use of bireactants (Fig. 6.9) toward Al clusters enables a comparison of their reactivity and gas-phase interaction. Comparing with alcohols, water dominates the competitive reaction with Al clusters and the O–H bond in water is readily activated to form aluminum hydroxide cluster products in the room-temperature fast-flow tube apparatus. The transition states to produce hydrogen refer to combination of the adsorbed H atoms, which is akin to a Tafel step in general HER mechanism. Furthermore, water was found to contribute to

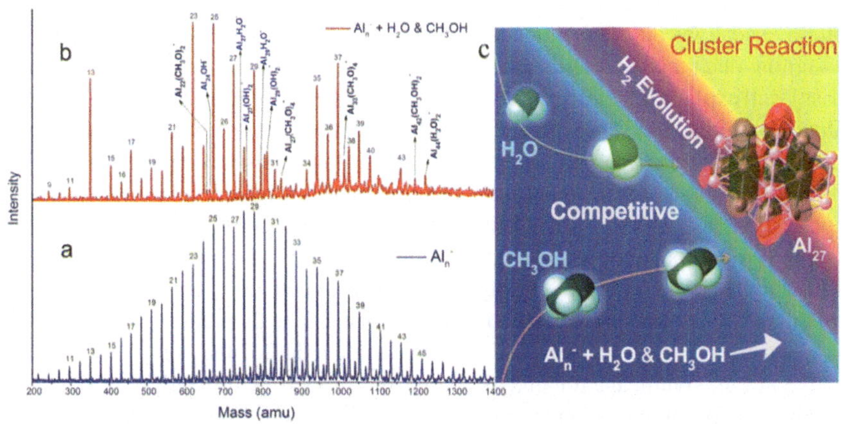

Fig. 6.9 Reaction of H_2O and CH_3OH (1:1 molar ratio) with Al clusters: **a** the original Al_n^- spectrum before the reaction, **b** the spectrum showing the products after simultaneous exposure to the reactants. **c** A sketch showing the competition and interaction between H_2O and CH_3OH when reacting with Al_n^-

the activation of the O–H bond in alcohols when reacting with Al cluster anions. These investigations help improve the understanding of the HER mechanism and indicate potential application for hydrogen generation [10]. Subsequent studies have demonstrated how partial atomic charges and bonding orbitals [93], and the doping of heteroatoms could affect the HER processes on clusters [38, 94–99], shedding light on an alternative Eley–Rideal and Langmuir–Hinshelwood mechanisms in the presence of two OH-group molecules [100].

Based on these HER investigations of –OH group molecules on Al clusters, the complementary-active-sites (CAS) mechanism [7] has been well established to explain the size-selective reactivity of metal clusters with polar molecules [5, 9, 29, 40, 42, 101, 102]. Recently, the reactions of vanadium clusters with water find a prominent hydrogen evolution reaction (HER) of single H_2O molecule for $V_{n\geq3}^+$ but no HER products were observed in the same condition for $n = 1, 2$. DFT-calculation results reveal that the wagging vibration of –OH group results in readily formed V–O–V intermediate states which allow the terminal hydrogen to interact with an adsorbed hydrogen atom giving rise to H_2 release. The presence of three vanadium atoms decreases the energy barrier of the rate-determine step transition state, resulting in effective H_2 production from a single water molecule. This mechanism is essentially different from the aformentioned reactivity of water with aluminum clusters by dissociative chemisorption of at least two water molecules at multiple surface sites followed by a Tafel step of recombination of the two adsorbed H atoms.

6.7 Summary

The understanding of size-selective metal reactivity towards polar molecules in gas phase shed light on complementary active sites. This established mechanism (by assigning one metal atom acts as a Lewis acid and a second Al atom acts as a Lewis base) well explained the contradiction for size-selectivity of Al_n^- reacting with water, alcohols, thiols, and some other molecular systems. It provides an insight into the origin of hydrogen evolution reactions. It is expected this mechanism will induce further understanding metal–organic reactivity and formation, as well as potential application in catalysis and industrial production. Recently Behrens et al. [103] reported a comprehensive experimental and theoretical analysis of the active site structure of a heterogeneous catalyst that is of crucial importance in industry to produce methanol. The active site of such catalyst can be thought of as the ensemble of atoms that directly catalyzes a reaction. Knowledge of the composition of the active site is significant for understanding the properties of catalysts [104, 105], and helping probing more challenging catalysts the interactions between the metals and the oxide supports are highly synergistic and sensitive to the environmental and reaction conditions [106, 107].

References

1. A.J. Pierik, W. Roseboom, R.P. Happe, K.A. Bagley, S.P.J. Albracht, J. Biol. Chem. **274**, 3331–3337 (1999)
2. L. Que, J. Chem. Soc., Dalton Trans. 3933–3940 (1997)
3. E.I. Solomon, D.E. Heppner, E.M. Johnston, J.W. Ginsbach, J. Cirera, M. Qayyum, M.T. Kieber-Emmons, C.H. Kjaergaard, R.G. Hadt, L. Tian, Chem. Rev. **114**, 3659–3853 (2014)
4. M.B. Abreu, C. Powell, A.C. Reber, S.N. Khanna, J. Am. Chem. Soc. **134**, 20507–20512 (2012)
5. A.C. Reber, S.N. Khanna, P.J. Roach, W.H. Woodward, A.W. Castleman Jr., J. Phys. Chem. A **114**, 6071–6081 (2010)
6. A.C. Reber, P.J. Roach, W.H. Woodward, S.N. Khanna, A.W. Castleman Jr., J. Phys. Chem. A **116**, 8085–8091 (2012)
7. P.J. Roach, W.H. Woodward, A.W. Castleman Jr., A.C. Reber, S.N. Khanna, Science **323**, 492–495 (2009)
8. W.H. Woodward, A.C. Reber, J.C. Smith, S.N. Khanna, A.W. Castleman Jr., J. Phys. Chem. C **117**, 7445–7450 (2013)
9. Z. Luo, A.W. Castleman Jr., S.N. Khanna, Chem. Rev. **116**, 14456–14492 (2016)
10. Z. Luo, J.C. Smith, W.H. Woodward, A.W. Castleman Jr., J. Phys. Chem. Lett. **3**, 3818–3821 (2012)
11. M.J. Danton, M.S. Coleman, Pediatr. Res. **19**, 751–751 (1985)
12. L. Dzantiev, L.J. Romano, Biochemistry **39**, 356–361 (2000)
13. D. Herschlag, T.R. Cech, Biochemistry **29**, 10159–10171 (1990)
14. D.C. Johnston, B.G. Silbernagel, M. Daage, R.R. Chianelli, Abstr. Pap. Am. Chem. Soc. **189**, 76-PETR (1985)
15. S. K. Khanna, Abstr. Pap. Am. Chem. Soc. **247** (2014)
16. D.H. Park, B.V. Plapp, J. Biol. Chem. **266**, 13296–13302 (1991)
17. T.M. Penning, W.R. Abrams, J.E. Pawlowski, J. Biol. Chem. **266**, 8826–8834 (1991)
18. T.M. Penning, W.R. Abrams, J.E. Pawlowski, FASEB J. **5**, A442–A442 (1991)
19. K.M. Peterson, D.K. Srivastava, Biochemistry **39**, 12678–12687 (2000)
20. D. Rennex, B.A. Hemmings, J. Hofsteenge, S.R. Stone, Biochemistry **30**, 2195–2203 (1991)
21. E. Schleicher, K. Hitomi, C.W.M. Kay, E.D. Getzoff, T. Todo, S. Weber, J. Biol. Chem. **282**, 4738–4747 (2007)
22. S.A. Strobel, L. Ortoleva-Donnelly, S.P. Ryder, J.H. Cate, E. Moncoeur, Nature Structural Biology **5**, 60–66 (1998)
23. J.-Y. Zhao, F.-Q. Zhao, S.-Y. Xu, X.-H. Ju, J. Phys. Chem. A **117**, 2213–2222 (2013)
24. Y. Liu, Y. Hua, M. Jiang, M. Xu, F. Yu, J. Chen, Eur. Phys. J. D **67**, 194 (2013)
25. S. Das, S. Pal, S. Krishnamurty, J. Phys. Chem. A **117**, 8691–8702 (2013)
26. M. Arakawa, K. Kohara, T. Ito, A. Terasaki, Eur. Phys. J. D **67**, 80 (2013)
27. Y. Liu, Y. Hua, M. Jiang, G. Jiang, J. Chen, J. Chem. Phys. **136**, 084703 (2012)
28. P.N. Day, K.A. Nguyen, R. Pachter, J. Chem. Theory Comput. **8**, 152–161 (2012)
29. S. Ohmura, F. Shimojo, R.K. Kalia, M. Kunaseth, A. Nakano, P. Vashishta, J. Chem. Phys. **134**, 244702 (2011)
30. W. Mou, S. Ohmura, A. Hemeryck, F. Shimojo, R.K. Kalia, A. Nakano, P. Vashishta, Aip Adv. **1**, 042149 (2011)
31. S.A. Krasnokutski, F. Huisken, J. Phys. Chem. A **115**, 7120–7126 (2011)
32. M. Garcia-Rates, P. Miro, J. Maria Poblet, C. Bo, J. Bonet Avalos, J. Phys. Chem. B, **115**, 5980–5992 (2011)
33. F. Shimojo, S. Ohmura, R.K. Kalia, A. Nakano, P. Vashishta, Phys. Rev. Lett. **104**, 126102 (2010)
34. L. Zhu, L. Li, S. D. Hao, X.Y. Lin, H.C. Liang, C.J. Yang, Theoretical investigation of molecular water adsorption on Al(111) surface (2005)
35. A.G. Stack, J.R. Rustad, W.H. Casey, Geochim. Cosmochim. Acta **69**, A48–A48 (2005)

36. B.M. Reinhard, G. Niedner-Schatteburg, J. Phys. Chem. A **106**, 7988–7992 (2002)
37. V.A. Mikhailov, P.E. Barran, A.J. Stace, Phys. Chem. Chem. Phys. **1**, 3461–3465 (1999)
38. F. Li, L. Sun, J. Zhao, F. Xu, H.-Y. Zhou, Q.-M. Zhang, F.-L. Huang, Int. J. Hydrogen Energy **38**, 6930–6937 (2013)
39. C.K. Siu, Z.F. Liu, J.S. Tse, J. Am. Chem. Soc. **124**, 10846–10860 (2002)
40. S. Álvarez-Barcia, J.R. Flores, Chem. Phys. **374**, 131–137 (2010)
41. S. Álvarez-Barcia, J.R. Flores, Chem. Phys. **382**, 92–97 (2011)
42. S. Álvarez-Barcia, J.R. Flores, J. Phys. Chem. A **116**, 8040–8050 (2012)
43. R. Hofmannsievert, A.W. Castleman Jr., J. Phys. Chem. **88**, 3329–3333 (1984)
44. W. H. H. Woodward, PhD, The Pennsylvania State University, 2011.
45. K. Sugawara, F. Sobott, A.B. Vakhtin, J. Chem. Phys. **118**, 7808–7816 (2003)
46. D.A. Daramola, G.G. Botte, Comput. Theo. Chem. **989**, 7–17 (2012)
47. G.K. Koyanagi, V. Kapishon, D.K. Bohme, X. Zhang, H. Schwarz, Eur. J. Inorg. Chem. 1516–1521 (2010)
48. A. Grubisic, X. Li, G. Gantefoer, K.H. Bowen, H. Schnoeckel, F.J. Tenorio, A. Martinez, J. Chem. Phys. **131**, 184305 (2009)
49. M. Oncak, Y. Cao, M.K. Beyer, R. Zahradnik, H. Schwarz, Chem. Phys. Lett. **450**, 268–273 (2008)
50. E.S. Kryachko, F. Remacle, J. Chem. Phys. **127**, 194305 (2007)
51. A. Martinez, J. Braz. Chem. Soc. **16**, 337–344 (2005)
52. K. Koszinowski, D. Schroder, H. Schwarz, Organometallics **23**, 1132–1139 (2004)
53. K. Koszinowski, M. Schlangen, D. Schroder, H. Schwarz, Int. J. Mass Spectrom. **237**, 19–23 (2004)
54. A. Antusek, M. Urban, A.J. Sadlej, J. Chem. Phys. **119**, 7247–7262 (2003)
55. M.F. Zhou, M.H. Chen, L.N. Zhang, H. Lu, J. Phys. Chem. A **106**, 9017–9023 (2002)
56. C. Lacaze-Dufaure, T. Mineva, N. Russo, J. Comput. Chem. **22**, 1557–1564 (2001)
57. K.A. Jackson, M. Knickelbein, G. Koretsky, S. Srinivas, Chem. Phys. **262**, 41–51 (2000)
58. S.E. Kooi, A.W. Castleman, Chem. Phys. Lett. **315**, 49–54 (1999)
59. M. Garcia-Hernandez, N. Lopez, I.D. Moreira, J.C. Paniagua, F. Illas, Surf. Sci. **430**, 18–28 (1999)
60. W.T. Chan, R. Fournier, Chem. Phys. Lett. **315**, 257–265 (1999)
61. L. Lian, S.A. Mitchell, P.A. Hackett, D.M. Rayner, J. Chem. Phys. **104**, 5338–5344 (1996)
62. A. Fahmi, R.A. vanSanten, Z. Phys. Chem. **197**, 203–217 (1996)
63. U.N. Andersen, G. Bojesen, Int. J. Mass Spectrom. Ion Processes **153**, 1–7 (1996)
64. B. J. Winter, T. D. Klots, E. K. Parks and S. J. Riley, *Z. Phys. D: At., Mol. Clusters*, 1991, **19**, 381–384.
65. K. Fuke, S. Nonose, N. Kikuchi, K. Kaya, Chem. Phys. Lett. **147**, 479–483 (1988)
66. L. Geng, C. Cui, Y. Jia, H. Wu, H. Zhang, B. Yin, Z.D. Sun, Z. Luo, J. Phys. Chem. A **124**, 5879–5886 (2020)
67. A.N. Pestryakov, V.V. Lunin, N. Bogdanchikova, O.N. Temkin, E. Smolentseva, Fuel **110**, 48–53 (2013)
68. M. Ichihashi, C.A. Corbett, T. Hanmura, J.M. Lisy, T. Kondow, J. Phys. Chem. A **109**, 7872–7880 (2005)
69. S. Keki, L. Nagy, G. Deak, M. Zsuga, L. Somogyi, A. Levai, J. Am. Soc. Mass Spectrom. **15**, 879–883 (2004)
70. Y.L. Cao, X.A. Zhao, B. Xin, S.X. Xiong, Z.C. Tang, J. Mol. Struct. Theochem **683**, 141–146 (2004)
71. S.H. Cai, K. Sohlberg, J. Mol. Catal. a-Chem. **193**, 157–164 (2003)
72. G.M. Koretsky, M.B. Knickelbein, R. Rousseau, D. Marx, J. Phys. Chem. A **105**, 11197–11203 (2001)
73. R. Rousseau, D. Marx, J. Chem. Phys. **112**, 761–769 (2000)
74. A.M.L. Oiestad, E. Uggerud, Chem. Phys. **262**, 169–177 (2000)
75. G. Dietrich, S. Kruckeberg, K. Lutzenkirchen, L. Schweikhard, C. Walther, J. Chem. Phys. **112**, 752–760 (2000)

76. R. Rousseau, G. Dietrich, S. Kruckeberg, K. Lutzenkirchen, D. Marx, L. Schweikhard, C. Walther, Chem. Phys. Lett. **295**, 41–46 (1998)
77. M.B. Knickelbein, G.M. Koretsky, J. Phys. Chem. A **102**, 580–586 (1998)
78. R.T. Yadav, M. Ichihashi, T. Kondow, J. Phys. Chem. A **108**, 7188–7192 (2004)
79. Z.X. Tian, X.P. Xing, Z.C. Tang, Rapid Commun. Mass Spectrom. **17**, 17–23 (2003)
80. E.E. Fileti, P. Chaudhuri, S. Canuto, Chem. Phys. Lett. **400**, 494–499 (2004)
81. M.W. Chase, *NIST-JANAF Thermochemical Tables, Fourth edition* (American Institute Of Phys, 1998)
82. S. Mori, M. Suginoya, Y. Tamai, Asle Transactions **25**, 261–266 (1982)
83. S.A. Claridge, A.W. Castleman, S.N. Khanna, C.B. Murray, A. Sen, P.S. Weiss, ACS Nano **3**, 244–255 (2009)
84. A. Guevara-Garcia, A. Martinez, J.V. Ortiz, J. Chem. Phys. **122**, 214309 (2005)
85. V. Bonacic-Koutecky, J. Burda, R. Mitric, M.F. Ge, G. Zampella, P. Fantucci, J. Chem. Phys. **117**, 3120–3131 (2002)
86. A. Sanchez, S. Abbet, U. Heiz, W.D. Schneider, H. Hakkinen, R.N. Barnett, U. Landman, J. Phys. Chem. A **103**, 9573–9578 (1999)
87. J.A. Anderson, M.F. Garcia, *Supported Metals in Catalysis* (Imperial College Press, 2005)
88. A.F. Carley, M.W. Roberts, *Surface Chemistry and Catalysis* (Kluwer Academic/Plenum Publishers, 2002)
89. R.E. McClean, H.H. Nelson, M.L. Campbell, J. Phys. Chem. **97**, 9673–9676 (1993)
90. S. Alvarez-Barcia, J.R. Flores, J. Chem. Phys. **131**, 174307 (2009)
91. C.X. Zhang, C.H. Chen, H.X. Dong, J.R. Shen, H. Dau, J.Q. Zhao, Science **348**, 690–693 (2015)
92. *US Patient*, US20190024216A1 (2019)
93. A.M.S. Pembere, X. Liu, W. Ding, Z. Luo, J. Phys. Chem. A **122**, 3107–3114 (2018)
94. H. Zhang, C. Cui, Z. Luo, J. Phys. Chem. A **123**, 4891–4899 (2019)
95. H.Z. Wang, D.Y.C. Leung, M.K.H. Leung, M. Ni, Renew. Sust. Energ. Rev. **13**, 845–853 (2009)
96. J.M. Woodall, J.T. Ziebarth, C.R. Allen, D.M. Sherman, J. Jeon, G. Choi, in *Materials Innovations in an Emerging Hydrogen Economy*, ed by G.G. Wicks, J. Simon, vol. 202 (2009), pp. 121–127
97. K. Cheng, R. Fu, D. Sang, L. Jing, Y. Li, Mater. Lett. **129**, 84–87 (2014)
98. K.N. Li, C.L. Yang, M.S. Wang, X.G. Ma, L.Z. Wang, Int. J. Hydrogen Energy **40**, 8911–8916 (2015)
99. K.N. Li, C.L. Yang, M.S. Wang, X.G. Ma, L.Z. Wang, Int. J. Hydrogen Energy **41**, 17858–17863 (2016)
100. J. Chen, Z. Luo, ChemPhysChem **20**, 499–505 (2019)
101. S. Álvarez-Barcia, J.R. Flores, J. Phys. Chem. C **115**, 24849–24857 (2011)
102. S. Alvarez-Barcia, J.R. Flores, J. Chem. Phys. **140**, 084313 (2014)
103. M. Behrens, F. Studt, I. Kasatkin, S. Kühl, M. Hävecker, F. Abild-Pedersen, S. Zander, F. Girgsdies, P. Kurr, B.-L. Kniep, M. Tovar, R.W. Fischer, J.K. Nørskov, R. Schlögl, Science **336**, 893–897 (2012)
104. G.A. Somorjai, *Introduction to Surface Chemistry and Catalysis* (Wiley, New York, 1994)
105. J. Greeley, J.K. Norskov, M. Mavrikakis, Annu. Rev. Phys. Chem. **53**, 319–348 (2002)
106. J.P. Greeley, Science **336**, 810–811 (2012)
107. P.L. Hansen, J.B. Wagner, S. Helveg, J.R. Rostrup-Nielsen, B.S. Clausen, H. Topsøe, Science **295**, 2053–2055 (2002)

Chapter 7
The Reactions with Monoxides for Pollution Removal

7.1 Introduction

In general, the main cluster types include naked clusters without stabilizing ligands. Besides, there are abundant investigations on ligand-protected clusters which serve as building blocks in numerous materials [1]. The stability of such clusters is often reconciled by ligands granting the cluster a closed electronic shell through covalent or ionic bonding [2]. For example, clusters especially gold usually employ a class of ligands including thiols [3–9] and phosphines [10–13] to improve the stabilization. In gas phase, typical clusters for main group elements can be stabilized by hydride ligands; while for transition metal clusters, typical stabilizing ligands have been found to be carbon monoxide (also halides, isocyanides, alkenes and hydrides, etc.) [1, 14–19]. These ligands undergo electronegative bonding with the metallic core of the cluster, perturbing the electronic structure of the cluster. In this chapter, we will present the curious influence of CO encountering metal clusters. In addition to exploring the stability of cluster species in protection of such ligands, equitable research interest in studying the cluster reactivity of metals and metal oxides with carbon monoxide has also been recognized of importance in catalysis. Transition metal oxides (molecules or clusters) provide good catalytic activity in oxidation reactions due to the potential advantage that the oxygen atoms are bound atomically to the transition metal atoms and there are no O–O bonds unless the oxygen to metal ratio is large [20].

It is well-known that CO is used to manufacture steel from iron ore in blast furnaces where the general reactions are "$Fe_2O_3 + 3CO = 2Fe + 3CO_2$, $Fe_3O_4 + 4CO = 3Fe + 4CO_2$" [21, 22]. However, as the reaction to provide CO by "$C + CO_2 = 2CO$" is reversible, there is always some CO included in the exhaust gas polluting air if no effective treatments. In addition to the exhaust gas in metal smelting, the environmentally harmful gas, CO, is also involved in automobile exhaust. In general, the oxidation of CO to CO_2 is an important environmental reaction to the abatement of harmful carbon monoxide gas. This reaction involves breaking the O–O bond of

Z. Luo and S. N. Khanna, *Metal Clusters and Their Reactivity*,
https://doi.org/10.1007/978-981-15-9704-6_7

the O_2 molecule and hence catalysts are generally used to overcome a large energy barrier. Finding more efficient and selective catalyst materials with lower operating temperatures for the oxidation of CO is beneficial. Recent years researchers have found that iron oxide clusters are important alternative species to effect the oxidation of CO at low temperature without the activation of strong O–O bonds [21]. Iron oxides are very practical in pollution abatement applications as they are abundant and inexpensive, and hence attracted reasonable research interest [23–35].

7.2 Transition Metal Clusters React with CO

Probing the unique size-dependent properties of small metal clusters and metal oxide clusters, CO-chemisorbed metal clusters, and especially their reactivity and catalysis towards CO has attracted extensive interest in this field [36–61].

7.2.1 Cobalt Clusters React with CO

Cobalt metal is a useful catalyst in many industrial processes, especially in C_1 chemistry or Fisher-Tropsch synthesis which refers to a collection of chemical reactions that converts a mixture of carbon monoxide and hydrogen into liquid hydrocarbons [62, 63]. It was also noted that cobalt is the best metal catalyst for the conversion of methane into large hydrocarbon molecules [43]. It is of potential importance to study the reactivity of cobalt clusters, so as to provide insights into the micro-processes which occur at the catalytic surface of cobalt metal, for instance, what kind or size of cobalt particles are expected to display greater catalytic activity [64, 65]. The reactivity of anionic cobalt clusters with CO has been studied by Kapiloff and Ervin [66], who noted the sequential addition of CO to the cobalt cluster anions leading to saturated species $Co_2(CO)_7^-$, $Co_3(CO)_{10}^-$, $Co_4(CO)_{12}^-$, $Co_5(CO)_{13}^-$, and $Co_6(CO)_{15}^-$ for which skeletal structures were proposed in accordance with electron-counting rules, as shown in Figs. 7.1 and 7.2. Different numbers of CO molecules add to the clusters but do not break the C-O bonds, and almost no fragmentation of the metal cores. The DFT calculations showed that there are small CO-adsorption energies which could be partially removed by collisions with the buffer gas before reaction or cluster fragmentation. It is worth mentioning that, comparing with smaller clusters, the larger clusters produced greater fragmentation versus addition products.

An investigation by Guo et al. [43] reported the reactions of mass-selected clusters Co_n^+ (n = 2–8) with CO in the gas phase using a selected ion drift tube affixed with a laser vaporization source (SIDT-LV) operated under well-defined thermal conditions. All reactions for "$CO + Co_n^+$ (n = 2–8)" were found to be association reactions although their absolute rate constants displayed a strong dependence on cluster size. Among these cobalt clusters, Co_4^+ and Co_5^+ display a higher reactivity toward the

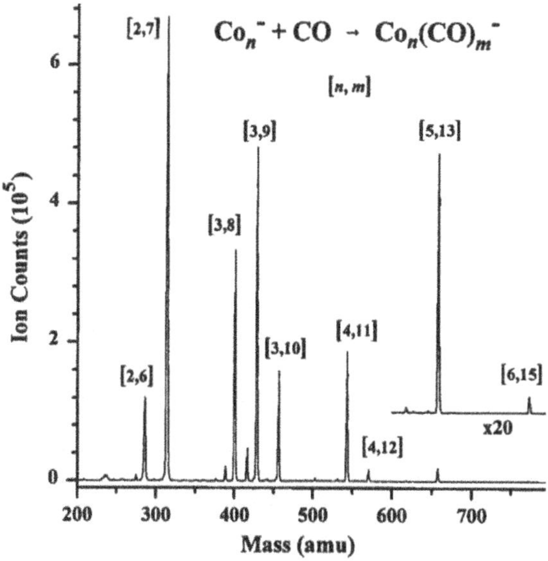

Fig. 7.1 A mass spectrum at high CO flow rates corresponding to 0.05 Torr pressure. $Co_n(CO)_m^-$ ions are identified by [n, m]. The small peaks are cobalt oxides ions with added carbonyls or other impurities. Reproduced with permission from Ref. [66]. Copyright 1997 American Chemical Society

CO molecules than do clusters of neighboring sizes. The multiple-collision conditions employed in their work enabled a determination of the maximum coordination number of CO molecules bound onto each Co_n^+ cluster, which were interpreted in terms of *Lauher's* calculation and the polyhedral skeletal electron pair theory. For example, the tetramer tends to bond 12 CO molecules, the pentamer bond 14 CO, while the hexamer could attach 16 CO molecules. However, for the trimer Co_3^+, the measured maximum coordination number is one CO less than the predicted value. These findings also helped to have determined the cobalt cluster structures, where tetramer cation Co_4^+ was interpreted to have a tetrahedral structure, the pentamer Co_5^+ a trigonal bipyramid, and the hexamer Co_6^+ an octahedral structure [43].

7.2.2 Ni_n^+ Clusters React with CO

A study in *Wöste* group [67] by using a triple quadrupole mass spectrometer with a sputter source demonstrated the reactions of nickel cation clusters Ni_n^+ with CO, as shown in Fig. 7.3. Interestingly, the size-selected nickel clusters were found to react with carbon monoxide and produce gas-phase nickel-carbonyl complexes of the type $Ni_n(CO)_k^+$, $Ni_nC(CO)_l^+$ and $Ni_{n-1}C(CO)_m^+$ where n ranges from 1 through 13, while k, l and m vary as a function of the cluster size n. Controlled syntheses of these homoleptic nickel carbonyls $Ni_x(CO)_y^+$ of which the stoichiometry can be related to the bonding models for organometallic cages and clusters. Many of the measured numbers were found to coincide with the theoretical prediction [68–72]. Individual CO ligands in multi-metallic complexes rapidly interchange positions

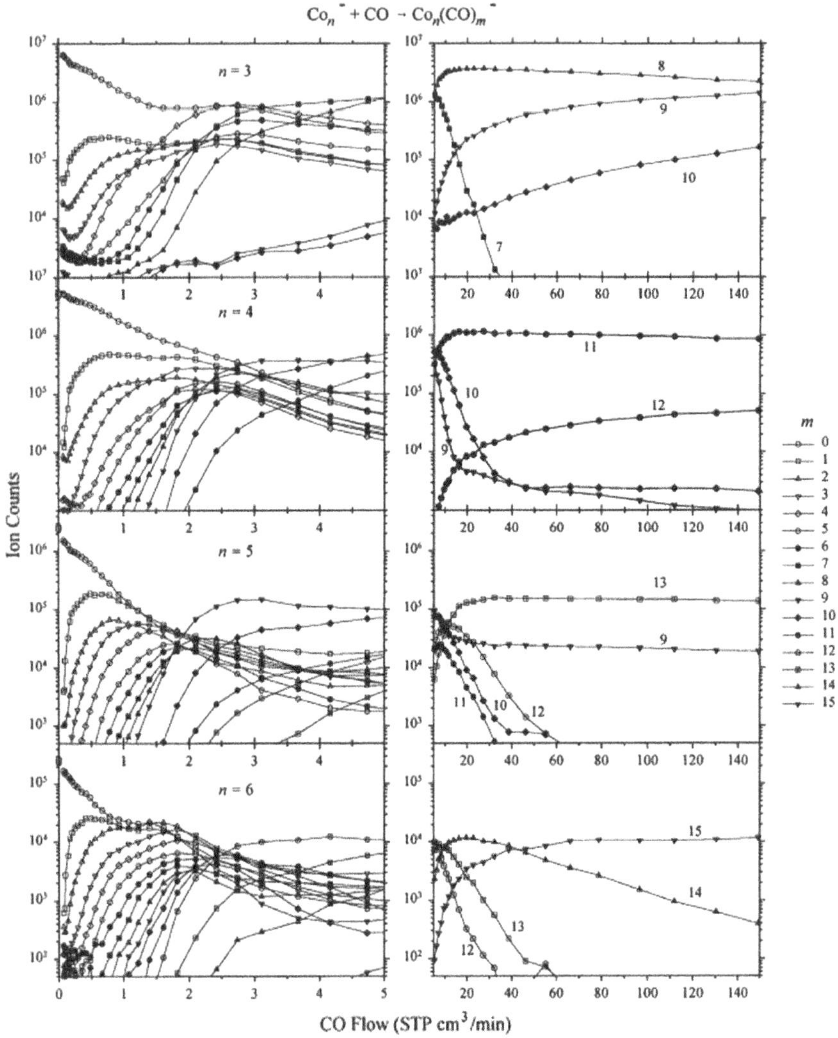

Fig. 7.2 Ion intensities of the anionic products from the reactions of Co_n^- with carbon monoxide as a function of CO flow rate. $Co_n(CO)_m^-$ ions are grouped by n from n = 3 to n = 6 (top to bottom panels), and m values are given by the symbols according to the legend at the right. The left panels show all products at low CO flow rates (0–5 STP cm^{3}min^{-1}), while the right panels show only the species corresponding to the last few carbonyls added up to the saturation limit (labelled by m values) at higher CO flow rates. Reproduced with permission from Ref. [66]. Copyright 1997 American Chemical Society

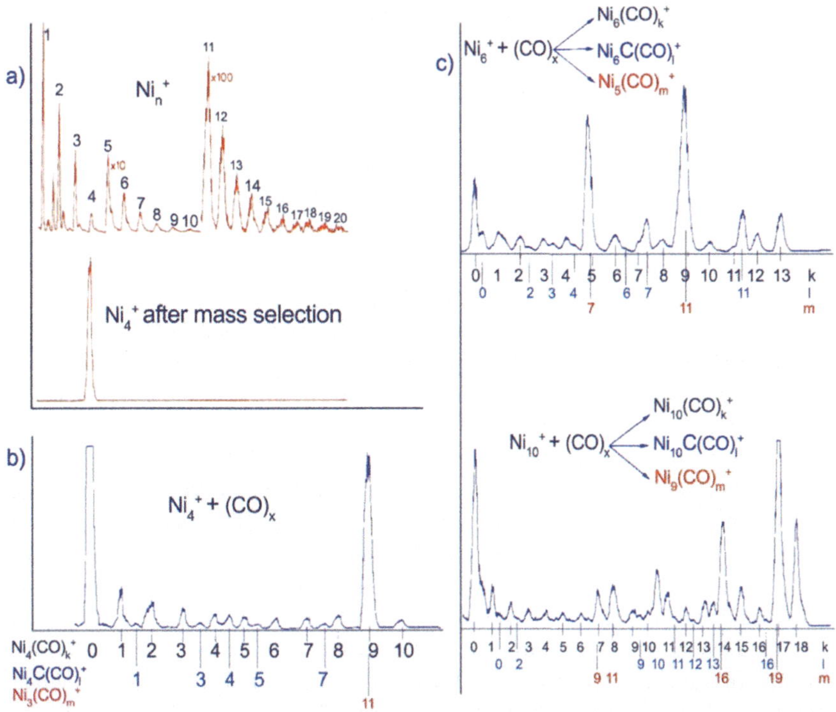

Fig. 7.3 **a** A mass spectrum of positively charged sputtered nickel clusters. Note the sensitivity change at Ni_5^+ and Ni_{11}^+. The monodispersed cluster beam was recorded by setting the first QMS in front of the ion drift tube on the mass of Ni_4^+, while the second spectrometer behind the drift tube was scanned. Products of the reaction of carbon monoxide with **b** Ni_4^+, **c** Ni_6^+, and **d** Ni_{10}^+ at a CO pressure of approximately 3 mbar. Reproduced with permission from Ref. [67]. Copyright 1987 American Chemical Society

intramolecularly. As ligands are interacting with the cluster as a whole and not just with individual metal atoms, it was proposed that there is no special arrangement of ligands around the central metal core which is greatly preferred energetically over all others [73, 74]. Lauher [68] has calculated the most favorable molecular geometry for any given transition metal cluster and predicted the bonding capabilities of such clusters; many of *Lauher's* predictions have been experimentally verified, [75] including the reactivity of Co_n^+ cationic clusters with CO as discussed above [43].

7.3 Reactivity of CO with Iron Oxides

7.3.1 Anionic Clusters $Fe_nO_m^-$

Castleman and Khanna [22] showed a synergistic investigation combining gas-phase experiments and theoretical first-principles calculations to study the structure, stability, and reactivity of $Fe_nO_m^-$ clusters towards CO. Collision-induced dissociation of these iron oxide species under xenon gas collisions showed that both FeO_3^- and FeO_2^- are stable building blocks in forming larger iron oxide clusters. Based on mass-selected experiments, the transfer of oxygen atoms from $Fe_nO_m^-$ to CO was seen as the dominant reaction pathway. Further, theoretical calculations demonstrated that the fragmentation patterns leading to the production of O or FeO_n fragments are governed both by the energetics of the overall process (as well as the number of intermediate states) and the changes in spin multiplicity [22].

Figure 7.4A shows the anionic iron oxide cluster distribution with both dissociated and molecular oxygen adsorbed at near thermal energy. The calculated optimized lowest-energy structures for these anionic FeO_{1-4}^- and $Fe_2O_{2-6}^-$ clusters are shown in Fig. 7.4B. For iron oxide clusters containing a single Fe atom, oxygen atoms bind directly to the metal with no molecular oxygen units, where the maximum coordination number of Fe is four with a tetrahedral form of the FeO_4^- as ground state structure. In comparison, an isomer of FeO_4^- possessing a molecular oxygen subunit exhibits a higher energy of 0.78 eV above the ground-state energy. All the Fe–O bond lengths in FeO_{1-4}^- are relatively similar (~1.65 Å); but the spin multiplicity changes from quartet in FeO_{1-3}^- species to doublet for FeO_4^-. Different from the structures of FeO_{1-4}^-, a basic ring structure composed of $Fe_2O_2^-$ with oxygen bridging each iron atom is formed for all the clusters $Fe_2O_{2-6}^-$. The species $Fe_2O_{3-6}^-$ have their additional oxygen atoms attached to iron outside the 1.85–1.79 Å within the ring (i.e., Fe–O–Fe) while 1.65–1.62 Å outside. Optimized calculations showed that the maximum coordination for the $Fe_2O_6^-$ cluster is two extra oxygen atoms bounding directly to each Fe atom. In addition, an anti-ferromagnetic spin coupling was noted for $Fe_2O_2^-$, $Fe_2O_3^-$, $Fe_2O_4^-$ and $Fe_2O_5^-$; however, it was found that the Fe sites in $Fe_2O_6^-$ are coupled ferromagnetically [22]. Note that the progression of the exchange coupling with oxidation coincides with the case of Cr_2O_n clusters [76].

In order to ascertain the stability of these iron oxide clusters, collision induced dissociation studies were undertaken with inert xenon gas under single (0.09 mTorr) and multiple (0.2 mTorr) collision conditions. Simultaneously, calculations were done on the energies required to remove an O, O_2, O^-, and O_2^- from FeO_n^- and $Fe_2O_n^-$ clusters, as well as the energies for fragmentation pathways leading to the production of Fe, FeO, FeO_2 and FeO_3 for the $Fe_2O_n^-$ clusters. The results of these investigations are plotted in Fig. 7.4C, which corresponded to the observed experimental products. It was noted that the loss of O_2 reaches an energetic minimum for $Fe_2O_6^-$, indicating the energy needed to break two Fe–O bonds possibly overcome by the exothermic formation of O_2 (which releases 6.21 eV energy). Such oxygen recombination was also observed in studies conducted with $V_2O_5^+$ [77]. In

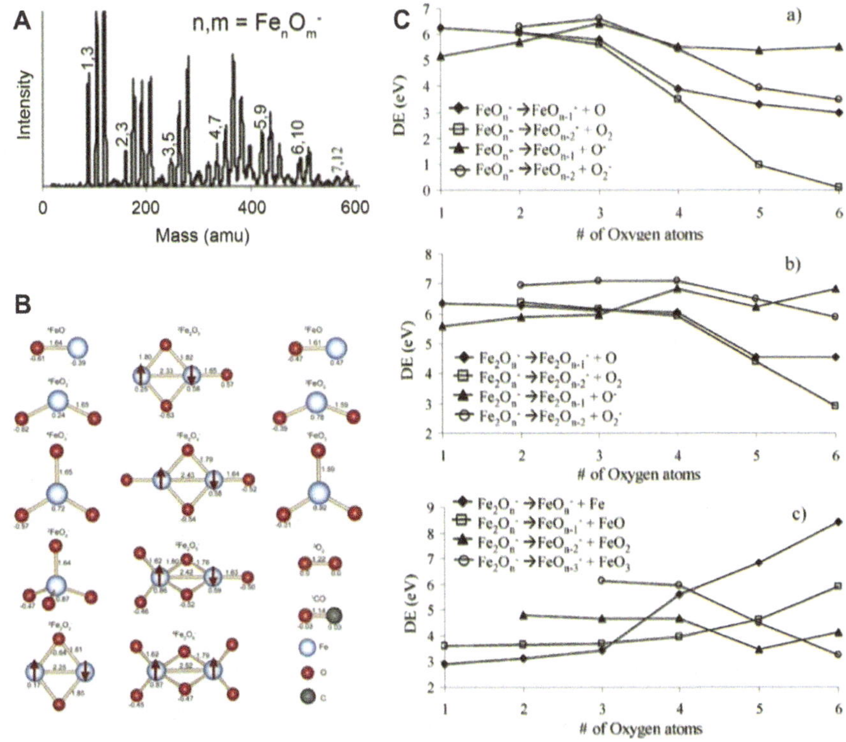

Fig. 7.4 **A** A typical mass distribution produced for iron oxide anionic clusters. The first iron oxide in each series is labelled according to $Fe_nO_m^-$, where (n, m) with subsequent peaks in the series have one additional oxygen atom. **B** The ground state geometries of O_2, CO, FeO_{1-3}, and $Fe_nO_m^-$ clusters. The bond lengths are given in Angstroms and the superscripts indicate the spin multiplicity. The arrows indicate the spin polarization at the Fe atoms for the $Fe_2O_m^-$ clusters. The Mulliken charges are marked below each atom. **C** Graphs of the dissociation energy associated with removing an O atom or O_2 subunit from (a) FeO_n^- and (b) $Fe_2O_n^-$. (c) Graph of the dissociation energy associated with removing Fe or FeO_n from $Fe_2O_n^-$ clusters

comparison, the breaking of Fe–Fe bonds in the dimer clusters $Fe_2O_n^-$ could mainly occurred at higher energies or under multiple collision conditions [22].

Figure 7.5 shows the reactivity of FeO_2^-, FeO_4^-, $Fe_2O_3^-$ and $Fe_2O_6^-$ with CO, where the reactant and product signals change as a function of increasing CO pressure. These cluster stoichiometries, especially FeO_2^- and $Fe_2O_3^-$ which both are composed of one more oxygen atom than the number of iron atoms, were found to be the most efficient iron oxide anions by following the CO oxidation reaction channel. Theoretical studies showed that the atomization energy of CO and CO_2 are 11.63 and 17.97 eV, respectively, that is, the formation of CO_2 is energetically feasible in cases where it takes less than 6.34 eV to remove an O atom from the cluster, providing no insurmountable reaction barriers to prevent the formation of CO_2. The presence of $Fe_nO_{m-1}^-$ products (mass-selected reaction from $Fe_nO_m^-$) suggested that oxygen

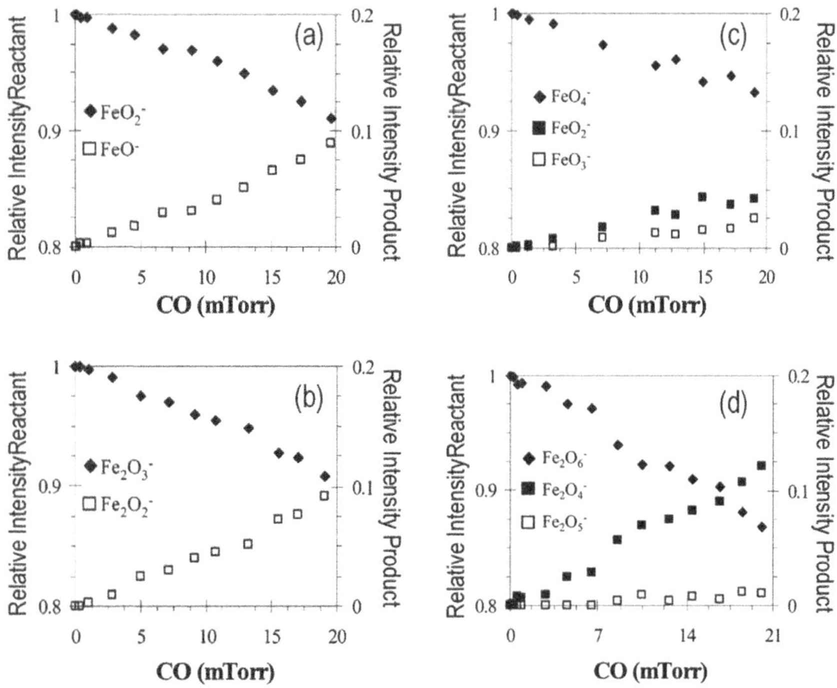

Fig. 7.5 Relative intensities of **a** FeO_2^- and FeO, **b** $Fe_2O_3^-$ and $Fe_2O_2^-$, **c** FeO_4^- and products species, and **d** $Fe_2O_6^-$ and product species, as functions of increasing CO pressure, respectively. The observed behavior shows that both FeO_2^- and $Fe_2O_3^-$ are effective for the oxygen atom transfer reaction. Note that both clusters take the dominant channel with molecular oxygen loss over atomic oxygen loss

is transferred from iron oxide clusters to CO and produce neutral CO_2 molecules. Further studies on the reaction pathways demonstrated that the reaction "$Fe_nO_m^- + CO \rightarrow Fe_nO_{m-1}^- + CO_2$" proceeds for those without barriers and follows a spin allowed path [22]. Calculation results showed that the first CO attaches to the Fe site or approach the O atom to form CO_2, and the more stable configuration corresponds to CO attached to the Fe site; and then the subsequent CO attaches to the O site forming CO_2. As revealed by the Mulliken population calculation, this result is consistent with the electron donating behavior of CO and the partial positive charge present on the Fe site [22, 78]. Besides, another reaction channel observed was the loss of molecular oxygen from the iron oxide clusters $Fe_nO_m^-$, as noted for FeO_4^- and $Fe_2O_6^-$ (Fig. 7.5c and d). The cluster intensities changed with increasing CO pressures and O_2 loss was identified as the dominant reaction channel followed by minor O atom loss [22].

7.3.2 Cationic Clusters $Fe_nO_m^+$ Reacting with CO

As is mentioned above, iron oxide clusters display practical applications in pollution abatement due to the effective reactivity towards CO, as well as the abundant and inexpensive iron ore resources [23–35]. Further insights into the reactivity of iron oxides with CO have also been studied involving small iron oxide cationic clusters $Fe_nO_m^+$ ($n = 1, 2, m = 1–5$), as displayed in Fig. 7.6. The reactivity in the presence of CO at near thermal energies was examined via a guided ion beam mass spectrometer.

First-principles calculations within the density functional theory framework were carried out to address the structures and energetics of small cationic clusters and to demonstrate the reaction pathways for "CO + $Fe_nO_m^+$ →". As results showing in Figs. 7.7 and 7.8, reaction channels including CO oxidation and oxygen replacement by CO were noted to be dependent on cluster size and stoichiometry. The reaction pathways were fully accounted for by the bond energies and the gain in energy when binding CO to the cluster. A detailed analysis of the reaction pathways for $Fe_nO_m^+$ clusters with CO was discussed showing the Fe–C attachment as the initial step of the reaction. Only for clusters with certain stoichiometries the CO oxidation is energetically feasible; and for oxygen-rich iron oxide clusters (e.g., $Fe_nO_m^+$ where $n \geq 3$), intermediate species with associated CO molecules were also observed (e.g.,

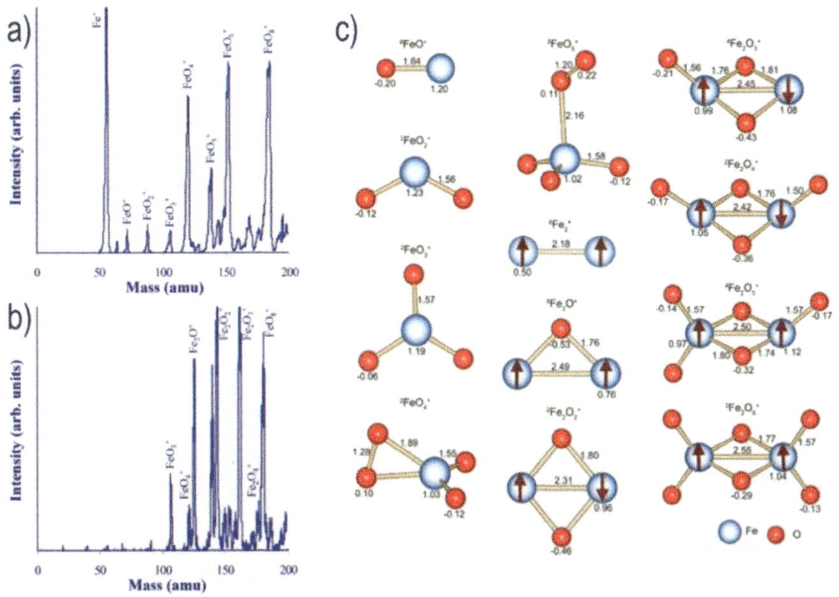

Fig. 7.6 Mass distribution of iron oxide cation clusters produced when employing **a** a 27 mm conical nozzle and **b** a 51 mm conical nozzle at the exit of the source; Ground-state geometries of $Fe_nO_m^+$ clusters. The bond lengths are given in angstroms, and the superscripts indicate the spin multiplicity. The arrows indicate the spin polarization at the Fe atoms for the $Fe_2O_m^+$ clusters. The Mulliken charges are marked below each atom

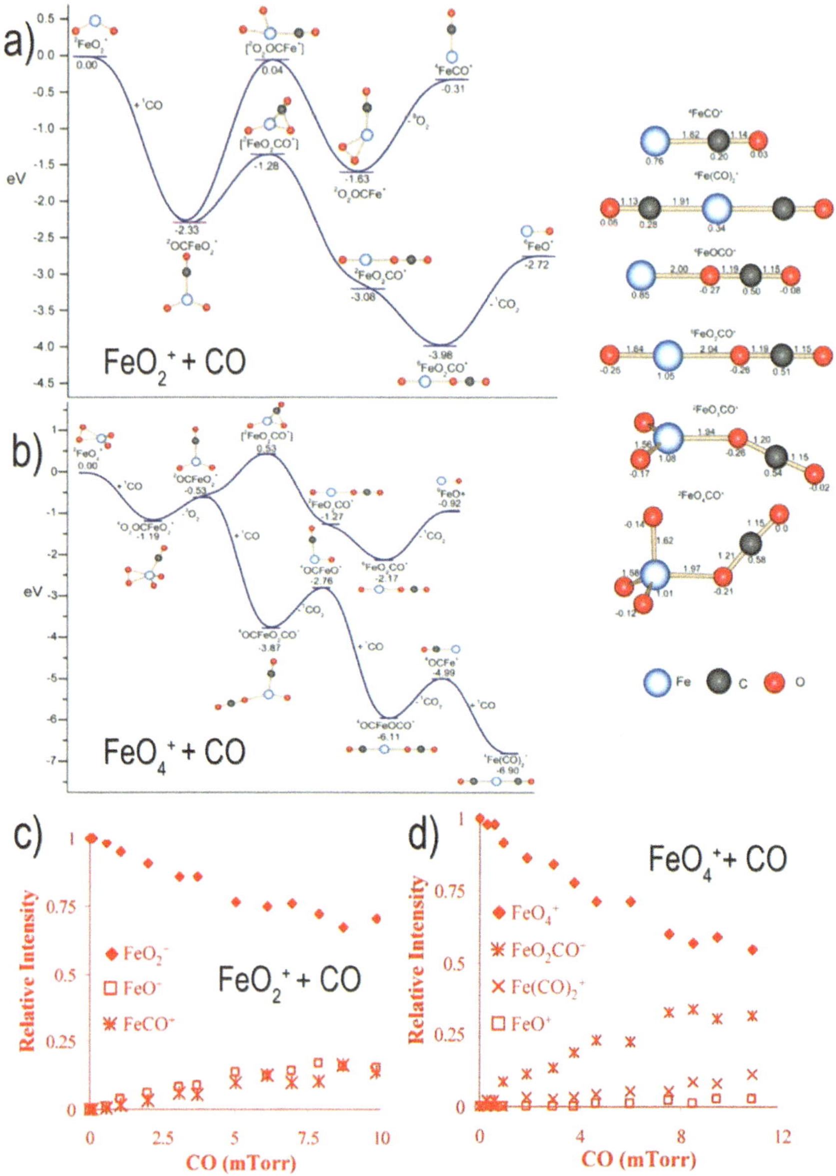

Fig. 7.7 **a/b** Change in energy (ΔE) for each step of the reaction pathway of FeO_2^+ and FeO_4^+ with CO. The superscripts indicate spin multiplicity; **c/d** ion intensity changes of FeO_2^+/FeO_4^+ and their respective observed products with CO as a function of increasing CO pressure

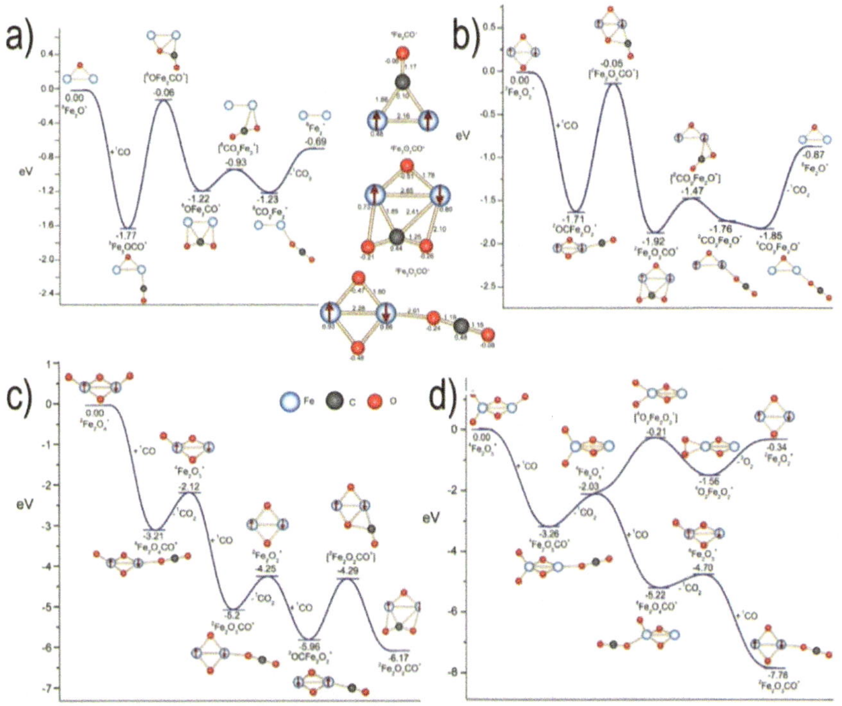

Fig. 7.8 Change in energy (ΔE) for each step of the reaction pathway of Fe_2O^+ (**a**), $Fe_2O_2^+$ (**b**), $Fe_2O_4^+$ (**c**), and $Fe_2O_5^+$ (**d**) with CO. The superscripts indicate spin multiplicity

$Fe[CO]_2^+$ and FeO_2CO^+ etc.). The relative ionic intensities of $Fe_nO_m^+$ and their correlated products *in* reacting with CO changed as a function of the increasing CO pressure. In conjunction of experimental findings with theoretical calculations, it was demonstrated a coherent picture about the trends in the dissociation energies which largely depend on the metal to oxygen ratio, and the nature of the structure–reactivity relationships in forming the selective products, as shown in the following Figs. 7.6, 7.7, and 7.8.

In brief, experimental results indicated that only $Fe_nO_m^+$ clusters with a "$m \leq n + 1$" stoichiometry (e.g., FeO^+, Fe_2O^+, and $Fe_2O_3^+$) showed oxygen atom transfer to CO hence producing CO_2 as a major reaction pathway. Therefore, iron oxides with a stoichiometry of three oxygen atoms or less are proposed to be the most important reactive centers in clusters containing one and two iron atoms. In comparison, higher oxides, $Fe_2O_m^+$ ($m > 3$) were not found to undertake such a dominant reaction channel with oxygen atom transfer to CO. For example, the major product for FeO_4^+ with CO was O_2 replacement by CO; the major reaction product for FeO_5^+ was O_2 release after CO collision; also the major reaction products for $Fe_2O_4^+$ were $Fe_2O_2CO^+$ and $Fe_2O_2^+$, but a very minor reaction pathway observed for $Fe_2O_4^+$ was oxygen atom transfer producing $Fe_2O_3^+$. Similarly, $Fe_2O_5^+$ reacts with CO to produce a major product $Fe_2O_3^+$; nevertheless, the detection of minor product species such as $Fe_2O_2^+$ and $Fe_2O_3CO^+$ in the reaction of $Fe_2O_5^+$ with CO still revealed the occurrence of a sequential oxidation process of CO. As shown in Fig. 7.8d, in the initial reaction step, a CO molecule attaches to the O atom bonded to Fe with the lowest coordination, with a binding energy at 3.26 eV; following that, an intermediate with CO_2 emanation and generating a $Fe_2O_4^+$ cluster requires 1.23 eV energy. Note the structural difference of this intermediate of $Fe_2O_4^+$ in a high-energy state where the two external O atoms are binding to one Fe atom, which differs from the ground state $Fe_2O_4^+$ and give rise to two subsequent reaction channels: (i) releases O_2 and proceeds through a transition state 1.82 eV higher in energy, resulting an energy gain of 1.35 eV to generate the $Fe_2O_2^+$ cluster; (ii) further attaches CO onto another O atom generating the $Fe_2O_3^+$ species after CO_2 release, and then $Fe_2O_3^+$ allows association with additional CO to form $Fe_2O_3CO^+$.

7.3.3 Neutral Clusters Fe_nO_m Reacting with CO

In addition to these cationic and anionic cluster ions, small neutral iron oxide clusters were also demonstrated to react with carbon monoxide in a comparable pathway. Xue et al. [79] reported the reactions of small neutral iron oxide clusters (FeO_{1-3} and $Fe_2O_{4,5}$) with carbon monoxide by a joint experimental and theoretical approach. Neutral Fe_nO_m clusters were generated by reaction of laser-ablation-generated iron plasma with O_2 in a supersonic expansion and were reacted with CO in a fast flow reactor. Detection of these neutral clusters was done through 118 nm VUV laser ionization together with time-of-flight mass spectrometry, as shown in Fig. 7.9.

Akin to the above results on Fe_nO_m cluster ions, it was demonstrated that FeO_2 and FeO_3 neutral clusters are also reactive toward CO, while Fe_2O_4, Fe_2O_5, and FeO are relatively less reactive. Furthermore, FeO_2 was noted to bear a higher reactivity than FeO_3 with CO. The reaction cross section σ or first order rate constant (k_1) in the fast flow reactor was estimated simply by using the equation:

$$I_{gas} = I_{He}exp(-\sigma nl) = I_{He}exp(-k_1 n\Delta t) \tag{7.1}$$

where I_{gas} and I_{He} are signal magnitudes of the clusters after reaction with the reactant gas CO (pure He involved as buffer gas); n is the molecular density of reactant gas which is simply estimated with the ideal gas law $P = nkT$ in which P is the pressure, k and T are the Boltzmann constant and the gas temperature; [79] l is the effective path length of the reactor, while Δt is the reaction time that can be estimated as $\Delta t = l/\upsilon$ (υ is the cluster beam velocity, e.g., 1 km/s in the mentioned study). Based on Eq. (7.1), it was evaluated that the independent quantities of cross section for "FeO_2 + CO" and "FeO_3 + CO" was ~$3 \times 10^{-17} cm^2$ and $1 \times 10^{-17} cm^2$ respectively.

The experimental observations were supported through density functional theory (DFT) calculations. The reaction pathways with negative or very slight overall

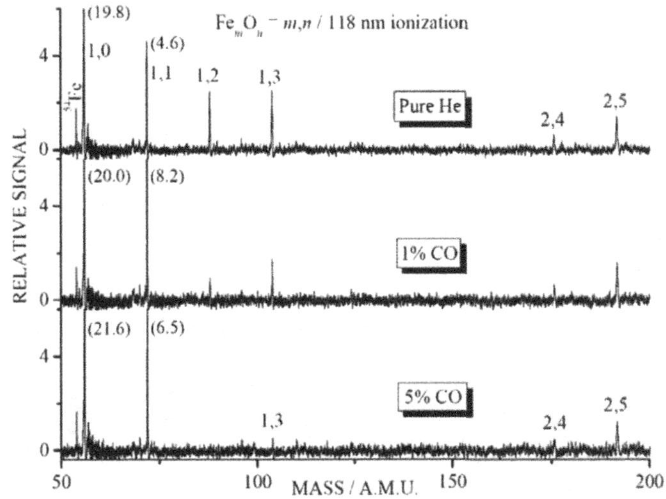

Fig. 7.9 TOF mass spectra for reaction of neutral iron oxide clusters with carbon monoxide in a fast flow reactor. CO concentrations are 0% (top trace), 1% (middle), and 5% (bottom) of the helium carrier gas. The relative signals of Fe and FeO are given in the parentheses. Reproduced with permission from Ref. [79]. Copyright 2008 American Chemical Society

barriers were identified for CO oxidation by FeO_2 and FeO_3. The lower reactivity of FeO_3 than FeO_2 was interpreted in a spin inversion process presented in the reaction of FeO_3 with CO. In comparison, significant reaction barriers were calculated for the reactions of FeO and Fe_2O_{4-5} with CO, which are coincident with the experimental observations.

7.3.4 Anionic and Cationic Co_nO_m Reacting with CO

Similar investigations have also been performed to examine the reactivity of both anionic and cationic Co_nO_m clusters with CO using guided-ion-beam mass spectrometry. As shown in Fig. 7.10, the anionic $Co_nO_m^-$ clusters display a size distribution dominated by CoO_2^-, CoO_3^-, $Co_2O_3^-$ and $Co_2O_3^-$ etc.; in contrast, the cationic species showing remarkable Co^+ and CoO_{10}^+. Mass-selected reaction of the cobalt oxide anions $Co_xO_y^-$ ($x = 1$–3, $y = 2$–6) and cations $Co_xO_y^+$ ($x = 1$, 2, $y = 1$–6) were studied. It was found that the anionic clusters having the stoichiometries $Co_2O_3^-$, $Co_2O_5^-$, $Co_3O_5^-$ and $Co_3O_6^-$ exhibited dominant products conforming to the transfer of a single oxygen atom to CO, forming CO_2. This reactivity closely resembles the above-mentioned reactions of $Fe_nO_m^-$ with the CO. The products resulted from the transfer of a single oxygen atom to CO forming CO_2 mostly according to the following Eqs. (7.2) and (7.3).

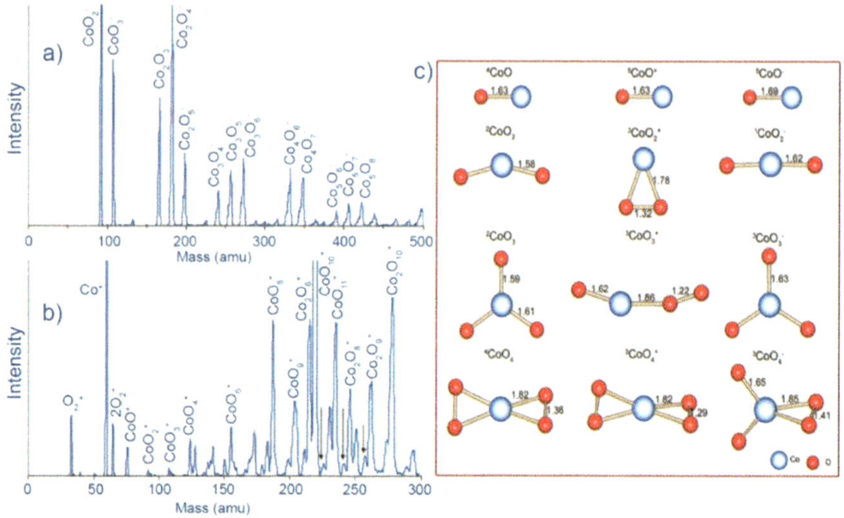

Fig. 7.10 Typical mass distribution of **a** anionic and **b** cationic cobalt oxide clusters obtained through laser vaporization; together with the calculated ground state geometries of neutral, cationic, and anionic CoO_y ($y = 1$–4) clusters (**c**). The bond lengths are given in angstroms and the superscripts indicate the spin multiplicity

$$Co_xO_y^- + CO \rightarrow Co_xO_{y-1}^- + CO_2 \quad (7.2)$$

$$Co_xO_y^- + CO \rightarrow Co_{x-1}O_{y-1}^- + Co + CO_2 \quad (7.3)$$

However, cationic clusters resulted in products with adsorption of CO onto the cluster accompanied by the loss of either O_2 molecules or cobalt oxide units, which contrasts with the products resulting from the chemical reactions for the mass selected $Co_xO_y^-$ anions. A comparison is listed in Table 7.1. For most of the $Co_xO_y^+$ species, their reactions with CO were demonstrated to mostly follow the equations:

$$Co_xO_y^+ + CO \rightarrow Co_xO_{y-2}(CO)^+ + O_2 \quad (7.4)$$

$$Co_xO_y^+ + CO \rightarrow Co_xO_{y-2}^+ + O_2 + CO \quad (7.5)$$

$$Co_xO_y^+ + CO \rightarrow Co_{x-1}O_{y-2} + Co^+ + O_2 + CO \quad (7.6)$$

First-principles calculations displayed the theoretical electronic structure within the density functional theory framework and showed that the enhanced reactivity of selective $Co_xO_y^-$ with CO is ascribed to the relatively minimal atomic oxygen dissociation energy which suggests the oxidation of CO energetically favorable. Also noted is that, for the cationic cobalt oxide clusters, calculation results showed that

Table 7.1 List of anionic and cationic cobalt oxide clusters, and the products resulting from the chemical reactions for the mass selected $Co_xO_y^-$ and $Co_xO_y^+$ with carbon monoxide [80]

$Co_xO_y^-(x, y)$	Products	$Co_xO_y^+(x, y)$	Products
1,2	1,1	1,1	Co^+
1,3	1,2	1,2	$CoCO^+$
2,3	2,2		Co^+
	1,2		$CoO(CO)_2^+$
2,4	2,3	1,3	$CoOCO^+$
2,5	2,4		CoO^+
	2,3		Co^+
	1,3		$CoO(CO)_2^+$
3,4	3,3	1,4	CoO_2CO^+
	2,4		$Co(CO)_2^+$
3,5	3,4		CoO_2^+
	3,3		$CoCO^+$
	2,3	2,2	$Co(CO)_2^+$
	2,4		CoO_2CO^+
3,6	3,5	2,4	$Co_2O_2^+$
	3,4		Co_2CO^+
	3,3		$Co(CO)_2^+$
	2,4		CoO_2CO^+
	2,3	2,6	$Co_2O_4^+$
			$Co_2O_2^+$
			$Co(CO)_2^+$
			$CoCO^+$

the oxygen binds preferentially in a form of less activated molecular O_2, as shown in Fig. 7.10c. Therefore, the displacement of weakly bound O_2 units through the exothermic adsorption of CO onto positively charged cobalt oxides is energetically favorable. In fact, CO adsorption energy was calculated to be larger for cationic clusters than for anionic species.

Further insight into the reactivity of the transition metal oxide clusters have been reported through a comparison of $MO_2^-/M_2O_3^-$ (M = Fe, Co, Ni, Cu) reacting with increasing pressure of CO. Figure 7.11 displays the reactivity of $NiO_2^-/Ni_2O_3^-$, $CuO_2^-/Cu_2O_3^-$, $FeO_2^-/Fe_2O_3^-$ and $CoO_2^-/Co_2O_3^-$ with increasing pressure of CO. It was revealed that these two series of anionic oxide clusters with the same number of metal atoms and stoichiometry but different elemental composition exhibit specific trends in relative oxidation reactivity with CO. Also found was that, the anionic MO_2^- and $M_2O_3^-$ clusters are more reactive for M = Fe and Cu than for M = Co and Ni. First-principles calculations indicated that the most reactive clusters $M_nO_m^-$ generally have relatively large initial binding energies of CO to the cluster which provide sufficient energy to overcome any subsequent barriers to oxidation [20]. On the other hand, the overall exothermicity of the reaction and spin multiplicity

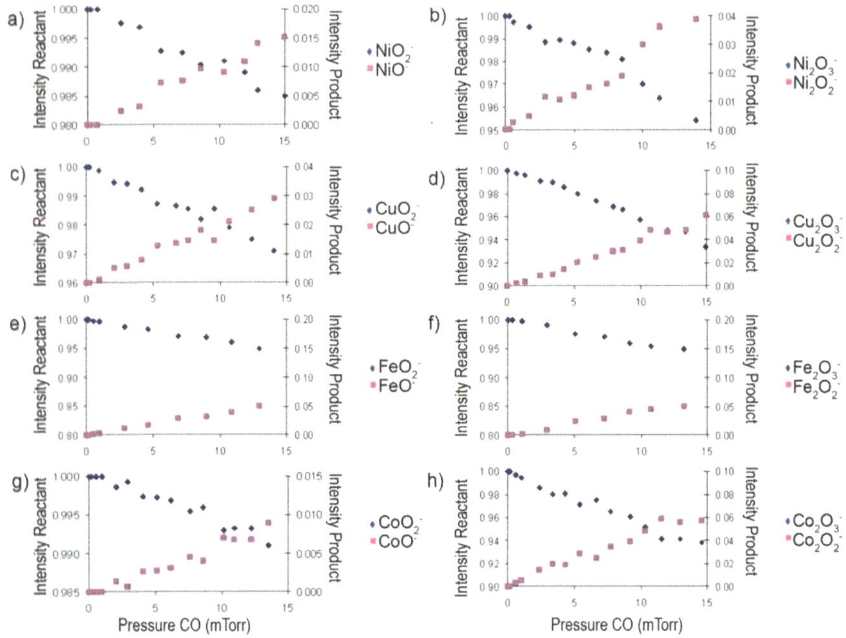

Fig. 7.11 The reactivity of $NiO_2^-/Ni_2O_3^-$, $CuO_2^-/Cu_2O_3^-$, $FeO_2^-/Fe_2O_3^-$ and $CoO_2^-/Co_2O_3^-$ with increasing pressure of CO. Normalized ion intensities of NiO_2^- and NiO^- (**a**), $Ni_2O_3^-$ and $Ni_2O_2^-$ (**b**), CuO_2^- and CuO^- (**c**), $Cu_2O_3^-$ and $Cu_2O_2^-$ (**d**), FeO_2^- and FeO^- (**e**), $Fe_2O_3^-$ and $Fe_2O_2^-$ (**f**), CoO_2^- and CoO^- (**g**), $Co_2O_3^-$ and $Co_2O_2^-$ (**h**)

also account for the variations in relative reactivity observed for the anionic MO_2^- and $M_2O_3^-$ clusters toward the oxidation of CO [20–22, 81, 82].

7.4 Reactivity of CO with $Ti_xO_y^+$ and $Zr_xO_y^+$

As mentioned above, the transition metal cluster oxides have been widely demonstrated to bear advantages within the oxidation of carbon monoxide used in air purification, pollution control, and fuel gas cleanup. Compared with molecular oxygen which bears a triplet ground state, atomic oxygen radical anions are important reactive intermediates; however, it is difficult to capture and characterize them for condensed phase systems. A recent investigation by Ma et al. [83] further developed this potential through the titanium and zirconium oxide cluster anions with dimensions up to nanoscale prepared by laser ablation method. Utilizing a fast flow reactor, they studied the reaction with CO through time-of-flight mass spectrometry together with DFT calculations, as shown in Fig. 7.12. During the reaction of titanium oxide clusters, the transfer of an oxygen atom from $(TiO_2)_nO^-$ (n = 3–25) to CO was observed leading to the formations of $(TiO_2)_n^-$ and CO_2 as products.

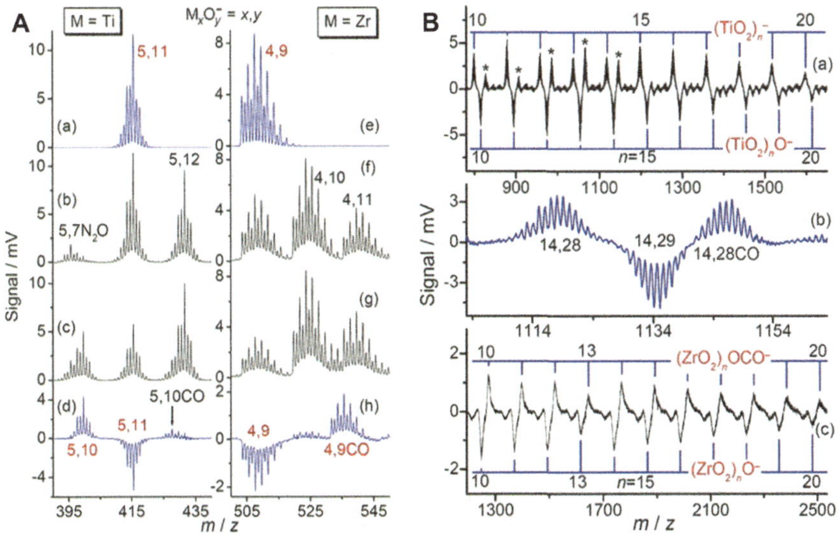

Fig. 7.12 **A** Selected TOF mass spectra for reactions of $Ti_5O_y^-$ (c) ad $Zr_4O_y^-$ (g) with 1.4 and 0.6 Pa CO in the reaction cell, respectively. The numbers x, y denote $M_xO_y^-$ in which M = Ti (left) or Zr (right). The reference spectra with N_2 in the reactor (b, f), the difference spectra (d = c − b, h = g − f), and the simulated $Ti_5O_{11}^-$ and $Zr_4O_9^-$ isotopomers (a, e) are shown. **B** The difference spectra for reactions of $Ti_{10\text{–}20}O_y^-$ (a) and $Zr_{10\text{–}20}O_y^-$ (c) with CO. Peaks marked with asterisks in spectrum a can be assigned as $(TiO_2)_nCO^-$ (n = 10–14). A portion of the spectrum in (a) is expanded and shown in (b). Reproduced with permission from Ref. [83]. Copyright 2013 American Chemical Society

Similar but different situation was noted for the reactions of $(ZrO_2)_nO^-$ (n = 3–25) with CO, where $(ZrO_2)_nOCO^-$ were observed as CO addition products. DFT calculation studies demonstrated that both $(TiO_2)_nO^-$ and $(ZrO_2)_nO^-$ clusters are atomic radical anion (O^-) bonded systems, but there is intense size-dependence on the energy for CO oxidation by O– radicals to form CO_2. The reactivity pattern of the O-bonded $(TiO_2)_nO^-$ and $(ZrO_2)_nO^-$ correlates very well with the aforementioned transition metal cluster oxides.

7.5 Similar Reactivity of CO and NO

In addition to chemical production, heterogeneous catalysts are also widely utilized in industry for the abatement of harmful atmospheric pollutants [84]. Recently it was found that Au particles supported on γ-Al_2O_3 exhibit enhanced catalytic activity for the oxidation of CO to CO_2, where the charge transfer to gold particles on γ-Al_2O_3 was demonstrated to be responsible for the enhanced oxidation activity of CO [85]. Alumina as a catalyst support showed advantages of its high mechanical strength and resistance to thermal degradation [86, 87]. A further insight was given to solve the fundamental questions whether aluminum oxides are reactive with CO and how an accumulation or deficiency of electron density influences their reactivity [88].

Employing guided-ion-beam mass spectrometry, Johnson et al. investigated the small aluminum oxide clusters regarding their reactivity toward CO [88]. As results, clusters with the same stoichiometry as bulk alumina, both cationic and anionic Al_2O_3, were found to exhibit atomic oxygen transfer at reacting with CO leading to the production of CO_2, as shown in Fig. 7.13. Cationic alumina clusters were more reactive than anionic species, and the cationic clusters exhibited additional products analogous to the transfer of two oxygen atoms and the loss of an aluminum atom, as described in Eqs. (7.7) and (7.8).

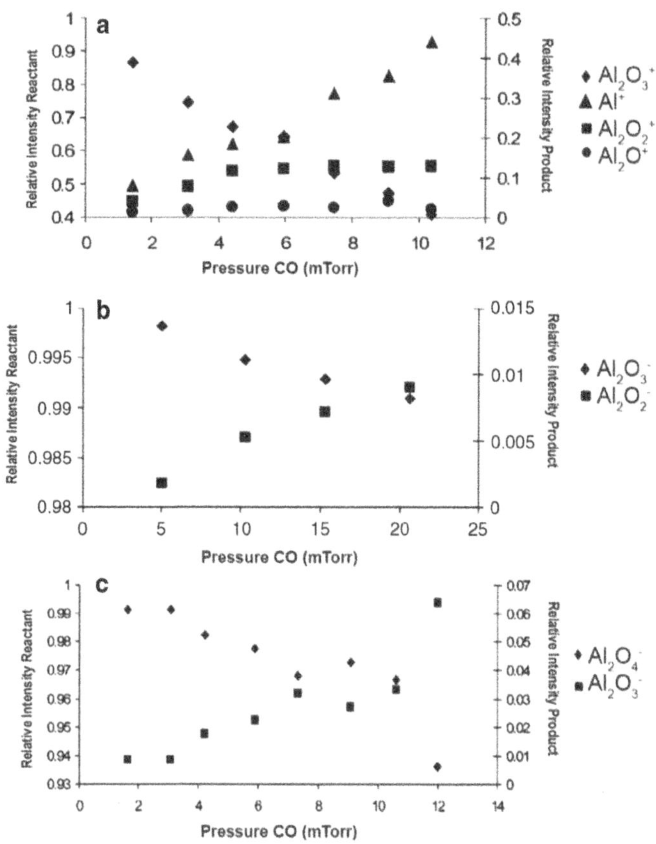

Fig. 7.13 Relative ion intensities of **A** $Al_2O_3^+$, **B** $Al_2O_3^-$ and (**A**) $Al_2O_4^-$ (**C**) and their reaction products with increasing pressure of CO. Notice the decrease in the reactant ion intensity and the concomitant increase in the O-atom transfer products. The reactant ion intensity is plotted on the left axis and the product is on the right. Reproduced with permission from Ref. [88]. Copyright 2008 American Chemical Society

$$Al_2O_4^- + CO \rightarrow Al_2O_3^- + CO_2^-;\ Al_2O_3^- + CO \rightarrow Al_2O_2^- + CO_2^- \quad (7.7)$$

$$Al_2O_3^+ + CO \rightarrow AlO_3CO + Al^+;\ Al_2O_3^+ + CO \rightarrow Al_2O_2^+ + CO_2;\ Al_2O_2^+ + CO \rightarrow Al_2O^+ + CO_2 \quad (7.8)$$

Similar interest was addressed on the reactivity on metal and metal oxide clusters reacting with NO [89–92]. Figure 7.14 displays four different types of distributions of nickel and nickel oxide cluster ions/cations which were formed by laser vaporization of a nickel rod. Several prominent peaks are labelled as Ni_x or Ni_xO_y, and the mass assignments of the unmarked peaks can be determined by adding or subtracting an oxygen atom (16 amu) from the peaks labelled in each spectrum.

The reactions of cationic nickel and nickel oxide clusters with NO were studied, as shown in Fig. 7.15, where the product distributions indicated that several different

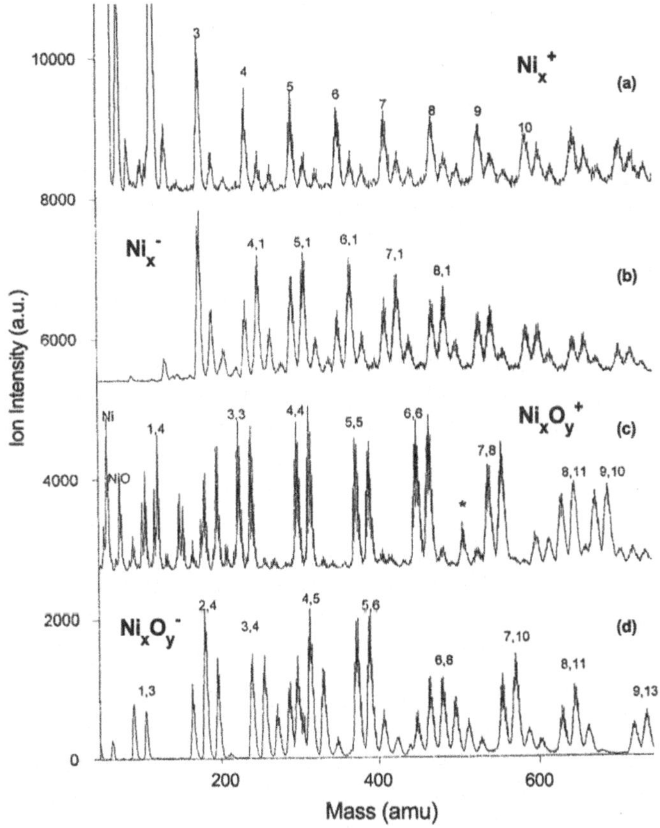

Fig. 7.14 Nickel and nickel oxide reactant ion cluster distributions, labeled as Ni_xO_y: **a** nickel cation clusters; **b** nickel anion clusters; **c** nickel oxide cation clusters; **d** nickel oxide anion clusters. Reproduced with permission from Ref. [89]. Copyright 1999 American Chemical Society

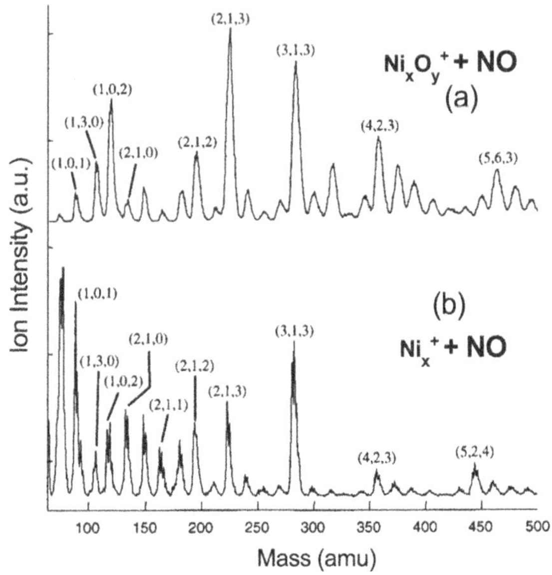

Fig. 7.15 Comparison of the reaction of **a** the nickel oxide clusters and **b** the bare nickel clusters with NO at 2 sccm. The species $Ni_xO_y(NO)_z^+$ are labeled as (x, y, z). These spectra show the similarity in reaction products formed. Reproduced with permission from Ref. [92]. Copyright 1999 American Chemical Society

reaction mechanisms occur between NO and $Ni_x^+/Ni_xO_y^+$. Together with an examination of pseudo-first-order bimolecular rate constants for these reactions, it was demonstrated that competing processes such as oxidation and replacement of oxygen with NO were involved. It was noted that the presence of a few magic peaks in the distributions indicated unusual stability of the size-selected product cluster species, such as $Ni_3O(NO)_3^+$ and $Ni_2O(NO)_3^+$ which were observed in the low-mass region of the product distributions for the reactions of NO with both Ni_x^+ and $Ni_xO_y^+$ [92].

Similarly, the fast flow reactor coupled with a quadrupole mass spectrometer was also used to study the gas-phase reactions of anionic nickel and nickel oxide clusters ($Ni_xO_y^-$, where x = 1–12 and y = 0–2) with nitric oxide, and the rate constants were examined for the initial reaction occurring between the Ni_x^- cluster anions and NO [91]. The results throwed light on three processes: (i) nickel and nickel oxide clusters are oxidized at the reaction with nitric oxide; (ii) addition products with the oxides are also formed; (iii) nitrogen dioxide and nitrogen trioxide are formed on nickel oxide clusters and subsequently released as anions (NO_2^- and NO_3^-). In comparison, N_2O^- was not observed because of its low electron affinity compared with NO_2 and nickel/nickel oxide clusters [93, 94].

References

1. M. Walter, J. Akola, O. Lopez-Acevedo, P.D. Jadzinsky, G. Calero, C.J. Ackerson, R.L. Whetten, H. Grönbeck, H. Häkkinen, Proc. Natl. Acad. Sci. U. S. A. **105**, 9157–9162 (2008)
2. M.B. Abreu, C. Powell, A.C. Reber, S.N. Khanna, J. Am. Chem. Soc. **134**, 20507–20512 (2012)

3. M. Moreno, F.J. Ibañez, J.B. Jasinski, F.P. Zamborini, J. Am. Chem. Soc. **133**, 4389–4397 (2011)
4. J. Akola, K.A. Kacprzak, O. Lopez-Acevedo, M. Walter, H. Grönbeck, H. Häkkinen, J. Phys. Chem. C **114**, 15986–15994 (2010)
5. O. Lopez-Acevedo, H. Tsunoyama, T. Tsukuda, H. Häkkinen, C.M. Aikens, J. Am. Chem. Soc. **132**, 8210–8218 (2010)
6. P.A. Clayborne, O. Lopez-Acevedo, R.L. Whetten, H. Gronbeck, H. Hakkinen, J. Chem. Phys. **135** (2011)
7. P.R. Nimmala, A. Dass, J. Am. Chem. Soc. **133**, 9175–9177 (2011)
8. M. Zhu, C.M. Aikens, F.J. Hollander, G.C. Schatz, R. Jin, J. Am. Chem. Soc. **130**, 5883–5885 (2008)
9. Y. Pei, Y. Gao, N. Shao, X.C. Zeng, J. Am. Chem. Soc. **131**, 13619–13621 (2009)
10. C.E. Briant, B.R.C. Theobald, J.W. White, L.K. Bell, D.M.P. Mingos, A.J. Welch, J. Chem. Soc., Chem. Commun. **0**, 201–202 (1981)
11. Y. Shichibu, Y. Negishi, T. Watanabe, N.K. Chaki, H. Kawaguchi, T. Tsukuda, J. Phys. Chem. C **111**, 7845–7847 (2007)
12. J.M. Pettibone, J.W. Hudgens, J. Phys. Chem. Lett. **1**, 2536–2540 (2010)
13. G. Shafai, S. Hong, M. Bertino, T.S. Rahman, J. Phys. Chem. C **113**, 12072–12078 (2009)
14. P. Braunstein, J. Rose, *Heterometallic Clusters for Heterogeneous Catalysis* (Wiley-VCH, New York, 1998)
15. A.K. Smith, J.M. Basset, J. Mol. Catal. **2**, 229–241 (1977)
16. Y. Shichibu, K. Suzuki, K. Konishi, Nanoscale **4**, 4125–4129 (2012)
17. D.-E. Jiang, M. Walter, Nanoscale **4**, 4234–4239 (2012)
18. D.E. Bergeron, P.J. Roach, A.W. Castleman Jr., N. Jones, S.N. Khanna, Science **307**, 231–235 (2005)
19. D.E. Bergeron, P.J. Roach, A.W. Castleman Jr., N.O. Jones, J.U. Reveles, S.N. Khanna, J. Am. Chem. Soc. **127**, 16048–16053 (2005)
20. J.U. Reveles, G.E. Johnson, S.N. Khanna, A.W. Castleman Jr., J. Phys. Chem. C **114**, 5438–5446 (2010)
21. N.M. Reilly, J.U. Reveles, G.E. Johnson, J.M. del Campo, S.N. Khanna, A.M. Koster, A.W. Castleman Jr., J. Phys. Chem. C **111**, 19086–19097 (2007)
22. N.M. Reilly, J.U. Reveles, G.E. Johnson, S.N. Khanna, A.W. Castleman Jr., J. Phys. Chem. A **111**, 4158–4166 (2007)
23. K.S.A. Halim, M.H. Khedr, M.I. Nasr, A.M. El-Mansy, Mater. Res. Bull. **42**, 731–741 (2007)
24. P.B. Armentrout, H. Koizumi, M. MacKenna, J. Phys. Chem. A **109**, 11365–11375 (2005)
25. B.V. Reddy, F. Rasouli, M.R. Hajaligol, S.N. Khanna, Chem. Phys. Lett. **384**, 242–245 (2004)
26. C.T. Liu, P.F. Shi, J.X. Zhang, Chin. J. Catal. **25**, 697–701 (2004)
27. D. Schroder, P. Jackson, H. Schwarz, Eur. J. Inorg. Chem. 1171–1175 (2000)
28. H.B. Wu, S.R. Desai, L.S. Wang, J. Am. Chem. Soc. **118**, 5296–5301 (1996)
29. R.H. Schultz, K.C. Crellin, P.B. Armentrout, J. Am. Chem. Soc. **113**, 8590–8601 (1991)
30. S.K. Loh, L. Lian, P.B. Armentrout, J. Chem. Phys. **91**, 6148–6163 (1989)
31. D.L. Hildenbrand, Chem. Phys. Lett. **34**, 352–354 (1975)
32. B.V. Reddy, S.N. Khanna, Phys. Rev. Lett. **93**, 068301 (2004)
33. P. Li, D.E. Miser, S. Rabiei, R.T. Yadav, M.R. Hajaligol, Appl. Catal. B: Environ. **43**, 151–162 (2003)
34. L. Guczi, K. Frey, A. Beck, B. Peto, C.S. Daroczi, N. Kruse, S. Chenakin, Appl. Catal. A: General **291**, 116–125 (2005)
35. W.X. Huang, W. Ranke, R. Schlogl, J. Phys. Chem. B **109**, 9202–9204 (2005)
36. P.C. Steinhardt, W.L. Gladfelter, A.D. Harley, J.R. Fox, G.L. Geoffroy, Inorg. Chem. **19**, 332–339 (1980)
37. K. Whitmire, D.F. Shriver, J. Am. Chem. Soc. **102**, 1456–1457 (1980)
38. M.O. Albers, N.J. Coville, E. Singleton, J. Organomet. Chem. **234**, C13–C15 (1982)
39. D.J.A. Fredeen, D.H. Russell, J. Am. Chem. Soc. **107**, 3762–3768 (1985)
40. M.D. Morse, M.E. Geusic, J.R. Heath, R.E. Smalley, J. Chem. Phys. **83**, 2293–2304 (1985)

41. K.H. Whitmire, K.S. Raghuveer, M.R. Churchill, J.C. Fettinger, R.F. See, J. Am. Chem. Soc. **108**, 2778–2780 (1986)
42. M.F. Jarrold, J.E. Bower, J. Am. Chem. Soc. **110**, 6706–6716 (1988)
43. B.C. Guo, K.P. Kerns, A.W. Castleman Jr., J. Chem. Phys. **96**, 8177–8186 (1992)
44. B.G. Ershov, E. Janata, A. Henglein, J. Phys. Chem. **98**, 7619–7623 (1994)
45. S. Aime, W. Dastru, R. Gobetto, J. Krause, E. Sappa, Organometallics **14**, 3224–3228 (1995)
46. H.G. Ang, G.K. Chuah, S. Jaenicke, W.L. Loh, J. Chem. Soc., Dalton Tran. 1243–1249 (1997)
47. E.K. Parks, G.C. Nieman, K.P. Kerns, S.J. Riley, J. Chem. Phys. **107**, 1861–1871 (1997)
48. R. Hernandez-Molina, A.G. Sykes, Coord. Chem. Rev. **187**, 291–302 (1999)
49. L. Piccolo, C. Becker, C.R. Henry, Eur. Phys. J. D **9**, 415–419 (1999)
50. G.L. Gutsev, C.W. Bauschlicher, L. Andrews, J. Chem. Phys. **119**, 3681–3690 (2003)
51. L.D. Socaciu, J. Hagen, J. Le Roux, D. Popolan, T.M. Bernhardt, L. Woste, S. Vajda, J. Chem. Phys. **120**, 2078–2081 (2004)
52. T.M. Bernhardt, L.D. Socaciu-Siebert, J. Hagen, L. Woste, Appl. Catal. A: General **291**, 170–178 (2005)
53. L.M. Han, Q.L. Suo, Y.B. Wang, J.H. Ye, N. Zhu, X.B. Leng, J. Sun, Polyhedron **24**, 759–764 (2005)
54. L. Jiang, Q. Xu, J. Phys. Chem. A **109**, 1026–1032 (2005)
55. H. Kajitani, H. Seino, Y. Mizobe, Organometallics **24**, 6260–6267 (2005)
56. M. Neumaier, F. Weigend, O. Hampe, M.M. Kappes, J. Chem. Phys. **125**, 104308 (2006)
57. J.B. Park, J.S. Ratliff, S. Ma, D.A. Chen, J. Phys. Chem. C **111**, 2165–2176 (2007)
58. R.B. Wyrwas, E.M. Robertson, C.C. Jarrold, J. Chem. Phys. **126**, 214309 (2007)
59. G. Lanzani, A.G. Nasibulin, K. Laasonen, E.I. Kauppinen, Nano Res. **2**, 660–670 (2009)
60. H. Noda, H. Uehara, M. Abe, T. Michi, M. Osawa, K. Uosaki, Y. Sasaki, Bull. Chem. Soc. Jpn. **82**, 1227–1231 (2009)
61. J. Cheng, M.J. Ferguson, J. Takats, J. Am. Chem. Soc. **132**, 2 (2010)
62. W. Wang, S. Wang, X. Ma, J. Gong, Chem. Soc. Rev. **40**, 3703–3727 (2011)
63. T. Seiyama, K. Tanale, *New Horizons in Catalysis* (Elsevier, New York, 1981)
64. P.J. Brucat, C.L. Pettiette, S. Yang, L.S. Zheng, M.J. Craycraft, R.E. Smalley, J. Chem. Phys. **85**, 4747–4748 (1986)
65. Y.H. Pan, K. Sohlberg, D.P. Ridge, J. Am. Chem. Soc. **113**, 2406–2411 (1991)
66. E. Kapiloff, K.M. Ervin, J. Phys. Chem. A **101**, 8460–8469 (1997)
67. P. Fayet, M.J. McGlinchey, L.H. Woeste, J. Am. Chem. Soc. **109**, 1733–1738 (1987)
68. J.W. Lauher, J. Am. Chem. Soc. **100**, 5305–5315 (1978)
69. D.M.P. Mingos, Acc. Chem. Res. **17**, 311–319 (1984)
70. D.M.P. Mingos, Chem. Soc. Rev. **15**, 31–61 (1986)
71. D.M.P. Mingos, T. Slee, Z.Y. Lin, Chem. Rev. **90**, 383–402 (1990)
72. D.M.P. Mingos, D.J. Wales, J. Am. Chem. Soc. **112**, 930–932 (1990)
73. F.A. Cotton, J.D. Jamerson, J. Am. Chem. Soc. **98**, 1273–1274 (1976)
74. R.E. Benfield, B.F.G. Johnson, J. Chem. Soc., Dalton Trans. 1743–1767 (1980)
75. B.F.G. Johnson, *Transition Metal Clusters* (John Wiely, New York, 1980)
76. B.V. Reddy, S.N. Khanna, Phys. Rev. Lett. **83**, 3170–3173 (1999)
77. R.C. Bell, K.A. Zemski, K.P. Kerns, H.T. Deng, A.W. Castleman Jr., J. Phys. Chem. A **102**, 1733–1742 (1998)
78. N.O. Jones, B.V. Reddy, F. Rasouli, S.N. Khanna, Phys. Rev. B **72**, 165411 (2005)
79. W. Xue, Z.-C. Wang, S.-G. He, Y. Xie, E.R. Bernstein, J. Am. Chem. Soc. **130**, 15879–15888 (2008)
80. G.E. Johnson, J.U. Reveles, N.M. Reilly, E.C. Tyo, S.N. Khanna, A.W. Castleman Jr., J. Phys. Chem. A **112**, 11330–11340 (2008)
81. N.M. Reilly, J.U. Reveles, G.E. Johnson, S.N. Khanna, A.W. Castleman Jr., Chem. Phys. Lett. **435**, 295–300 (2007)
82. G.E. Johnson, N.M. Reilly, A.W. Castleman Jr., Int. J. Mass Spectrom. **280**, 93–100 (2009)
83. J.-B. Ma, B. Xu, J.-H. Meng, X.-N. Wu, X.-L. Ding, X.-N. Li, S.-G. He, J. Am. Chem. Soc. **135**, 2991–2998 (2013)

84. E.L. Muetterties, Science **196**, 839 (1977)
85. Y.-F. Han, Z. Zhong, K. Ramesh, F. Chen, L. Chen, J. Phys. Chem. C **111**, 3163–3170 (2007)
86. S. Feyel, J. Döbler, R. Höckendorf, M.K. Beyer, J. Sauer, H. Schwarz, Angew. Chem. Int. Ed. **47**, 1946–1950 (2008)
87. N.N. Greenwood, A. Earnshaw, *Chemistry of the Elements* (Elsevier, Amsterdam, 1997)
88. G.E. Johnson, E.C. Tyo, A.W. Castleman Jr., J. Phys. Chem. A **112**, 4732–4735 (2008)
89. W.D. Vann, A.W. Castleman Jr., J. Phys. Chem. A **103**, 847–857 (1999)
90. W.D. Vann, R.L. Wagner, A.W. Castleman Jr., J. Phys. Chem. A **102**, 1708–1718 (1998)
91. W.D. Vann, R.L. Wagner, A.W. Castleman Jr., J. Phys. Chem. A **102**, 8804–8811 (1998)
92. W.D. Vann, R.C. Bell, A.W. Castleman Jr., J. Phys. Chem. A **103**, 10846–10850 (1999)
93. D.G. Hopper, A.C. Wahl, R.L.C. Wu, T.O. Tiernan, J. Chem. Phys. **65**, 5474–5494 (1976)
94. K.M. Ervin, J. Ho, W.C. Lineberger, J. Phys. Chem. **92**, 5405–5412 (1988)

Chapter 8
Energetic Reactions with Hydrocarbons

8.1 Introduction

Cluster science has undergone an explosive growth in reactivity during the past years, prompted both by basic chemistry to which studies of clusters may provide new insight, and a vast array of applied areas to which clusters relate. Elucidating the differences and similarities in the properties and reactivity of matter in the gaseous compared to condensed state from a molecular (/cluster) point of view has been an overriding theme of abundant investigations. Investigations of the chemical properties and the reaction kinetics have been particularly important in the subject of phase transition where progress has been impeded for lack of fundamental data for comparison with molecular theories. Among the investigations involving the scattering of high energy neutral and ionic particles from surfaces, clusters are often the observed reaction products, related to studies of condensed phases and surfaces. Determining factors that affect cluster size, stability and the mechanism of its formation provide a basis for explaining the products of such reactions.

In this chapter we summarize the advances in studying the reactivity of metal clusters and their oxides, mainly including transition metals of which the potential effectiveness, both as catalysts and catalytic supports, has undergone increased scrutiny in recent decades due to their wide range of applications [1, 2]. Among the extensive experimental and theoretical studies which were undertaken on transition metals in catalyzing the chemical reactions of various organic molecules, the selective activation of chemical bonds such as C–C, C–H or C–O etc. plays a significant role in optimizing synthetic schemes. The examination of gas phase reactions, by avoiding the complications arising from solvent environments and crystalline forces, has potential for elucidating details of transition metal catalytic activity and changes in the reaction pathways as a function of cluster sizes [3–5].

Z. Luo and S. N. Khanna, *Metal Clusters and Their Reactivity*,
https://doi.org/10.1007/978-981-15-9704-6_8

8.2 C–C Bond Cracking—Reactivity of Group V Metal Oxides

As outlined in the above section how all-metal cluster reactivity with oxygen has been a continuing topic of interest over the past twenty years; it is of particular interest for such clusters to enable potential use as energy density materials and energy storage capability. On the other hand, reactions conducted under endothermic conductions also have notable implications, such as the ability to acquire large thrust without undue stress being placed on the aircraft engines and related components, nor the need to exchange engines on a frequent basis. Moreover, with well selected conditions the engines operate in conditions of higher efficiency, acquiring better burning rates and selective reactivity. Among desired conditions, two particularly impact engine temperatures, namely the use of fuel to act as a coolant and, for the fuel to undergo desired cracking thereby absorbing energy.

During the last several years *Castleman* group [2] have been engaged in studying related problems and during the course of investigations they have identified some reactions that do display endothermic conditions, finding that in some cases it is the metal center that plays the dominant role in effecting the reaction class. For example, Bell et al. [6] showed an investigation into the reactivity and collision-induced dissociation of vanadium oxide clusters using a triple quadrupole mass spectrometer coupled with a LaVa-source. As shown in Fig. 8.1a, the dominant peaks in the mass distribution correspond to $(VO_2)_n(V_2O_5)_m(O_2)_x^+$. Studies on collision-induced dissociation of $V_2O_{4-6}^+$, $V_3O_{6-9}^+$, $V_4O_{8-10}^+$, $V_5O_{11-13}^+$, $V_6O_{13-15}^+$ and $V_7O_{16-18}^+$ indicated that VO_2, VO_3 and V_2O_5 units were the core building blocks for these clusters. Further investigation on the reactivity for these vanadium oxide clusters towards hydrocarbons showed that the reaction pathways include molecular association, cracking, dehydration and oxygenation of the neutral hydrocarbons. For example, the reaction of $V_3O_7^+$ with 1-butene (C_4H_8) displays predominant C–C cracking in forming $V_3O_7C_2H_4^+$, while its reaction with 1,3-butadiene (C_4H_6) displays dehydration of the association product, as shown in Fig. 8.1b and c [2].

The variation of C–C cracking and dehydration has been thoroughly investigated on the VB group metal cluster oxides [3, 6–24]. Figure 8.2 shows the mass distributions of vanadium, niobium, and tantalum oxide cluster cations, and the reactions of several classes of small organic molecules with cluster oxides composed of vanadium, niobium and tantalum are displayed in Fig. 8.3 [2]. Note that the n-butene (C_4H_8) displays negligible, small, and predominant C–C cracking for $V_nO_m^+$, $Nb_nO_m^+$, and $Ta_nO_m^+$, respectively. Similar selectivity was observed for their reactivities with 1, 3-butadiene [2]. It was found that 1, 3-butadine (C_4H_6) displays virtually no C–C cracking with vanadium oxides, a trivial amount with niobium oxide clusters, while considerable cracking with Ta oxide clusters.

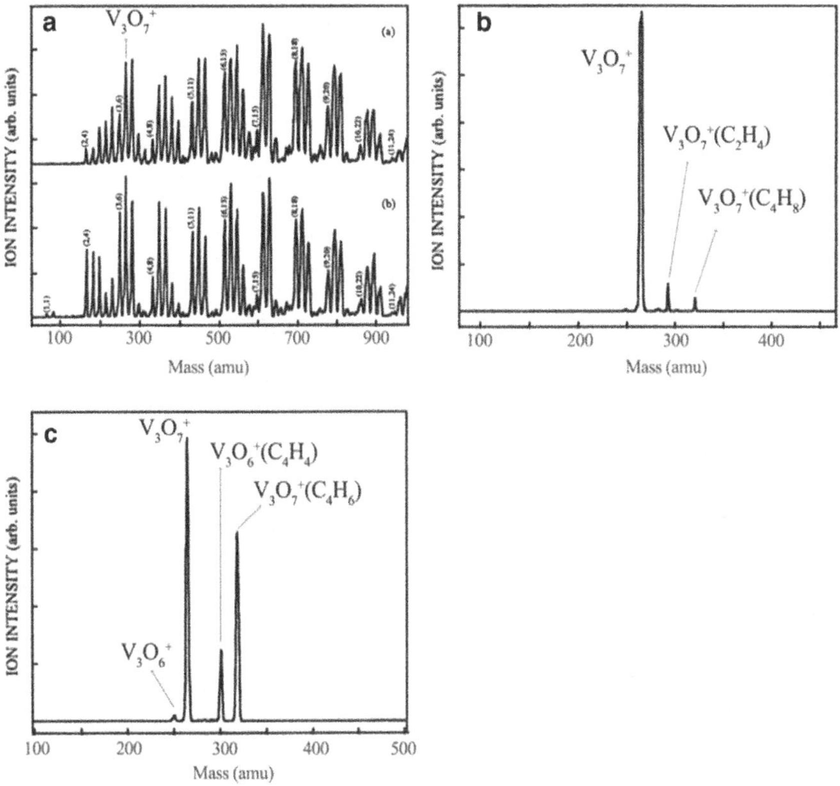

Fig. 8.1 **a** Total ion mass distribution of vanadium oxide cluster cations. Spectra in (a) display the mass distribution from the laser plasma reactions of vanadium with 10% oxygen seeded in the helium carrier gas. The same conditions were used to generate the cluster distribution in (b); however, 0.4 mTorr of krypton was added to the collision cell to determine the most stable cations. The numbers in parentheses, (x, y), denote the number of vanadium and oxygen atoms in the cluster $V_xO_y^+$; the remaining peaks correspond to masses with an additional oxygen atom as a series progresses. **b** Spectra of reaction $V_3O_7^+$ with 0.2 mTorr 1-butene displays predominant C–C cracking. **c** Spectra of reaction of $V_3O_7^+$ with 0.2 mTorr of 1, 3-butadiene displays predominant dehydration of the association product. Reproduced with permission from Ref. [6]. Copyright 1998 American Chemical Society

8.3 C–H Bond Activation

Extensive investigations on the interesting C–H bond activation by ionic gas phase clusters have been reported [25–34], in particular the C–H bond activation in methane and other small alkanes due to the well-known industrial application interest. These C–H bond activation studies can be achieved by various gas phase cationic oxide clusters, such as FeO^+ [30, 32], $(MoO_3)_{1-2}^+$ [35, 36], OsO_4^+ [37], $(V_2O_5)_{1-5}^+$ [35, 38, 39], MgO^+ [40–42], SO_2^+ [43, 44], $P_4O_{10}^+$ [45–47], CuO^+ [48], GeO^+/SnO^+/PbO^+ [49], $Al_nO_m^+$ [50–52], and all the early transition-metal dioxide cations MO_2^+ (M

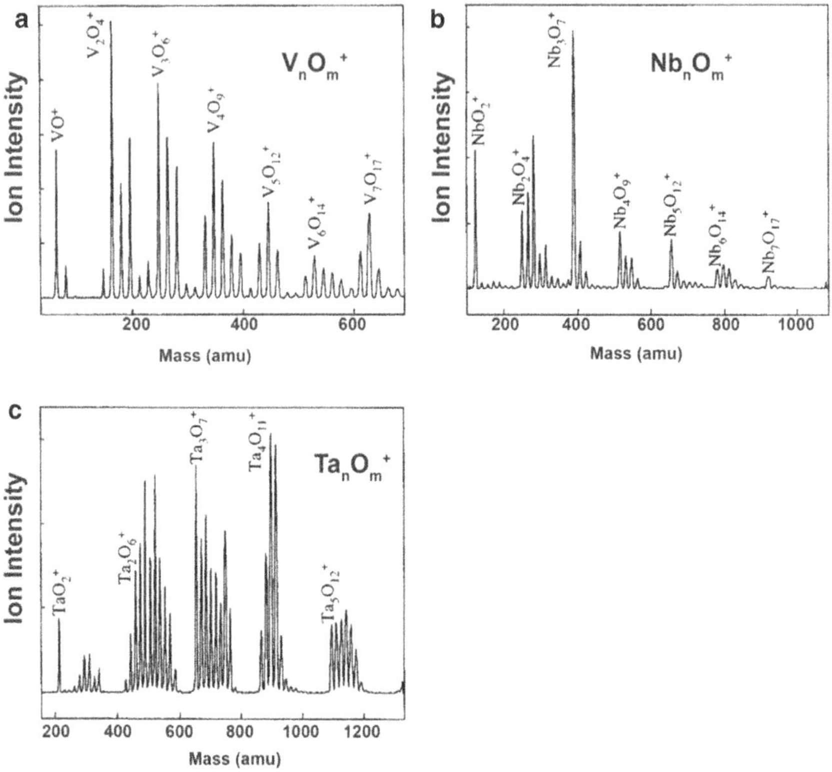

Fig. 8.2 Mass distributions of **a** vanadium, **b** niobium, and **c** tantalum oxide cluster cations. Reproduced with permission from Ref. [11]. Copyright 2000 American Chemical Society

= Ti, V, Zr, Nb) [35, 53] including $(TiO_2)_{1\text{-}5}^+$, $(ZrO_2)_{1\text{-}4}^+$, $(HfO_2)_{1\text{-}2}^+$, $(Nb_2O_5)_{1\text{-}3}^+$, $(Ta_2O_5)_{1,2}^+$, and $Re_2O_7^+$. On the other hand, several anionic oxide clusters such as $ScO_{3,4}^-$ [54], $Sc_3O_6^-$ [55], $(La_2O_3)_{1\text{-}3}O^-$ [56], and $Zr_2O_8^-$ [57], etc. [58–69]. These cluster systems can serve as a more detailed molecular approach for better understanding of the active sites in catalytic systems. A few interesting and distinctive aspects on the C–H bond activation of metal clusters with certain hydrocarbons are highlighted below.

With a certain similarity to H–H and N–N bonds in hydrogen and nitrogen molecules, the C–H bonds (1.09 Å, 413 kJ/mol) in small organic molecules also embodies a covalent bond, that is, carbon shares its outer valence electrons with hydrogen atoms giving rise to both-filled outer shells and reasonable stability. Note that, the electronegativity between C (2.55) and H (2.2) atoms is close to each other on a basis of Pauling's scale, with small differences enough to be regarded as being non-polar [70–72], especially for high symmetrical hydrocarbons such as CH_4, C_2H_2 and C_2H_4. The C–H bonds are very strong and usually unreactive, however, depending

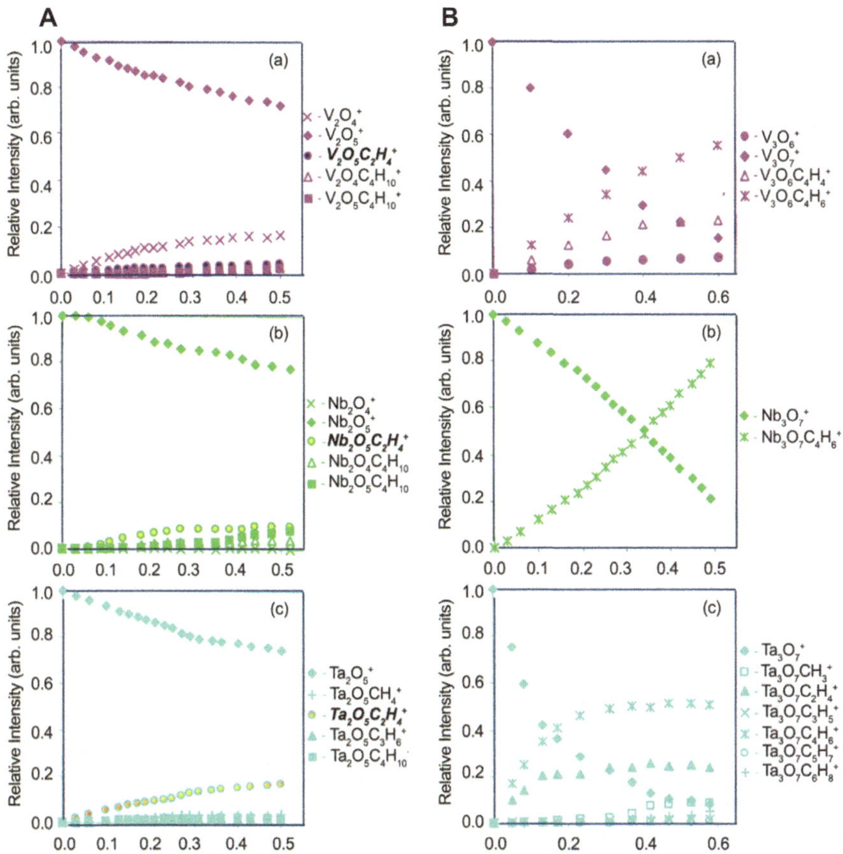

Fig. 8.3 **A** Relative product branching ratios for metal oxides with n-butane: (a) $V_2O_5^+$, oxygen transfer dominates, *negligible cracking*; (b) $Nb_2O_5^+$, small oxygen transfer and *cracking channel*; (c) $Ta_2O_5^+$, negligible oxygen transfer, *cracking reaction dominates.* **B** Relative product branching ratios for metal oxides with 1, 3-butadiene. Note various cracking channels which differ with the metal type and organic composition. Reproduced with permission from Ref. [2]. Copyright 2002 American Chemical Society

on active electron transfer and proton transfer, they participate in radical substitution or allow C–H bond activation by well-designed reactants/catalysts. The homo- and heterolytic C–H bond cleavage (as well as functionalization) is well known for its significance in chemistry and is regarded as a longstanding central challenge [28, 73–85].

Metal cluster reactivity shows its own novelty in this topic. For example, considerable investigations have been conducted upon the activation of methane by gas-phase palladium model systems [86–92]. Experimental observations showed that neutral clusters Pd_x ($x \leq 24$) tend to adsorb methane with a few exceptions (e.g., Pd_3 and Pd_4). While in contrast, the reactivity of small palladium oxides toward methane finds size

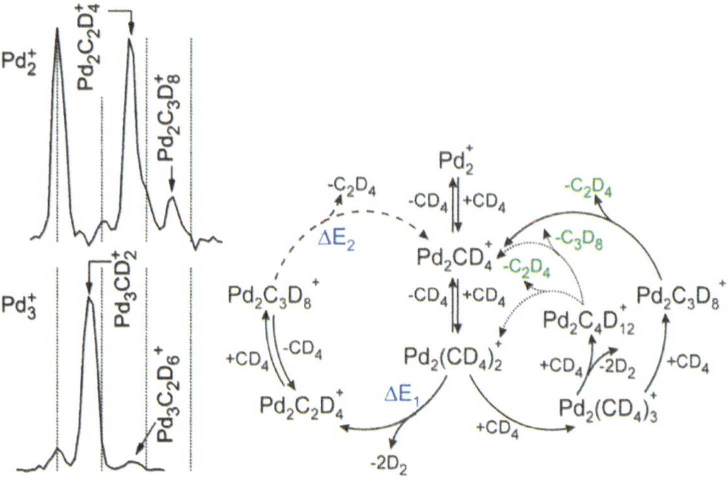

Fig. 8.4 (Left) Ion mass distributions obtained after reaction of palladium clusters Pd_2^+ (upper) and Pd_3^+ (bottom) with CD_4 at room temperature ($t_R = 0.1$ s). (Right) The proposed reaction mechanism for Pd_2^+ with CD_4. Reproduced from Ref. [93]. Copyright 2013 American Chemical Society

dependence for methane dehydrogenation [93, 94]. Bernhardt and coworkers [93] showed an interesting study on the reactivity of methane with Pd_x^+ ($x = 2$–4) clusters. Mass spectrometric and reaction kinetic studies under different temperature conditions have elucidated the intrinsic propensity of palladium clusters in reacting with methane molecules, as depicted in Fig. 8.4 [93]. The mass spectrum recorded in the case of Pd_2^+ displayed four signal peaks corresponding to the bare unreacted Pd_2^+, the association complex $Pd_2CD_4^+$ (weak), the methane-activated products $Pd_2C_2D_4^+$ (strong) and $Pd_2C_3D_8^+$ (weak). Similarly, Pd_3^+ reacted with CD_4 resulting in a main product $Pd_3CD_2^+$, together with a byproduct $Pd_3C_2D_6^+$ pointing to the additional adsorption of a second methane molecule. Tetramer Pd_4^+ was not observed to exhibit apparent reactivity under the room-temperature condition. With Pd_2^+ as a typical example, theoretical investigations addressed the activation of a first CH_4 on Pd atoms, and demonstrated strong dependence of C–H bond cleavage to form metal-hydride-methyl complexes (H–Pd_x–CH_3) [87–89, 91, 95–98]. Such studies provide valuable information for homologous metal and metal oxides used as important catalysts in industrial processes [99], and are helpful in understanding the pivotal parameters and elementary reaction mechanisms involved in catalysis.

8.3.1 Iso-Valence of ZrO and Pd

The element Pd has filled d shells and is widely utilized as catalysts due to their high activity for a number of reactions [100–104]. However, as limited quantities of this precious metal exist on the earth, attempts to identify inexpensive replacement catalysts with comparable efficiency are motivated. It has been proposed that by incorporating non-metallic elements (e.g., carbon or oxygen) into the early transition metal, the surface reactivity of the metal can be moderated to produce an effective catalyst [105]. The presence of nonmetal element changes the electron density of the metal and modulates the position of the d states near the Fermi energy, and the coordination of nonmetal atoms around metal atoms has a large influence upon catalytic activity. Studying the reactivity of isolated gas-phase ions/clusters with simple alkanes provide opportunities to identify microscopic origins of the reactivity [106, 107].

Early thermodynamical studies have suggested the reactivity of C_2H_6 and C_3H_8 activated by Pd^+, which revealed the existence of channels for the cleavage of C–C and C–H bonds at high kinetic energies [108, 109]. Figure 8.5 shows the reactivity of cationic Pd^+ and ZrO^+ with propane (C_3H_8) utilizing a guided-ion-beam mass spectrometer [110–112]. Major products observed for the reaction "$Pd^+ + C_3H_8$" (Fig. 8.5a) are CH_3Pd^+, $C_2H_3Pd^+$, and $C_3H_5Pd^+$. Similarly, the reaction "$ZrO^+ +$ C_3H_8" (Fig. 8.5b) presents products as CH_3ZrO^+, $C_2H_3ZrO^+$ and $C_3H_5ZrO^+$. Note that the intensities of the products from the two reactions display the consistent ordering. By conducting the reactions in both single and multiple collision conditions assists and together with first-principles calculations, a few reaction pathways were ascertained with similarity between Pd^+ and ZrO^+, indicating C–C and C–H bond breaking [110].

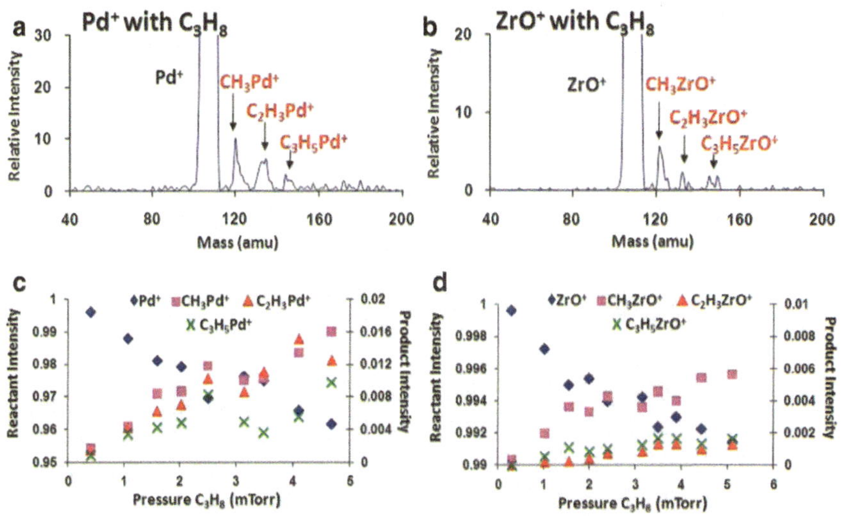

Fig. 8.5 Spectra for the interaction of **a** Pd^+ with 3.50 mTorr of C_3H_8 and **b** ZrO^+ with 3.90 mTorr of C_3H_8 both occurring at 20 eV in the lab frame. **c** Pd^+ and **d** ZrO^+ have been reacted with 0–5 mTorr of C_3H_8 at 20 eV in the lab frame. Branching ratios display the decrease in reactant intensity with the concomitant rise in product intensity as the pressure of gas in the reaction cell is increased

Figure 8.6 presents the reaction of Pd^+ and ZrO^+ with 0–5 mTorr of C_2H_6 at 20 eV in the lab frame, where the branching ratios display the decrease in reactant intensity with the concomitant rise in product intensity as the pressure of gas in the reaction cell is increased. Reaction pathways for the C–C bond cracking are shown in Fig. 8.6c and d; while the channels for the C–H bond to be cleaved with a loss of H_2 are shown in Fig. 8.6e and f, respectively. The energy required to break the first C–H bond with Pd^+ is 0.65 eV, while it is 1.38 eV with ZrO^+ and the intermediary state with the C–H bond cleaved by ZrO^+ is barely a local minimum in the energy landscape. Instability of intermediates versus products is a positive trait for catalysts; otherwise, trapping the complex in a stable intermediate may poison the catalyst [110].

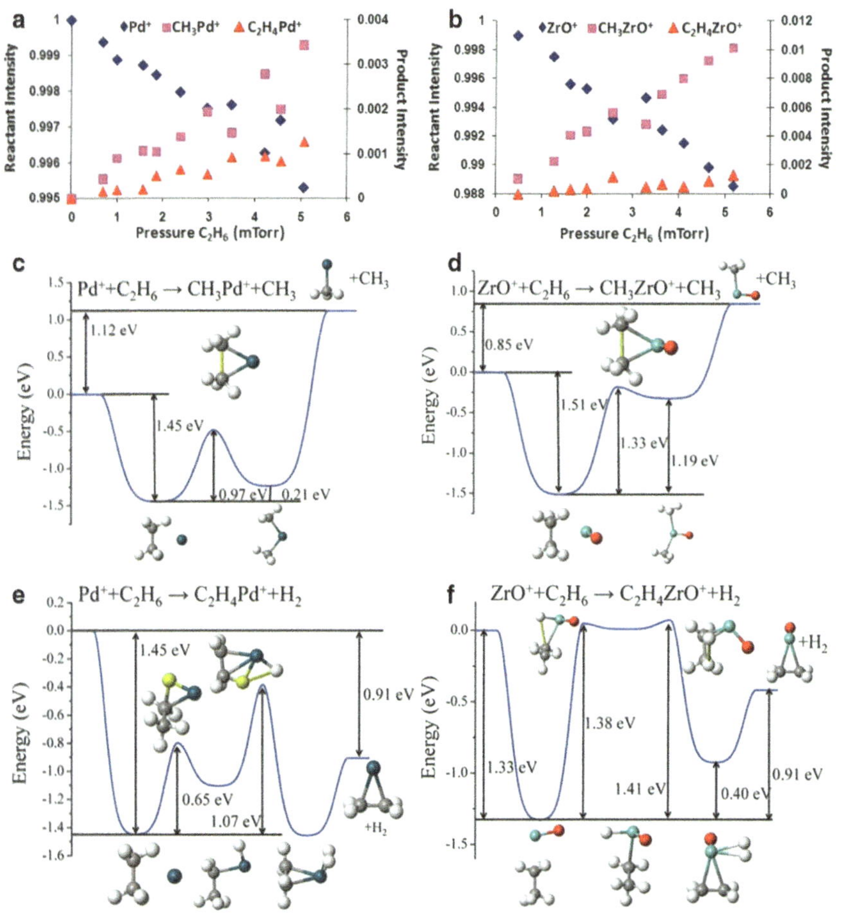

Fig. 8.6 **a** Pd^+ and **b** ZrO^+ reacted with 0–5 mTorr of C_2H_6 at 20 eV in the lab frame. **c**–**f** Energy profiles for the reaction of Pd^+ and ZrO^+ with ethane to form CH_3Pd^+ and CH_3ZrO^+, $C_2H_4Pd^+$, and $C_2H_4ZrO^+$, respectively

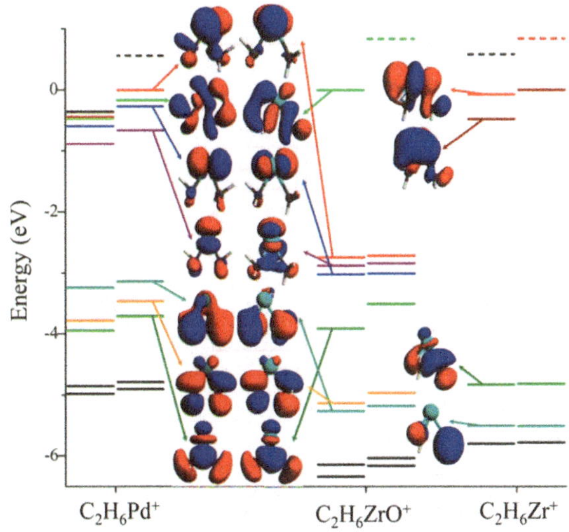

Fig. 8.7 Electronic structures of $C_2H_6Pd^+$, $C_2H_6ZrO^+$, and $C_2H_6Zr^+$ at the transition state for the cleavage of the C−C bond. The color coding corresponds to analogous orbitals. Solid lines are occupied, and dashed lines are unoccupied orbitals

The similar reactivity between Pd^+ and ZrO^+ with hydrocarbons reveals the chemical mimics for oxides like ZrO in place of the precious metals like Pd. Additional insight into such origins has been noted on the electronic properties of them. Figure 8.7 displays the electronic structures of Pd^+, ZrO^+ and Zr^+, plotted at the transition state for C–C bond cleavage in ethane (i.e., $C_2H_6Pd^+$, $C_2H_6ZrO^+$, and $C_2H_6Zr^+$). Primarily as a result of hybridization between Zr and O orbitals, molecular orbitals with 4d Zr components in ZrO are occupied, which is similar to Pd with a fully filled 4d manifold. By comparing with the transition state for Zr^+ of which the electronic structure is quite different, it was demonstrated that the addition of O to Zr populates the molecular orbitals with significant 4d components bringing the similarity to that of Pd, and hence moderates the reactivity [110].

In addition to the finding of the analogous reactivity between ZrO^+ and Pd^+, there are a few other anionic series on which the electronic state correlation between the elements and their isoelectronic molecular counterparts were examined by photoelectron spectroscopy studies, including MoC^-/Ru^-, WC^-/Pt^-, TiO^-/Ni^-, and ZrO^-/Pd^- [105, 106, 113–117]. These couples revealed comparable electronic transitions and orbital symmetry; further, on the experimental and theoretical basis as shown in Fig. 8.8, it was noted that ZrO^-, WC^-, and TiO^- can be viewed as the superatomic form of Pd^-, Ni^-, and Pd^- [106]. For Ni^- versus TiO^- system (Fig. 8.8a), surface plots of the highest occupied 9σ and 1δ molecular orbitals of TiO^- (from ab initio calculations) appear to resemble the associated 3d and 4 s atomic orbitals of Ni^-. Similarly, for ZrO^- in Fig. 8.8b, peaks C' and the $3\Delta_1 \leftarrow {}^2\Sigma^-$ component of D' appear as unresolved shoulders to more intense transitions; transitions beyond peak H' access the $v = 1$ level of the ${}^1\Delta_2$ excited state. For Pt^- versus WC^- system in Fig. 8.8c, the 16σ molecular orbital of WC^- is constructed of $c_s^2 = 0.19$, $c_p^2 = 0.29$, $c_d^2 = 0.52$ atomic orbital coefficients. Considering the relativistic radial contraction

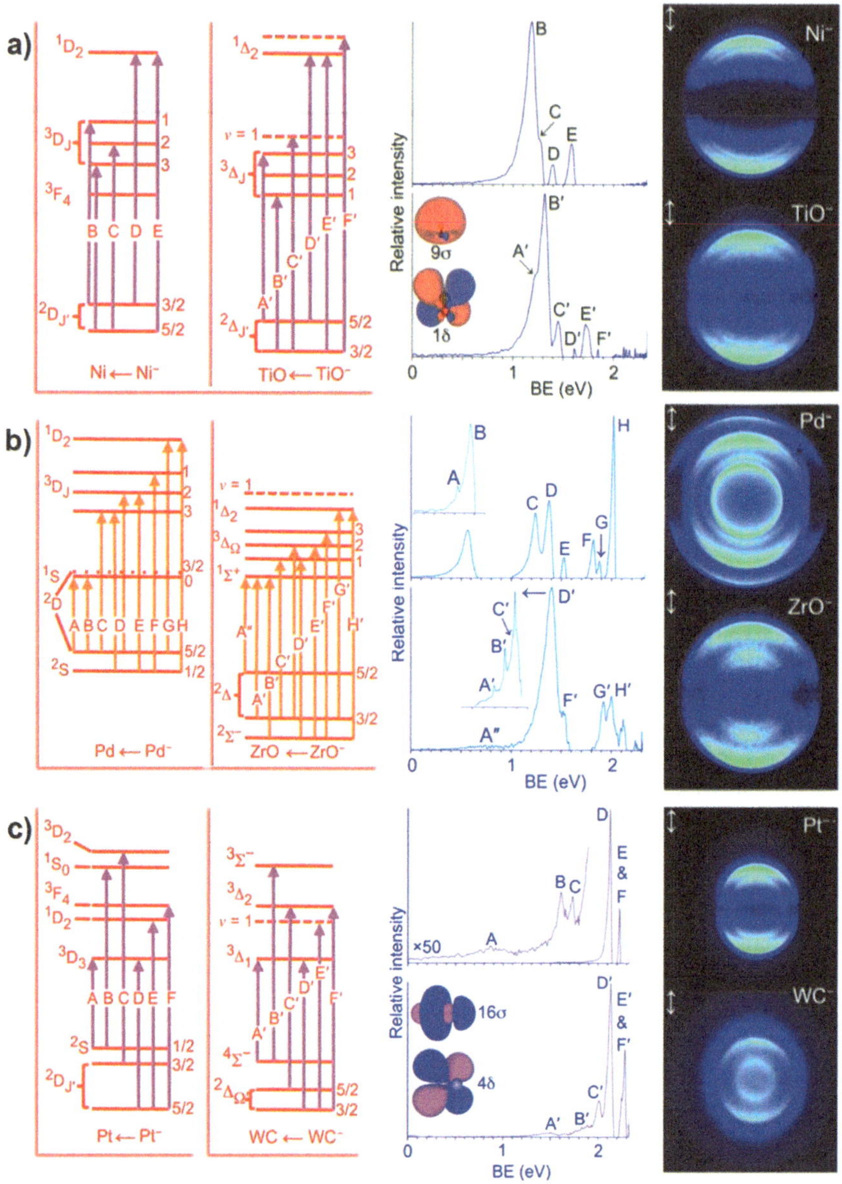

Fig. 8.8 Energy level diagrams (Left), binding energy spectra (Middle), and raw photoelectron images (Right) at a photon energy of 2.33 eV (532 nm) for TiO^-/Ni^- (**a**), ZrO^-/Pd^- (**b**), and WC^-/Pt^- (**c**). The laser polarization is vertical in the plane of the images (indicated by the double headed arrow). The insets of the binding energy spectra display the highest occupied molecular orbitals. Reproduced with permission from Ref. [119]. Copyright 2011 Springer Nature

of the 6 s orbital of Pt^- is ~20% [118], the constructive overlap of the 16σ bonding region is a mimic of the Pt 6 s orbital contraction, manifested by the $c_s^2 = 0.19$ coefficient [106].

8.3.2 Reactivity of VIII Group Metal Cluster Ions

There are abundant investigations concerning the reactivity of group VIII metals (i.e., Fe, Co, Ni series) with hydrocarbons [120–127]. The gas-phase reactivity of such metal clusters leading to the activation of C–H bonds (producing hydrogen) attracts reasonable research interest involving a variety of interesting questions [9]. For example, early studies showed that the Mn_2^+ and Co_2^+ dimers do not directly react with alkanes [128], but $Co_2(CO)^+$ reacts with butane (C_4H_{10}) to form $Co_2(CO)C_4H_8^+$ with a loss of hydrogen [129]. Similarly, $Re_3(CO)_n^+$, $Re_4(CO)_n^+$, and $Ir_4(CO)_n^+$ were also observed to support dehydrogenation at the presence of cyclohexane (C_6H_{12}) providing n was not too large, and a critical value of n for each case was rationalized in terms of frontier orbital theory [130, 131].

Figure 8.9 lays out the rate constants for reaction of $Co_n(CO)_n^+$ and $Ir_n(CO)_n^+$ with cyclohexane. For bare metal Co clusters, their reactions toward cyclohexane (C_6H_{12}) were demonstrated to follow:

$$Co_n^+ + C_6H_{12} \rightarrow Co_n(C_6H_{12-2m})^+ + mH_2 \tag{8.1}$$

This reaction pathway was obviously seen for Co clusters in the case n = 1 & m = 1–3, and n = 3–4 & m = 2–3. Note that $Con(C_6H_{12\text{-}2m})^+$ may undergo subsequent reactions. In comparison, the reactivity for Ir clusters with cyclohexane was noted as:

$$Ir_n^+ + C_6H_{12} \rightarrow Ir_n(C_6H_6)_m^+ + 3mH_2 \tag{8.2}$$

where the Ir_n^+ with n = 2–4 & m = 1–3 were found to be highly reactive in this way, with a total rate constant up to 7.2, 10.1, and $11.8 \times 10^{-10}\ cm^3s^{-1}$ respectively [120].

Also clusters Pt_n (n = 2–8) and Nb_n (n = 4–13) were found to dehydrogenate small alkanes in a fast flow [132–135], and showed a trend that the extent of dehydrogenation and the number of molecules attached increases with the cluster size [9]. Sputtered ionic clusters of Cu_n, Pt_n, Pd_n, and Ni_n were also found to dehydrogenate small alkanes [136, 137], and it appeared that Cu clusters are more reactive than Ag clusters, while Pt clusters are more reactive than Pd clusters which are more reactive than Ni clusters [9].

These reactions aim at remarkable dehydrogenation reactions which generally need an initial oxidative addition of the C–H bond onto the metal. In atomic metal cations, the empty and low-energy s orbitals facilitate the oxidative addition owing to

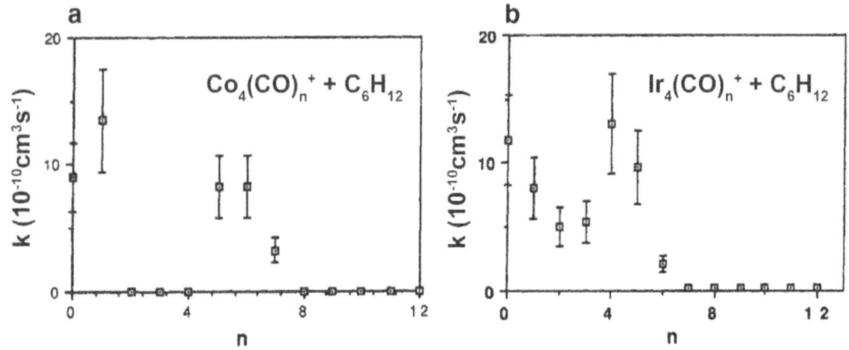

c

reactant ion	neutral products	total rate constant,[a] 10^{-10} cm^3 s^{-1}
$M_n(CO)_m^+$	$3H_2$, $2H_2$, H_2	
Ir_2^+	1.0, 0.0, 0.0	7.2
Ir_3^+	1.0, 0.0, 0.0	10.1
Ir_4^+	1.0, 0.0, 0.0	11.8
Co^+	0.1, 0.3, 0.3[b]	8.0
Co_2^+	products not observed	<0.08
Co_3^+	0.5, 0.4, 0.0	6.0
Co_4^+	0.9, 0.1, 0.0	9.2
$Co_4(CO)^+$	0.4, 0.6, 0.0	13.8
$Co_4(CO)_{2,3,4}^+$	products not observed	<0.2
$Co_4(CO)_5^+$	1.0, 0.0, 0.0	8.4
$Co_4(CO)_6^+$	1.0, 0.0, 0.0	8.4
$Co_4(CO)_7^+$	1.0, 0.0, 0.0	3.3
$Co_4(CO)_{8-12}^+$	products not observed	<1.0

Fig. 8.9 **a/b** Rate constants for reaction of $Co_4(CO)_n^+$ and $Ir_4(CO)_n^+$ with cyclohexane versus n, the number of CO ligands in the cluster. **c** A table listing the rate constants for all reactions of $M_n(CO)_m^+$ with cyclohexane (C_6H_{12}). Reproduced with permission from Ref. [120]. Copyright 1991 American Chemical Society

the ability to accept electron density from a σ bonding of C–H. It is noted in cationic clusters the molecular orbitals play an analogous role to the atomic s orbitals. The decisive factor for such metal cluster reactivity can be taken as the presence or absence of a vacant s-band orbital with sufficient orbital electron affinity to act as a good acceptor. Simple calculations by molecular orbital theory, although a number of empirical parameters and assumptions, can provide a basis to show the unreactive species may lack such an orbital. The predominance of 4 s orbitals in the bonding of late-first-row transition elements has been well-established by such calculation results; in comparison, d orbitals were found to be more significant in bonding for second- and third-row elements and early transition elements [9].

In addition, several alloy metal species such as MFe^+ (M = Co, V, Cu) were found to not react with alkanes [138–140], but $LaFe^+$ and Co_2Fe^+ react [141, 142]. In view of a $(3d)^6(4s)^2$ ground state of Fe and a $(5d)^2$ ground state of La^+, the ground state of $LaFe^+$ displays eight "d" electrons and two "s" electrons. It is expected of a high-energy antibonding σ* LUMO would result as in the first-row transition metal dimers such as Ni_2. As there are only eight "d" electrons (not ten), there will be at least one vacant d orbital even if they are not spin paired. The d orbitals are more critical in bonding in early transition metal and in second- and third-row transition metals. In fact, the d and s orbitals in La^+ are very close in energy, where two $5d^1 6s^1$ states are lower than the J = 4 state of the $5d^2\,{}^3F$ ground state. The La^+ could use one vacant s-d hybrid to bond to Fe while the other vacant s-d hybrid slightly perturbed from the energy of the LUMO in the free ion, hence allowing the dimer to be reactive with alkanes. The dimer ions with fewer than ten d electrons generally behave differently than dimers with more than ten d electrons, just as an atomic ion with fewer than five d electrons behaving differently compared to those with more than five d electrons [143].

In comparison, the failure of Mn_2^+ to react was explained in a similar argument [128]. The ground-state configurations of Mn^+ and Mn, i.e., $3d^5 4s^1$ and $3d^5 4s^2$, result in a $3d^{10} 4s\sigma^2 4s\sigma^{*1}$ configuration for Mn_2^+. As a partially occupied $4s\sigma^*$ orbital is an effective acceptor even if it is at low energy, the alert reactivity of Mn_2^+ is reasonable. On the other hand, it was found that Mn_2^+ can react with alcohols to form $Mn(ROH)^+$ together with a breaking of the metal-to-metal bond [144]. Therefore, a singly occupied antibonding orbital not only reduces the reactivity but also weakens the metal-to-metal bond. This is in sharp contrast with Co_2^+ which reacts with alcohols to form olefin complexes with the dimer ion, due to the difference of the bonding on Mn_2^+ and Co_2^+ [120]. In comparison, $FeCo^+$, FeV^+, and $FeCu^+$ clusters react to dehydrogenate olefins which are more energetic reagents than alkanes [138–140]; olefins have relatively weak allylic C−H bonds hence good donor $HOMO_S$ that they can interact with the high-energy LUMOs of the metal dimers.

8.3.3 Reactivity of Neutral Metal Oxide Clusters

Moreover, reactions of neutral vanadium oxide clusters with small hydrocarbons, namely C_2H_6, C_2H_4, and C_2H_2, have been investigated by experiments and DFT calculations [145]. Both approaches of single photon ionization through extreme ultraviolet (46.9 nm, 26.5 eV) and vacuum ultraviolet (118 nm, 10.5 eV) laser were used to detect the neutral cluster distributions and reaction products. Under the two ionization conditions, the results of mass spectrometry in the presence and absence of C_2H_6, C_2H_2, and C_2H_4 are displayed in Fig. 8.10 respectively. A few stable vanadium oxide clusters, VO_2, $V_2O_{4,5}$, and $V_3O_{6,7}$ etc. were observed with notable intensity in the presence of these reactants. While C_2H_6 is stable toward reaction with neutral vanadium oxide clusters, there were peaks observed due to an attachment of the reactant forming correlative products $V_mO_nC_2H_4$ and $V_mO_nC_2H_4$ respectively in the cases of C_2H_4 and C_2H_2. Besides the dominant association products, certain oxygen-rich clusters $VO_3(V_2O_5)_{n=0,1,2\ldots}$, (e.g., VO_3, V_3O_8, and V_5O_{13}) were found to react with C_2H_4 molecules and cause a cleavage of the C=C bond of C_2H_4 to

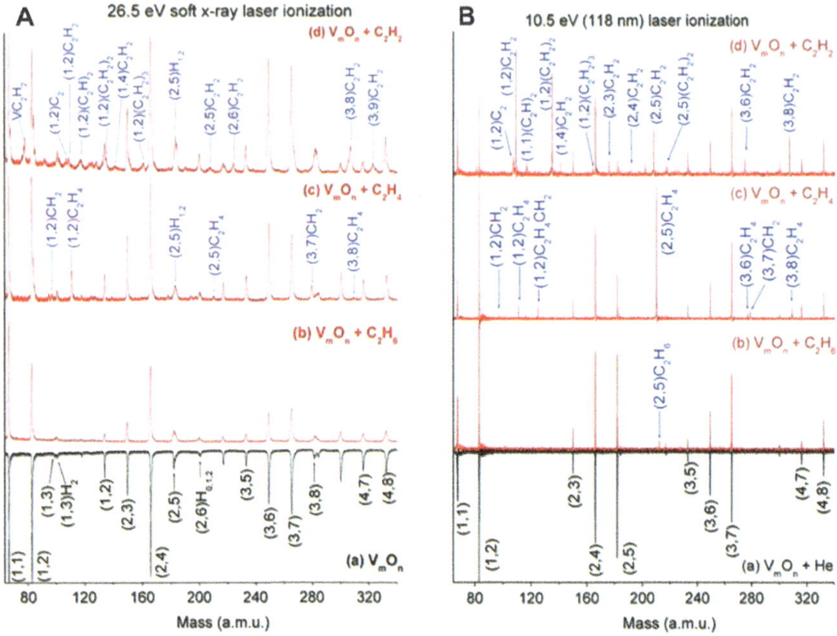

Fig. 8.10 **A** Reactions of V_mO_n clusters with C_2H_x studied by 26.5 eV soft X-ray laser ionization. New products of the reaction "$V_mO_n + C_2H_x$" are detected. (a) V_mO_n cluster distribution generated with 0.5% O_2/He expansion gas. Reactant gases, (b) Pure C_2H_6 (c) Pure C_2H_4. (d) Pure C_2H_2, are added to the flow tube reactor, respectively. **B** Reactions of V_mO_n clusters with C_2H_x studied by 118 nm (10.5 eV) laser ionization. (a) V_mO_n cluster distribution generated with a 0.5% O_2/He expansion gas, and added He gas in the flow tube reactor. Reactant gases, (b) Mixed 5% C_2H_6/He, (c) Mixed 5% C_2H_4/He, and (d) Mixed 5%C_2H_2/He, are added into flow tube reactor, respectively. The products detected by 10.5 eV laser are similar to those detected by 26.5 eV soft X-ray laser ionization. Reproduced with permission from Ref. [145]. Copyright 2008 American Chemical Society

produce $(V_2O_5)_nVO_2CH_2$ clusters. In contrast, in the reactions with C_2H_2, no C=C bond cleavage products were observed; instead, a dehydration reaction (i.e., C–H bond activation) was noted for "$VO_3 + C_2H_2$" in forming VO_2C_2.

Together with first-principle calculations which were employed to have investigated the association reactions for "V_mO_n+ C_2H_x", the dominant association reaction pathways are written as:

$$V_mO_n + C_2H_2 \rightarrow V_mO_nC_2H_2 \quad (8.3)$$

$$V_mO_n + C_2H_4 \rightarrow V_mO_nC_2H_4 \quad (8.4)$$

Besides, further studies showed the reactivities of C_2H_2 and C_2H_4 toward a simple oxygen-rich vanadium oxide cluster VO_3 using the DFT calculated binding energies, as displayed in Fig. 8.11. DFT calculations for "VO_3+ C_2H_4" and "VO_3+ C_2H_2" indicated the feasible reaction pathways that are thermodynamically favorable and overall barrierless at room temperature, which is in good agreement with the results of experimental observations. They are:

$$VO_3 + C_2H_4 \rightarrow VO_2CH_2 + CH_2O, \quad \Delta H = -0.25\ \text{eV} \tag{8.5}$$

$$\rightarrow VO_2 + CH_4O, \quad \Delta H = -0.83\ \text{eV} \tag{8.6}$$

$$\rightarrow VO_2 + C_2H_4O, \quad \Delta H = -0.35\ \text{eV} \tag{8.7}$$

$$VO_3 + C_2H_2 \rightarrow VO_2C_2 + H_2O, \quad \Delta H = -0.59\ \text{eV} \tag{8.8}$$

$$\rightarrow VO_2C_2 + CH_2CO, \quad \Delta H = -1.6\ \text{eV} \tag{8.9}$$

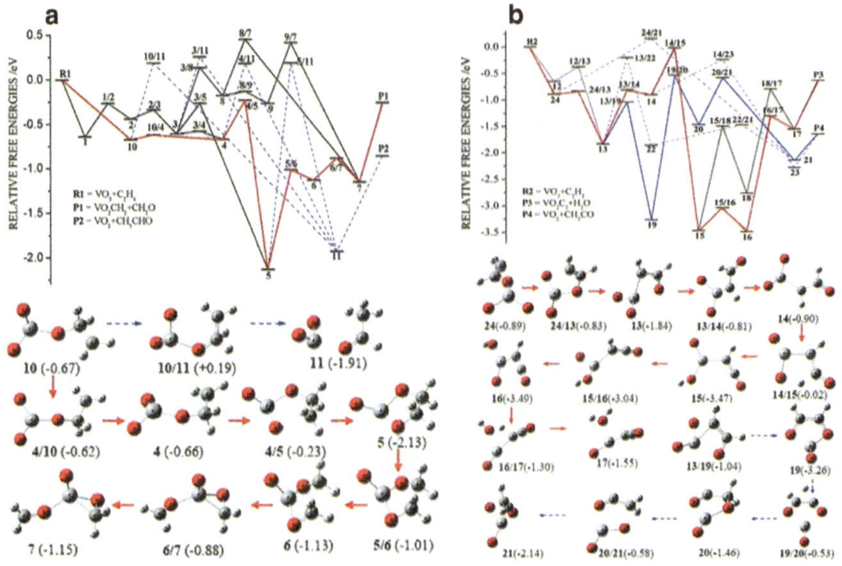

Fig. 8.11 **a** DFT calculated relative Gibbs free energies at 298 K for overall reactions $VO_3 + C_2H_4 \rightarrow VO_2CH_2 + CH_2O$ (P1) and $VO_3 + C_2H_4 \rightarrow VO_2 + CH_3CHO$ (P2). Structures were DFT optimized geometries of the reaction intermediates and the transition states in the lowest reaction pathway for generation of products P_1 and P_2. **b** DFT calculated relative Gibbs free energies at 298 K for reaction channels $VO_3 + C_2H_2 \rightarrow VO_2C_2 + H_2O$ (P3) and $VO_3 + C_2H_2 \rightarrow VO_2 + CH_2CO$ (P4). The structures were the optimized geometries of the reaction intermediates and transition states of two lowest energy pathways for the generation of products P3 and P4, respectively. All the values (in eV) in parentheses below each geometry are Gibbs free energies at 298 K. Reproduced with permission from Ref. [145]. Copyright 2008 American Chemical Society

These values of the enthalpies of reactions given above were obtained for the lowest energy structures of the reactants and products in their singlet or doublet spin states at room-temperature 298 K. It is worth mentioning that the reactions which are thermodynamically allowed may still be limited due to dynamic constraints, providing a barrier is unsurmountable, e.g., in the case of the oxygen/hydrogen transfer and/or structural transformation, etc. [145].

Similar to the above case, recently Wang et al. [59] studied the reactivity of the butane (C_4H_{10}) with neutral cluster V_2O_5 in the gas phase, as given in Fig. 8.12A. The experimental results showed that neutral V_2O_5 can react with *n*-butane (C_4H_{10}) to generate $V_2O_5H_2$, indicating double hydrogen atom transfer from C_4H_{10} to V_2O_5 to produce C_4H_8, which is coincident with the previous findings on cationic cluster system [13, 14]:

$$V_2O_5 + C_4H_{10} \rightarrow V_2O_5H_2 + C_4H_8 \tag{8.10}$$

This reactivity has also been further ascertained by substantiating C_4H_{10} with C_4D_{10}, for which the deuterated product peak $V_2O_5D_2$ was clearly observed [59].

Figure 8.12B plots the reaction profile for the reaction of C_4H_{10} with ${}^3V_2O_5$ (the lowest energy triplet state). It is shown that the formation of ${}^3V_2O_5H \cdot C_4H_9$ is facile and straightforward; and the first H atom transfer process from one β-H of C_4H_{10} to the spin located O_t of ${}^3V_2O_5$ is energetically allowed. Such H-atom transfer reactions have been reported as a spin driven process in ionic reaction systems [28, 38, 146].

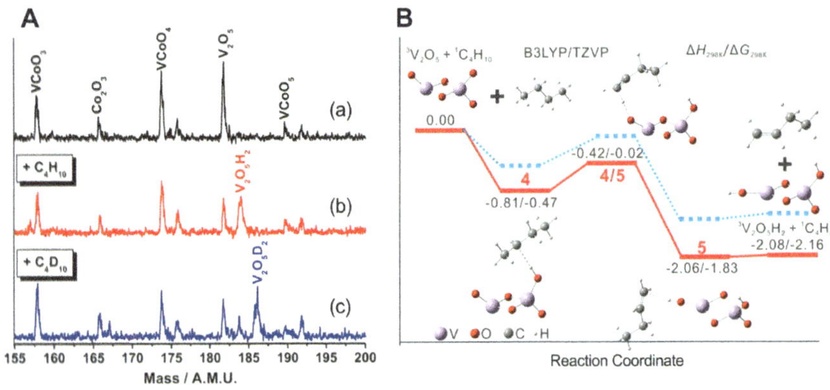

Fig. 8.12 **A** Gas phase V_2O_5 neutral cluster reacted with butane and deuterated butane: (a) distribution without any reactant; (b) C_4H_{10} added into the reaction cell; (c) C_4D_{10} added into the reaction cell. A V–Co target was employed to generate the signals. **B** Calculated energy profile of the reaction: lowest energy triplet ${}^3V_2O_5$ with C_4H_{10} (red solid lines) based on energy differences between the stationary and transition states at the B3LYP/TZVP level of theory is plotted. The energy values are relative to the entrance channel, denoted as reaction thermal enthalpy ($\Delta H_{298\ K}$) and Gibbs free energy ($\Delta G_{298\ K}$) and given in eV. The blue dotted lines denote the similar reaction of lowest energy triplet ${}^3V_2O_5$ with C_2H_6. Reproduced with permission from Ref. [59]. Copyright 2012 American Chemical Society

From the DFT calculations, both 1-butene and 2-butene are likely formed on the basis of a similar reaction mechanism [59].

The cluster reactivity towards hydrocarbons can be classified to a few aspects, such as: (i) direct association, (ii) C−H bond activation, (iii) C−C bond or C=C cracking, and (iv) O-atom transfer. These reactivities are coincident with calculated reaction coordinates via first-principles theory. By combining photoelectron spectral studies, comparable electronic transitions and orbital symmetry were found for WC versus Pt; Ni versus TiO, Pd versus ZrO, revealing commonality in their electronic spectra. On the basis of these experimental findings, it was proposed that ZrO can be viewed as the superatomic form of Pd, or at least they have similar electronic structures within 2.33 eV of the highest occupied molecular orbital (HOMO) of the anion. Similarly, TiO is mimic of Ni; as well as Pt versus WC.

References

1. A.W. Castleman Jr., S. Wei, Annu. Rev. Phys. Chem. **45**, 685–719 (1994)
2. K.A. Zemski, D.R. Justes, A.W. Castleman Jr., J. Phys. Chem. B **106**, 6136–6148 (2002)
3. D. Paul, K. Hong, T.K. Kim, J.-S. Oh, K.-W. Jung, Int. J. Mass Spectrom. **315**, 15–21 (2012)
4. M.T. Bowers, *Gas Phase Ion Chemistry* (Academic Press, New York, 1979)
5. J.F. Garvey, W.R. Peifer, M.T. Coolbaugh, Acc. Chem. Res. **24**, 48–54 (1991)
6. R.C. Bell, K.A. Zemski, K.P. Kerns, H.T. Deng, A.W. Castleman Jr., J. Phys. Chem. A **102**, 1733–1742 (1998)
7. R.C. Bell, K.A. Zemski, A.W. Castleman Jr., J. Phys. Chem. A **102**, 8293–8299 (1998)
8. R.C. Bell, A.W. Castleman Jr., D.L. Thorn, Inorg. Chem. **38**, 5709–5715 (1999)
9. R.C. Bell, K.A. Zemski, A.W. Castleman Jr., J. Phys. Chem. A **103**, 1585–1591 (1999)
10. R.C. Bell, K.A. Zemski, A.W. Castleman Jr., J. Phys. Chem. A **103**, 2992–2998 (1999)
11. K.A. Zemski, R.C. Bell, A.W. Castleman Jr., J. Phys. Chem. A **104**, 5732–5741 (2000)
12. R.C. Bell, K.A. Zemski, D.R. Justes, A.W. Castleman Jr., J. Chem. Phys. **114**, 798–811 (2001)
13. K.A. Zemski, R.C. Bell, D.R. Justes, A.W. Castleman, Jr., *A brief summary of reactions of group V transition metal oxide cluster ions with butane* (2001)
14. K.A. Zemski, D.R. Justes, R.C. Bell, A.W. Castleman Jr., J. Phys. Chem. A **105**, 4410–4417 (2001)
15. K.A. Zemski, D.R. Justes, A.W. Castleman Jr., J. Phys. Chem. A **105**, 10237–10245 (2001)
16. R.C. Bell, A.W. Castleman Jr., J. Phys. Chem. A **106**, 9893–9899 (2002)
17. D.R. Justes, A.W. Castleman Jr., R. Mitric, V. Bonacic-Koutecky, Europ. Phys. J. D **24**, 331–334 (2003)
18. D.R. Justes, R. Mitric, N.A. Moore, V. Bonacic-Koutecky, A.W. Castleman Jr., J. Am. Chem. Soc. **125**, 6289–6299 (2003)
19. J.R. Stairs, S.J. Peppernick, K.M. Davis, A.W. Castleman Jr., Isr. J. Chem. **44**, 223–228 (2004)
20. N.A. Moore, R. Mitric, D.R. Justes, V. Bonacic-Koutecky, A.W. Castleman Jr., J. Phys. Chem. B **110**, 3015–3022 (2006)
21. C. Berg, T. Schindler, G. Niednerschatteburg, V.E. Bondybey, J. Chem. Phys. **102**, 4870–4884 (1995)
22. C. Berg, M. Beyer, U. Achatz, S. Joos, G. Niedner-Schatteburg, V.E. Bondybey, J. Chem. Phys. **108**, 5398–5403 (1998)
23. Y. Xie, S.G. He, F. Dong, E.R. Bernstein, J. Chem. Phys. **128**, 044306 (2008)
24. K.A. Zemski, R.C. Bell, A.W. Castleman Jr., Int. J. Mass Spectrom. **184**, 119–128 (1999)
25. A. Božović, S. Feil, G.K. Koyanagi, A.A. Viggiano, X. Zhang, M. Schlangen, H. Schwarz, D.K. Bohme, Chem. Eur. J. **16**, 11605–11610 (2010)

26. A. Shayesteh, V.V. Lavrov, G.K. Koyanagi, D.K. Bohme, J. Phys. Chem. A **113**, 5602–5611 (2009)
27. D.K. Böhme, H. Schwarz, Angew. Chem. Int. Ed. **44**, 2336–2354 (2005)
28. H. Schwarz, Angew. Chem. Int. Ed. **50**, 10096–10115 (2011)
29. S. Yin, E.R. Bernstein, Int. J. Mass Spectrom. **321–322**, 49–65 (2012)
30. D. Schröder, H. Schwarz, Angew. Chem. Int. Ed. **29**, 1433–1434 (1990)
31. H. Schwarz, Angew. Chem. Int. Ed. **30**, 820–821 (1991)
32. D. Schröder, H. Schwarz, D.E. Clemmer, Y. Chen, P.B. Armentrout, V.I. Baranov, D.K. Böhme, Int. J. Mass Spectrom. Ion Processes **161**, 175–191 (1997)
33. X. Zhao, A.C. Hopkinson, D.K. Bohme, ChemPhysChem **9**, 873–881 (2008)
34. A. Bozovic, D.K. Bohme, Phys. Chem. Chem. Phys. **11**, 5940–5951 (2009)
35. Y.-X. Zhao, X.-N. Wu, Z.-C. Wang, S.-G. He, X.-L. Ding, Chem. Commun. **46**, 1736–1738 (2010)
36. I. Kretzschmar, A. Fiedler, J.N. Harvey, D. Schröder, H. Schwarz, J. Phys. Chem. A **101**, 6252–6264 (1997)
37. K.K. Irikura, J.L. Beauchamp, J. Am. Chem. Soc. **111**, 75–85 (1989)
38. S. Feyel, J. Döbler, D. Schröder, J. Sauer, H. Schwarz, Angew. Chem. Int. Ed. **45**, 4681–4685 (2006)
39. T. Wende, J. Döbler, L. Jiang, P. Claes, E. Janssens, P. Lievens, G. Meijer, K.R. Asmis, J. Sauer, Int. J. Mass Spectrom. **297**, 102–106 (2010)
40. D. Schröder, J. Roithová, Angew. Chem. Int. Ed. **45**, 5705–5708 (2006)
41. D. Schröder, J. Roithová, E. Alikhani, K. Kwapien, J. Sauer, Chem. Eur. J. **16**, 4110–4119 (2010)
42. K. Kwapien, M. Sierka, J. Döbler, J. Sauer, ChemCatChem **2**, 819–826 (2010)
43. G. de Petris, A. Cartoni, A. Troiani, V. Barone, P. Cimino, G. Angelini, O. Ursini, Chem. Eur. J. **16**, 6234–6242 (2010)
44. G. de Petris, A. Troiani, M. Rosi, G. Angelini, O. Ursini, Chem. Eur. J. **15**, 4248–4252 (2009)
45. N. Dietl, M. Engeser, H. Schwarz, Chem. Eur. J. **16**, 4452–4456 (2010)
46. N. Dietl, M. Engeser, H. Schwarz, Angew. Chem. Int. Ed. **48**, 4861–4863 (2009)
47. N. Dietl, M. Engeser, H. Schwarz, Chem. Eur. J. **15**, 11100–11104 (2009)
48. N. Dietl, C. van der Linde, M. Schlangen, M.K. Beyer, H. Schwarz, Angew. Chem. Int. Ed. **50**, 4966–4969 (2011)
49. K. Chen, Z.-C. Wang, M. Schlangen, Y.-D. Wu, X. Zhang, H. Schwarz, Chem. Eur. J. **17**, 9619–9625 (2011)
50. Z.-C. Wang, N. Dietl, R. Kretschmer, J.-B. Ma, T. Weiske, M. Schlangen, H. Schwarz, Angew. Chem. Int. Ed. **51**, 3703–3707 (2012)
51. S. Feyel, J. Döbler, R. Höckendorf, M.K. Beyer, J. Sauer, H. Schwarz, Angew. Chem. Int. Ed. **47**, 1946–1950 (2008)
52. Z.-C. Wang, T. Weiske, R. Kretschmer, M. Schlangen, M. Kaupp, H. Schwarz, J. Am. Chem. Soc. **133**, 16930–16937 (2011)
53. J.N. Harvey, M. Diefenbach, D. Schröder, H. Schwarz, Int. J. Mass Spectrom. **182–183**, 85–97 (1999)
54. L.-H. Tian, Y.-X. Zhao, X.-N. Wu, X.-L. Ding, S.-G. He, T.-M. Ma, ChemPhysChem **13**, 1282–1288 (2012)
55. Y.-X. Zhao, J.-Y. Yuan, X.-L. Ding, S.-G. He, W.-J. Zheng, Phys. Chem. Chem. Phys. **13**, 10084–10090 (2011)
56. B. Xu, Y.-X. Zhao, X.-N. Li, X.-L. Ding, S.-G. He, J. Phys. Chem. A **115**, 10245–10250 (2011)
57. J.-B. Ma, X.-N. Wu, Y.-X. Zhao, X.-L. Ding, S.-G. He, J. Phys. Chem. A **114**, 10024–10027 (2010)
58. Z.-C. Wang, S. Yin, E.R. Bernstein, J. Phys. Chem. Lett. **3**, 2415–2419 (2012)
59. Z.-C. Wang, S. Yin, E.R. Bernstein, J. Phys. Chem. A **117**, 2294–2301 (2013)
60. J.-B. Ma, Z.-C. Wang, M. Schlangen, S.-G. He, H. Schwarz, Angew. Chem. Int. Ed. **51**, 5991–5994 (2012)

61. Z.-C. Wang, N. Dietl, R. Kretschmer, T. Weiske, M. Schlangen, H. Schwarz, Angew. Chem. Int. Ed. **50**, 12351–12354 (2011)
62. N. Dietl, R.F. Höckendorf, M. Schlangen, M. Lerch, M.K. Beyer, H. Schwarz, Angew. Chem. Int. Ed. **50**, 1430–1434 (2011)
63. Z.G. Zhang, H.G. Xu, X.Y. Kong, W.J. Zheng, J. Phys. Chem. A **115**, 13–18 (2011)
64. Z.-G. Zhang, H.-G. Xu, Y. Zhao, W. Zheng, J. Chem. Phys. **133**, 154314 (2010)
65. Z.-C. Wang, X.-N. Wu, Y.-X. Zhao, J.-B. Ma, X.-L. Ding, S.-G. He, Chem. Eur. J. **17**, 3449–3457 (2011)
66. Z.-Y. Li, Y.-X. Zhao, X.-N. Wu, X.-L. Ding, S.-G. He, Chem. Eur. J. **17**, 11728–11733 (2011)
67. Y.-X. Zhao, X.-N. Wu, J.-B. Ma, S.-G. He, X.-L. Ding, J. Phys. Chem. C **114**, 12271–12279 (2010)
68. Z.-C. Wang, X.-N. Wu, Y.-X. Zhao, J.-B. Ma, X.-L. Ding, S.-G. He, Chem. Phys. Lett. **489**, 25–29 (2010)
69. J.-B. Ma, X.-N. Wu, X.-X. Zhao, X.-L. Ding, S.-G. He, Phys. Chem. Chem. Phys. **12**, 12223–12228 (2010)
70. J. March, *Advanced Organic Chemistry: Reactions, Mechanisms, and Structure* (John Wiley & Sons Inc., New York, 1986)
71. J.M. Bollinger, J.B. Broderick, Curr. Opin. Chem. Biol. **13**, 51–57 (2009)
72. Y.-R. Luo, J.-P. Cheng, *CRC Handbook of Chemistry and Physics*, ed. W.M. Haynes (CRC Press, 2015)
73. R.H. Crabtree, J. Organomet. Chem. **689**, 4083–4091 (2004)
74. S.M. Lang, A. Frank, T.M. Bernhardt, Catal. Sci. Technol. **3**, 2926–2933 (2013)
75. G.P. Petrova, G.N. Vayssilov, N. Roesch, Catal. Sci. Technol. **1**, 958–970 (2011)
76. F.-M. Li, H.-Q. Yang, T.-Y. Ju, X.-Y. Li, C.-W. Hu, Comput. Theo. Chem. **994**, 112–120 (2012)
77. Y.-Y. Liu, Z.-Y. Geng, Y.-C. Wang, J.-L. Liu, X.-F. Hou, Comput. Theo. Chem. **1015**, 52–63 (2013)
78. M.J. Manard, P.R. Kemper, M.T. Bowers, Int. J. Mass Spectrom. **249**, 252–262 (2006)
79. N.J. Barnett, L.V. Slipchenko, M.S. Gordon, J. Phys. Chem. A **113**, 7474–7481 (2009)
80. V.J.F. Lapoutre, B. Redlich, A.F.G. van der Meer, J. Oomens, J.M. Bakker, A. Sweeney, A. Mookherjee, P.B. Armentrout, J. Phys. Chem. A **117**, 4115–4126 (2013)
81. J. Li, E. Croiset, L. Ricardez-Sandoval, J. Phys. Chem. C **117**, 16907–16920 (2013)
82. S. Motozuka, M. Tagaya, T. Ikoma, M. Morinaga, T. Yoshioka, J. Tanaka, J. Phys. Chem. C **117**, 16104–16118 (2013)
83. P. Maity, S. Takano, S. Yamazoe, T. Wakabayashi, T. Tsukuda, J. Am. Chem. Soc. **135**, 9450–9457 (2013)
84. T.-Y. Ju, H.-Q. Yang, F.-M. Li, X.-Y. Li, C.-W. Hu, Theor. Chem. Acc. **132**, 1387 (2013)
85. Z. Cheng, N.A. Fine, C.S. Lo, Top. Catal. **55**, 345–352 (2012)
86. Y.-M. Chen, M.R. Sievers, P.B. Armentrout, Int. J. Mass Spectrom. Ion Processes **167–168**, 195–212 (1997)
87. M.R.A. Blomberg, P.E.M. Siegbahn, M. Svensson, J. Phys. Chem. **98**, 2062–2071 (1994)
88. M.R.A. Blomberg, P.E.M. Siegbahn, M. Svensson, J. Phys. Chem. **96**, 5783–5789 (1992)
89. J.J. Carroll, K.L. Haug, J.C. Weisshaar, M.R.A. Blomberg, P.E.M. Siegbahn, M. Svensson, J. Phys. Chem. **99**, 13955–13969 (1995)
90. J.J. Carroll, J.C. Weisshaar, J. Am. Chem. Soc. **115**, 800–801 (1993)
91. M.R.A. Blomberg, P.E.M. Siegbahn, M. Svensson, J. Am. Chem. Soc. **114**, 6095–6102 (1992)
92. W. Zhang, L. Wang, Comput. Theo. Chemistry **963**, 236–244 (2011)
93. S.M. Lang, A. Frank, T.M. Bernhardt, J. Phys. Chem. C **117**, 9791–9800 (2013)
94. S.M. Lang, T.M. Bernhardt, Faraday Discuss. **152**, 337–351 (2011)
95. Q. Cui, D.G. Musaev, K. Morokuma, J. Chem. Phys. **108**, 8418–8428 (1998)
96. A. Diefenbach, F.M. Bickelhaupt, J. Chem. Phys. **115**, 4030–4040 (2001)
97. M. Schlangen, H. Schwarz, Angew. Chem. Int. Ed. **46**, 5614–5617 (2007)
98. E. Broclawik, R. Yamauchi, A. Endou, M. Kubo, A. Miyamoto, Int. J. Quantum Chem. **61**, 673–682 (1997)

99. B.C. Gates, *Catalytic Chemistry* (John Wiley & Sons Inc., New York, Singapore, 1992)
100. T. Engel, G. Ertl, J. Chem. Phys. **69**, 1267–1281 (1978)
101. K. Judai, S. Abbet, A.S. Worz, U. Heiz, C.R. Henry, J. Am. Chem. Soc. **126**, 2732–2737 (2004)
102. M.S. Chen, D. Kumar, C.W. Yi, D.W. Goodman, Science **310**, 291–293 (2005)
103. W.E. Kaden, T. Wu, W.A. Kunkel, S.L. Anderson, Science **326**, 826–829 (2009)
104. R. Robles, S.N. Khanna, Phys. Rev. B **82**, 085428 (2010)
105. B. Delmon, P. Grange, P.A. Jacobs, G. Poncelet, *Preparation of Catalysis II* (Elsevier, Amsterdam, 1978)
106. S.J. Peppernick, K.D.D. Gunaratne, A.W. Castleman Jr., Proc. Natl. Acad. Sci. U. S. A. **107**, 975–980 (2010)
107. S.J. Peppernick, K.D.D. Gunaratne, A.W. Castleman Jr., Chem. Phys. Lett. **489**, 1–11 (2010)
108. Y.M. Chen, M.R. Sievers, P.B. Armentrout, Int. J. Mass Spectrom. **167**, 195–212 (1997)
109. M.R. Sievers, P.B. Armentrout, Organometallics **22**, 2599–2611 (2003)
110. E.C. Tyo, A.W. Castleman, A.C. Reber, S.N. Khanna, J. Phys. Chem. C **115**, 16797–16802 (2011)
111. S.M. Lang, T.M. Bernhardt, R.N. Barnett, U. Landman, Angew. Chem. Int. Ed. **49**, 980–983 (2010)
112. G.E. Johnson, R. Mitric, M. Noessler, E.C. Tyo, V. Bonacic-Koutecky, A.W. Castleman Jr., J. Am. Chem. Soc. **131**, 5460–5470 (2009)
113. R.B. Levy, M. Boudart, Science **181**, 547–549 (1973)
114. B.N. Kuznetsov, Y.I. Yermakov, M. Boudart, J.P. Collman, J. Mol. Catal. **4**, 49–57 (1978)
115. J.S. Lee, M. Boudart, Catal. Lett. **8**, 107–114 (1991)
116. J.A.J. Rodrigues, G.M. Cruz, G. Bugli, M. Boudart, G. DjegaMariadassou, Catal. Lett. **45**, 1–3 (1997)
117. E.C. Weigert, A.L. Stottlemyer, M.B. Zellner, J.G. Chen, J. Phys. Chem. C **111**, 14617–14620 (2007)
118. P. Pyykko, Chem. Rev. **88**, 563–594 (1988)
119. A.W. Castleman, Catal. Lett. **141**, 1243–1253 (2011)
120. Y.H. Pan, K. Sohlberg, D.P. Ridge, J. Am. Chem. Soc. **113**, 2406–2411 (1991)
121. W.Y. Lu, R.B. Huang, S.H. Yang, J. Phys. Chem. **99**, 12099–12102 (1995)
122. O. Gehret, M.P. Irion, Chem. Phys. Lett. **254**, 379–383 (1996)
123. R. Liyanage, X.G. Zhang, P.B. Armentrout, J. Chem. Phys. **115**, 9747–9763 (2001)
124. M.N. Yi, K.J. Fisher, I.G. Dance, Int. J. Mass Spectrom. **216**, 155–168 (2002)
125. K. Koszinowski, D. Schroder, H. Schwarz, Inorg. Chim. Acta **357**, 1881–1885 (2004)
126. D.M. Niu, S.D. Zhang, X.Y. Zhang, H.Y. Li, Chin. J. Chem. Phys. **17**, 116–120 (2004)
127. W. Xue, Z.-C. Wang, S.-G. He, Y. Xie, E.R. Bernstein, J. Am. Chem. Soc. **130**, 15879–15888 (2008)
128. R.B. Freas, D.P. Ridge, J. Am. Chem. Soc. **102**, 7129–7131 (1980)
129. R.B. Freas, D.P. Ridge, J. Am. Chem. Soc. **106**, 825–826 (1984)
130. J. Wronka, R.A. Forbes, F.H. Laukien, D.P. Ridge, J. Phys. Chem. **91**, 6450–6452 (1987)
131. Y.H. Pan, D.P. Ridge, J. Phys. Chem. **93**, 3375–3378 (1989)
132. D.J. Trevor, R.L. Whetten, D.M. Cox, A. Kaldor, J. Am. Chem. Soc. **107**, 518–519 (1985)
133. R.J. Stpierre, E.L. Chronister, M.A. Elsayed, J. Phys. Chem. **91**, 5228–5234 (1987)
134. R.J. Stpierre, M.A. Elsayed, J. Phys. Chem. **91**, 763–765 (1987)
135. M.R. Zakin, D.M. Cox, A. Kaldor, J. Phys. Chem. **91**, 5224–5228 (1987)
136. R.B. Freas, J.E. Campana, J. Am. Chem. Soc. **107**, 6202–6204 (1985)
137. T.F. Magnera, D.E. David, J. Michl, J. Am. Chem. Soc. **109**, 936–938 (1987)
138. R.L. Hettich, B.S. Freiser, J. Am. Chem. Soc. **107**, 6222–6226 (1985)
139. D.B. Jacobson, B.S. Freiser, J. Am. Chem. Soc. **107**, 1581–1589 (1985)
140. E.C. Tews, B.S. Freiser, J. Am. Chem. Soc. **109**, 4433–4439 (1987)
141. D.B. Jacobson, B.S. Freiser, J. Am. Chem. Soc. **106**, 5351–5352 (1984)
142. Y.Q. Huang, B.S. Freiser, J. Am. Chem. Soc. **110**, 387–392 (1988)

143. P.J. Ausloos, S.G. Lias, *Structure/Reactivity and Thermochemistry of Ions* (Reidel, Boston, 1987)
144. B.S. Larsen, R.B. Freas, D.P. Ridge, J. Phys. Chem. **88**, 6014–6018 (1984)
145. F. Dong, S. Heinbuch, Y. Xie, J.J. Rocca, E.R. Bernstein, Z.-C. Wang, K. Deng, S.-G. He, J. Am. Chem. Soc. **130**, 1932–1943 (2008)
146. X.-L. Ding, X.-N. Wu, Y.-X. Zhao, S.-G. He, Acc. Chem. Res. **45**, 382–390 (2012)

Chapter 9
Carbon-Carbon Cross-Coupling Reactions

9.1 Introduction

Chemical processes have played a vital role in the evolution of human civilization. Catalysts have been known for centuries even before mankind understood the chemical processes. From the fermentation of wine to vinegar, conversion of starch to sugar, to the initial industrial application of catalysts to produce sulphuric acid, catalytic industry has revolutionized our world. They are used in numerous processes including petrochemical industry for polymerization, oxidation, hydrogenation; production of fertilizers, catalytic convertors in cars, green chemistry; also, production of drugs in pharmaceutical industry. In this chapter, we will try to describe the later application that is revolutionizing the health care industry.

Ongoing demand for new drugs and the need to produce drugs at cheaper prices have stimulated tremendous efforts to develop novel catalysts in pharmaceutical industry [1]. The diverse materials employed as catalysts include metal oxide, metal complexes, organic and inorganic polymers, as well as biocatalysts and photocatalysts. An important class of reactions are the transition-metal catalyzed cross-coupling reactions that have wide-ranging applications such as in producing active pharmaceutical ingredients (API) [2]. As an example, Singulair is an important drug used for allergies and asthma and is marketed by Merck. Its active ingredient montelukast sodium requires a Heck reaction using aryl-halide catalyzed by a Pd catalyst [3–5]. In fact, Pd-catalyzed cross-coupling reactions account for almost 40% of cross-coupling reactions involving C–C bond formation in chemistry and pharmaceutical industry [6–49]. It is interesting to point out that nickel was used in cross-coupling reactions in early 1900's whereas the use of Pd started in 1950's. One of the objectives of this chapter is to highlight how recent efforts have led to novel Pd catalysts used for production of pharmaceuticals and what kind of attempts have been conducted to replace palladium by cheaper metals. Fundamental information on the mechanisms underlying metal cluster catalysis is critical to these developments.

Z. Luo and S. N. Khanna, *Metal Clusters and Their Reactivity*,
https://doi.org/10.1007/978-981-15-9704-6_9

9.2 Cross-Coupling Reactions by Metal Catalysts

Transition metal catalysts have long been used in organic and organometallic industry. Their use for coupling reaction can be traced back to the late 19th century [50]. One of the early reactions was Ullmann reaction that converts aryl halides to biphenyl species via copper catalysis [51–53]. While early work used copper catalysts, a copper/palladium catalyst was introduced to the coupling reaction by Heck in 1968 [54]. During the 70s' and 80s', several Pd catalyzed cross-coupling processes had been discovered and a summary of the reactions is given in Fig. 9.1 [55–62]. These are important developments that were recognized in the 2010 Nobel Prize in chemistry awarded to *Heck*, *Suzuki*, and *Negishi* [63].

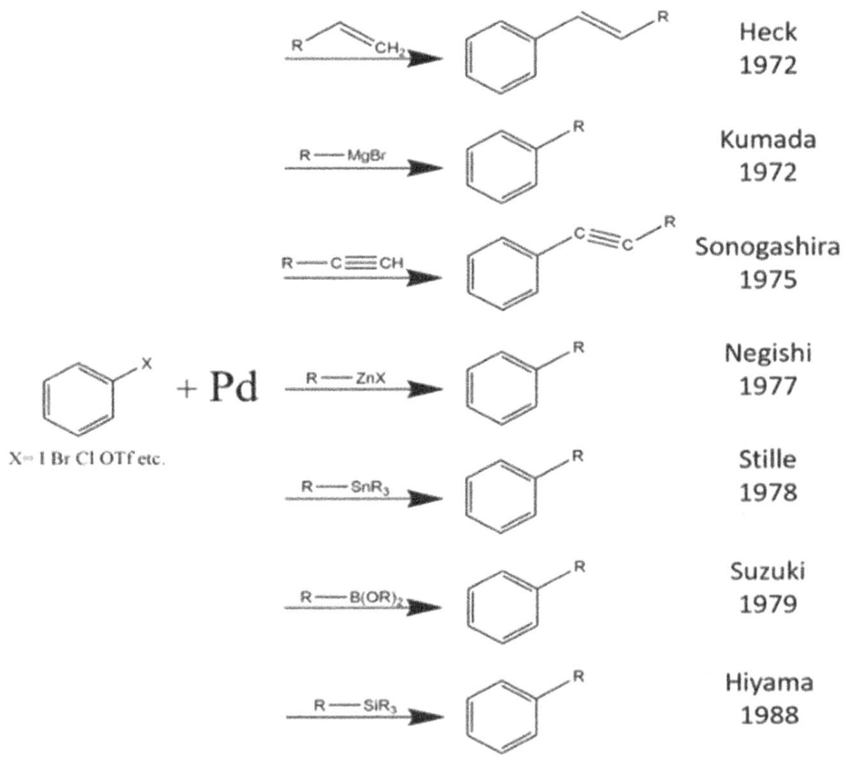

Fig. 9.1 Various cross-coupling reactions catalyzed by Palladium catalysts

9.3 Palladium Clusters Catalyse Cross-Coupling Reactions

As mentioned above, the development of Pd-catalyzed cross-coupling reactions represents one of the most significant advancements in contemporary organic synthesis, and these reactions are of strategic importance in the assembly of highly functionalized organic molecules [63, 64]. These developments in chemistry have increased the accessibility to molecules of great chemical complexity, particularly in the area of pharmaceutical drug discovery and development. The Pd-catalyzed cross-coupling reactions are typically performed under homogenous conditions, utilizing ligands to enhance activity and selectivity, and are extensively used in the assembly of API. However, the process leads to residual metal that contaminates the reaction product [65, 66]. This is particularly a major issue in pharmaceutical applications where this chemistry is extensively used, since palladium compounds can be highly toxic. Therefore, posttreatment coupled with the inability to recycle the platinum group metal, as well as the ligand, results in a significant cost component in API applications [4]. The development of supported catalysts that could reduce/eliminate leaching and sintering, could enhance the performance, and that could be recycled, would be a giant step in lowering the cost of the synthesis of chemicals and drugs. Such a development could open a pathway to "Drugs when Needed". Understanding the factors controlling the different reaction steps in the supported catalysis including the role of support in lowering the reaction barriers is the first step towards optimizing the catalysts. Such an understanding could also provide pathways to replace Palladium by less expensive and more effective metals. Following we describe some recent advances where joint experimental/theory efforts are providing new directions towards these objectives.

Several problems including residual metal contamination are germane to the homogeneous catalysts as the metal in solution is used to catalyze the reaction. One way to overcome the limitation is then to perform the catalysis by particles bound to supports so that they would prevent leaching of the catalyst into the solutions. These have led to the evaluation of palladium metal clusters with a wide range of traditional catalyst support systems employing a variety of synthetic techniques [65, 67–70]. However, several efforts in this area suggest that deposited Pd nanoparticles on solid supports merely serve as a reservoir for active and small soluble Pd_n species that catalyse cross-coupling reactions via a leaching/redeposition mechanism [71–76]. For the few examples of supported Pd catalysts that attest to function via a heterogeneous pathway, the dispersity of metal nanoparticles appears to play a major role in the mode of action.

It was recently discovered that palladium clusters/particles supported on reduced graphene oxide could represent a high-performance heterogeneous catalyst. These studies focused on a model Suzuki reaction using 4-bromobenzoic acid and phenylboronic acid as reagents. The reaction follows a three-step pathway of oxidative addition, trans-metalation and reductive elimination that could be recycled for multiple times, indicative of overcoming the leaching and recyclability issues [77]. In these studies, Pd nanoparticles supported on graphene oxide were synthesized by impregnating Pd precursor with graphene oxide followed by hydrazine and microwave-heating assisted co-reduction. Structural investigations using STM and other probes indicated that such a method generates vacancy defect sites/voids in the graphene sheet with Pd_n particles strongly bound to these vacancies/voids. It was interesting that the resulting catalysts exhibited remarkable catalytic activity (Table 9.1) as compared to other support systems with the turn-over frequency (TOF) being orders of magnitude higher than the other catalysts. Furthermore, negligible metal leaching was observed when these materials were used in Suzuki cross-coupling reactions (<300 ppb Pd in the reaction mixture) and the catalysts could be recovered and recycled multiple times without measurable loss of catalytic activity. The enhanced catalytic activity along with the apparently strong binding of Pd clusters

Table 9.1 Turnover frequency (TOF) of a variety of solid supported Pd catalyst for Suzuki reactions involving aryl bromides and phenylboronic acid

Solid Support	TOF (hr^{-1})	Reaction Condition
Defected Graphene [78]	230,000	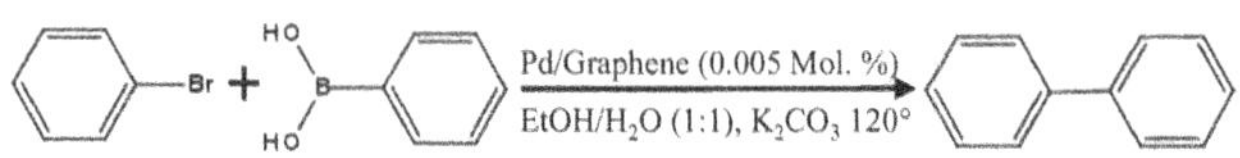
Graphite Oxide [79]	39,000	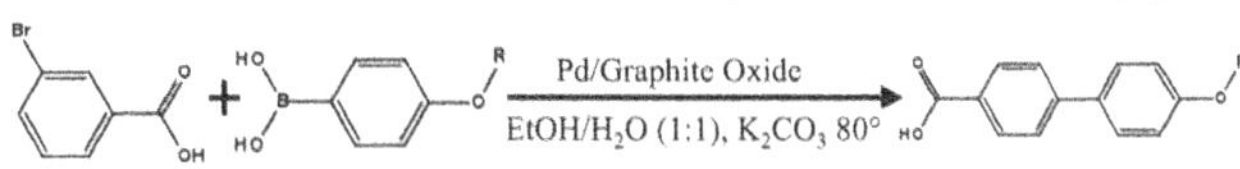
Silica [80]	25,000	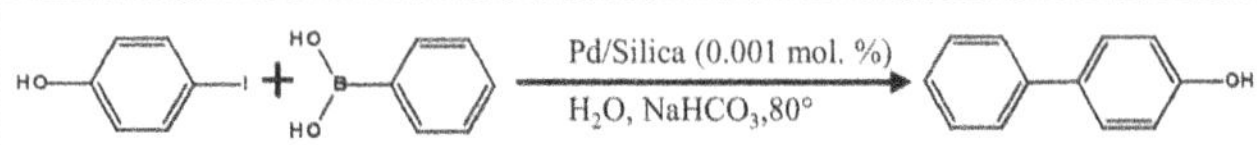
Activated Carbon [81]	16,600	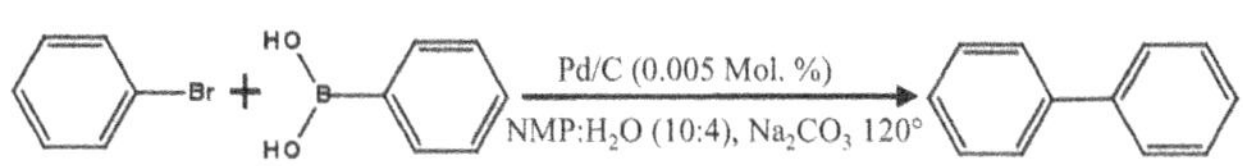
Al_2O_3 [81]	9,600	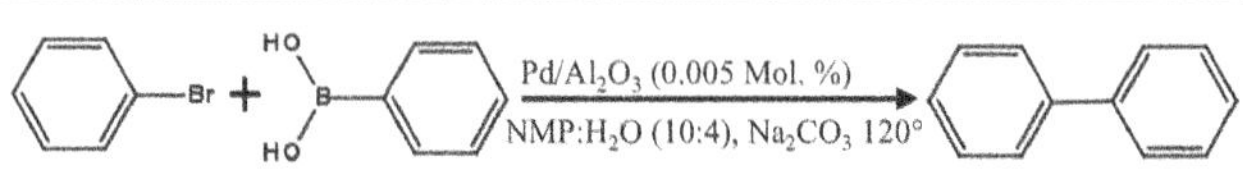
TiO_2 [81]	9,700	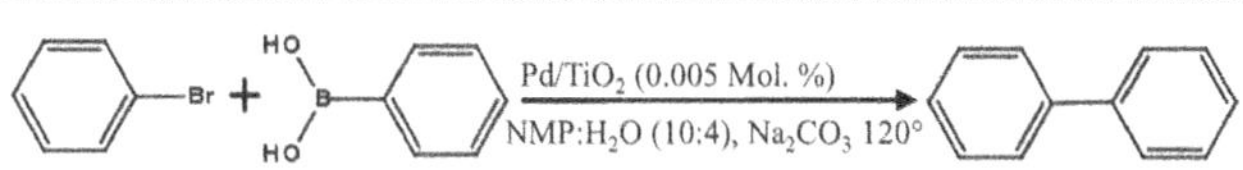
CeO_2 [81]	4,100	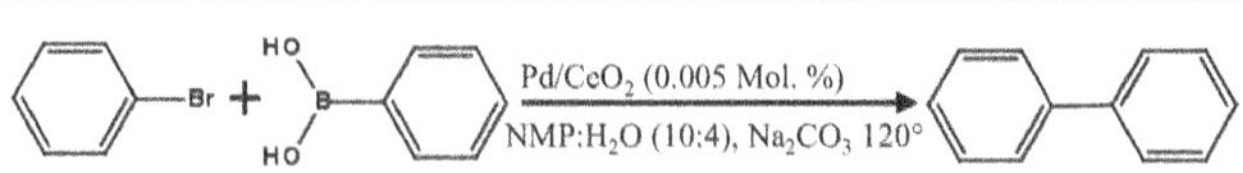
NaY [81]	4,100	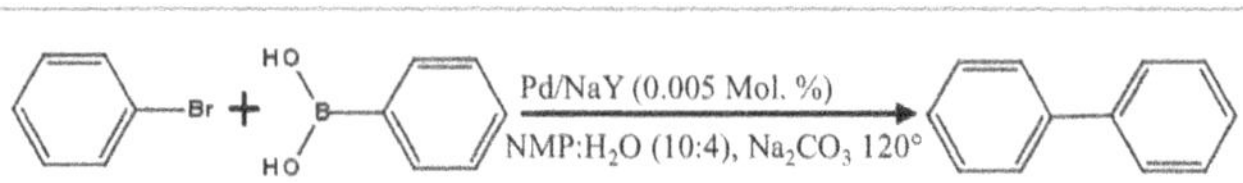

to the graphene support suggested that a unique interfacial interaction was occurring between Pd nanoclusters and defected graphene.

The remarkable activity coupled with minimal leaching raised two questions: (1) what is the microscopic reaction mechanism that makes these supports so active? (2) Can one use this information to optimize the catalysis and identify additional active supports?

To answer some of these questions, Yuan et al. [81]. Recently undertook a series of synergistic studies combining kinetic experiments on various palladium-based catalysts and coupled it to first-principles theory to determine the activation energy in each step of the catalytic cycle of Suzuki reaction for both free and supported Pd_n species [34]. The studies examined the effects of the support, active-site structure, Pd cluster size, and the electron rich ligands. To understand the effect of particle size and the particle-support interactions on the catalytic activity, Pd nanoparticles supported on graphene oxide were synthesized by three different methods: (i) impregnation of Pd precursor with graphene oxide followed by hydrazine and microwave-heating co-reduction (CO(MW)), [77, 82] (ii) strong electrostatic adsorption (SEA) [83] of Pd precursor on graphene oxide followed by hydrogen reduction (SEA-H) in tube furnace, and (iii) SEA followed by solvent-free microwave heating reduction (SEA-MW). In the first CO(MW) and third SEA-MW methods, microwave irradiation was used to induce vacancy defect sites that can strongly anchor Pd nanoparticles [84]. In comparison, the second method (i.e., SEA-H method) immobilizes the cationic Pd precursors to deprotonated oxygen functional groups and results in uniform Pd nanoparticles, but no vacancy defects are formed. TEM images of the resulting catalysts, as shown in Fig. 9.2, confirmed that the microwave irradiation changes the metal-support interaction.

Further, they evaluated the three supported catalysts and a commercially available supported catalyst along with two homogenous Pd-based catalysts. For each catalyst system, the activation energy and TOF were determined for the model Suzuki reaction using 4-bromobenzoic acid and phenyl boronic acid as reagent. The findings are shown in Fig. 9.2d. The Pd/G CO and Pd/G SEA-H catalysts possess similar particle size distributions (avg. ~2 nm), allowing a size independent comparison between catalysts of different metal-support interactions. The activation energy of the Pd/G SEA-H was four times larger than that of Pd/G CO, and the TOF is 8,000 h^{-1} versus 114,000 h^{-1} respectively. This demonstrated that the catalytic activity is highly sensitive to the catalyst-support interaction. The activation energy of Pd/G CO catalyst was also found to be significantly lower than that of the homogenous catalysts Pd/PPh_3 and Pd/CyJP. The activation energy of Pd/G SEA-MW was about three times lower than that of Pd/G SEA-H. It was proposed that the reduction in activation energy was caused by the microwave heating and induced defect sites. The activation energy of Pd/G SEA-MW was slightly higher than that of the Pd/G CO catalyst which was most likely due to the larger nanoparticle size reducing the ratio of lower coordination Pd active sites. In fact, the activation energy of the Pd/G SEA-H was found to be similar to the activation energy of commercially available Pd/C catalyst, which is believed to catalyze the cross-coupling reaction via a leaching induced homogenous mechanism

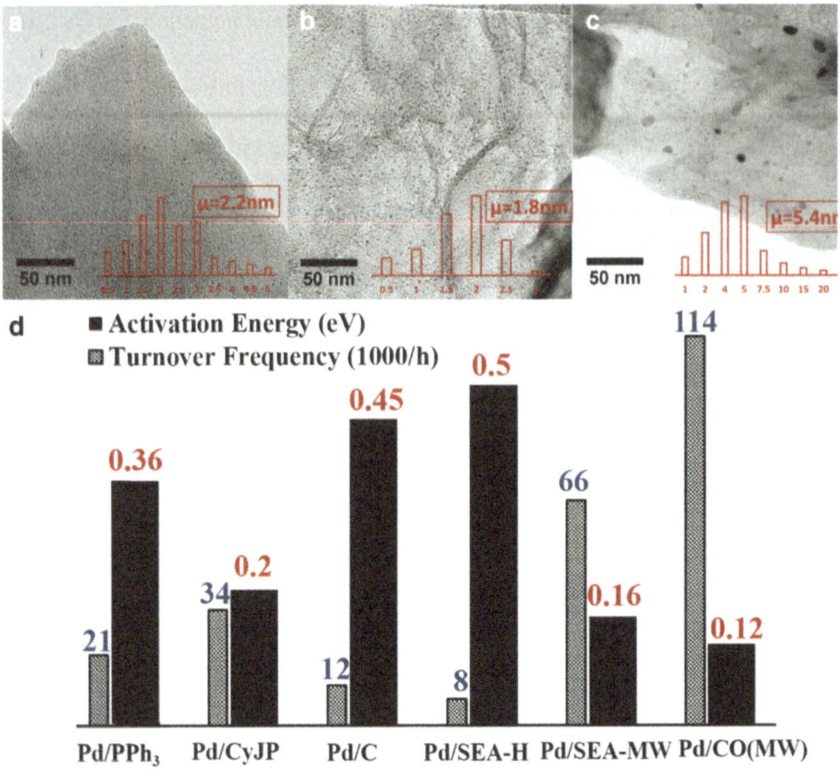

Fig. 9.2 TEM image and particle size distribution of Pd/graphene catalyst synthesized by **a** microwave assisted co-reduction (CO (MW)), **c** strong electrostatic adsorption and hydrogen reduction (SEA-H), and **c** strong electrostatic adsorption, and solvent-free microwave irradiation (SEA-MW). **d** Activation energy and TOF of the catalysts in Suzuki reaction

[67]. To summarize, catalysts formed by using defect-inducing microwave radiation exhibited the highest TOF and lowest activation energy.

9.4 Microscopic Catalytic Mechanism of Supported Palladium Clusters

Yuan et al. proceeded to examine the underlying microscopic mechanism for the exceptional performance of supported Pd_n clusters using theoretical methods based on a density functional approach. The first step was examined if the support binds the clusters strongly to prevent leaching and Ostwald ripening. To answer this, they began by studying the geometries of the pure Pd_n clusters. While bulk Pd is paramagnetic, it is known that small Pd_n clusters are marked by finite spin moments. Consequently, one must look at the structures and spin multiplicity. Figure 9.3 shows the ground

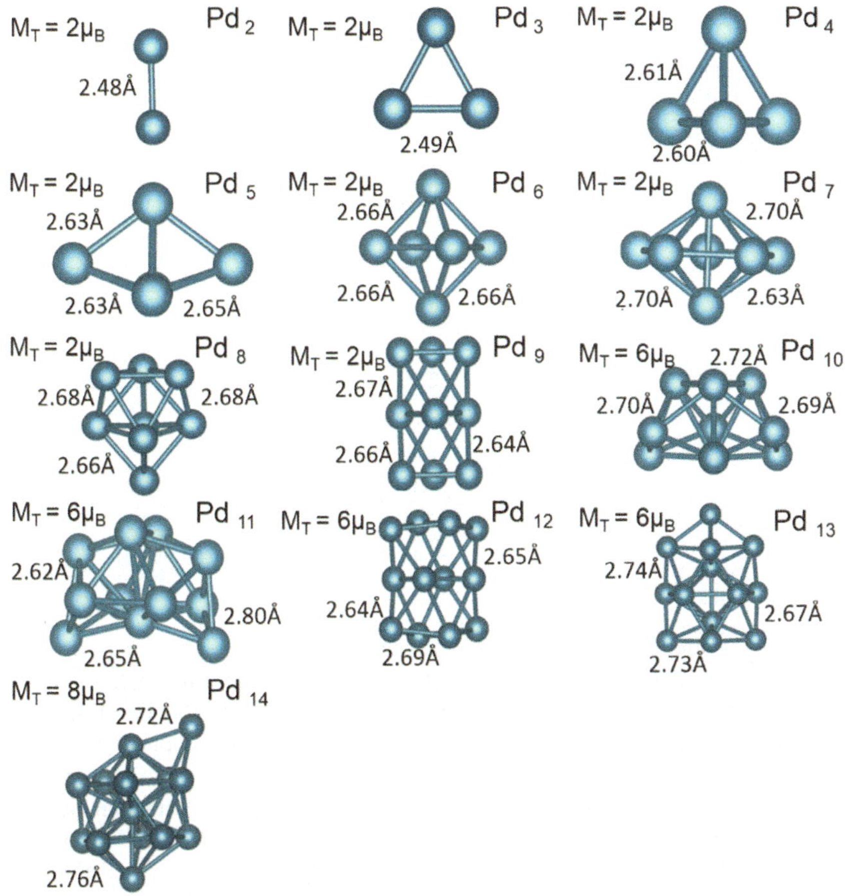

Fig. 9.3 The ground state structures, spin magnetic moment and selected bond length of free Pd_{2-14} clusters

states of $Pd_{2\text{-}14}$ clusters. To examine the stability, the average binding energies (ABE) per atom were calculated using this expression,

$$\mathrm{BE}\big/\,\mathrm{atom} = (\mathrm{n} \cdot \mathrm{E(Pd)} - \mathrm{E(Pd_n)})\big/\,n \qquad (9.1)$$

Here $E(Pd_n)$ is the total energy of the cluster of n atoms, and E(Pd) is the atomic energy. The authors also calculated the vertical ionization potential (VIP) representing the energy required to ionize a cluster without changing the geometry. The VIP represents the energy difference between the ground state of the neutral and that of the cation. The results of different binding energies are shown in Table 9.2.

As expected, the binding energy increases with cluster size. An examination of the increase in binding energy as a successive Pd atom is added shows that Pd_4, Pd_6 and Pd_1 clusters were more stable than the neighbouring sizes. Another interesting

Table 9.2 A list showing the average binding energy, spin multiplicity and the vertical ionization potential of Pd_{1-14} clusters

Pd cluster	Avg. binding energy (eV)	Spin multiplicity	Ionization potential (eV)
1	–	1	8.80
2	0.65	3	7.77
3	1.27	3	7.81
4	1.68	3	6.78
5	1.81	3	6.62
6	1.89	3	6.41
7	2.01	3	6.37
8	2.11	3	6.18
9	2.20	3	6.15
10	2.24	7	6.19
11	2.29	7	6.22
12	2.35	7	6.21
13	2.41	7	5.95
14	2.43	9	5.91

feature of the studies was that the ground state structure of Pd_{13} that has a complete coordination shell is not a regular icosahedral or cuboctahedral structure but a bilayer structure. From the catalysis point of view, the key issue was whether the clusters are more stable as they are supported on reduced graphene oxide. Graphene, generally obtained by the reduction of graphene oxide, can contain defects of various types and sizes [85] These defect sites have the potential to strongly bind palladium clusters. Theoretical studies were undertaken to investigate the properties of various defects including defects obtained by removing varying number of C atoms as well as Stone-Wales defects that do not involve the removal of atoms. Figure 9.4 shows the spin magnetic moment and the nature of the polygons around the defected site. Note that while the smaller defects have singlet ground states, multiplicities of 7 and higher emerge as the size of the defect increases beyond 6.7Å. The defects are labelled by the size of the polygons surrounding the defected region. The binding energy of the Pd atom and Pd_n clusters to the defected site show dependence on the size of the void [86].

Figure 9.5 shows the atomic structure of the ground state of the Pd site to the defect. For all the defects, the ground state is singlet. This is particularly interesting for DV(18) as the isolated vacancy had a spin multiplicity of 7 and the addition of Pd quenches the spin moment even through the Pd atom is linked to only two carbon sites. The Pd binding energy also shows large variations with the nature of the defect. One can broadly classify the binding sites into two categories: (1) defect sites where the graphene sheet heals to fill the void and forming three, five, or seven member rings and (2) the cases where the relaxed structure has a larger central hole that can accommodate an individual Pd atom or associated within a clusters. For the former which has a defect size no larger than 3.5 Å, the Pd binding energy ranges from 1.72

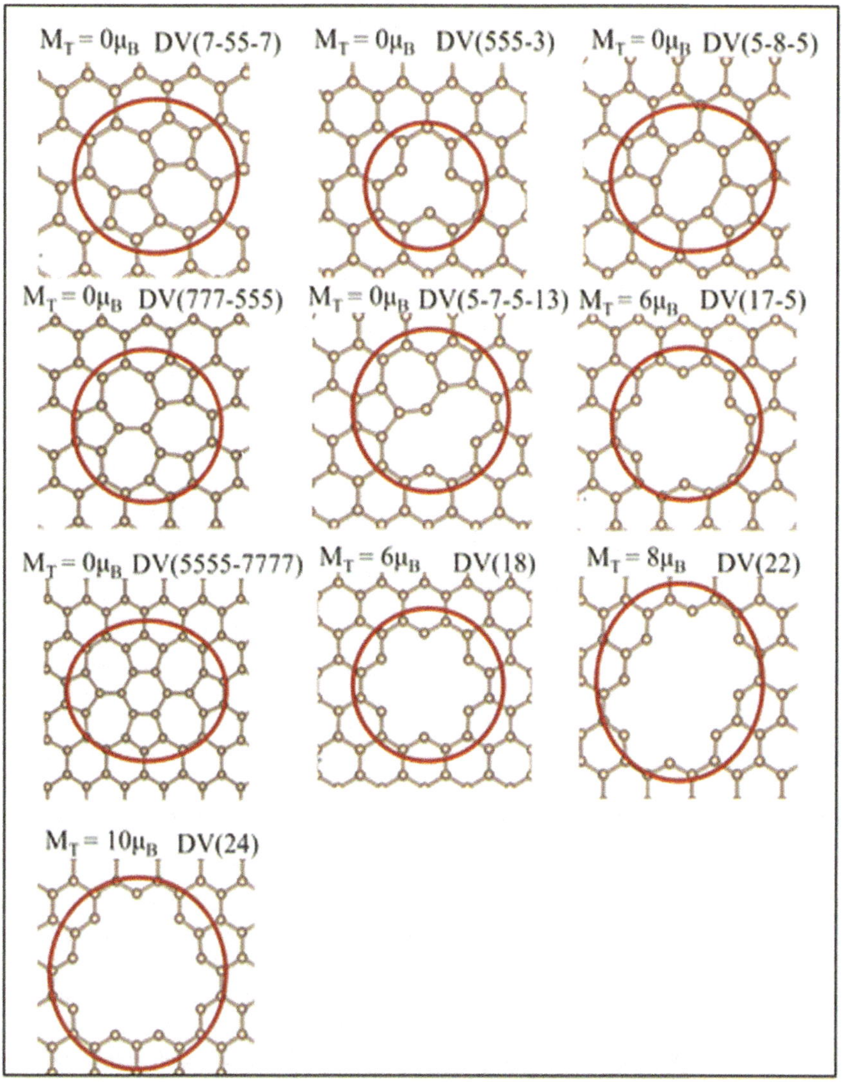

Fig. 9.4 Spin magnetic moment and the nature of the polygons around the defected site

to 2.13 eV compared to 1.26 eV for the pristine graphene. This shows that even the Stone-Wales defects bind Pd more strongly than the pristine graphene. The binding energy is however considerably enhanced to 7.28 eV as the defect size increased to 6.7 Å as in DV(18). This is clearly seen in Fig. 9.5 that shows the binding energy of the atom on different defect sizes.

As the large defect site can accommodate more Pd atoms, one can calculate the binding energy of multiple anchored Pd_n clusters to the defect. It was found that by

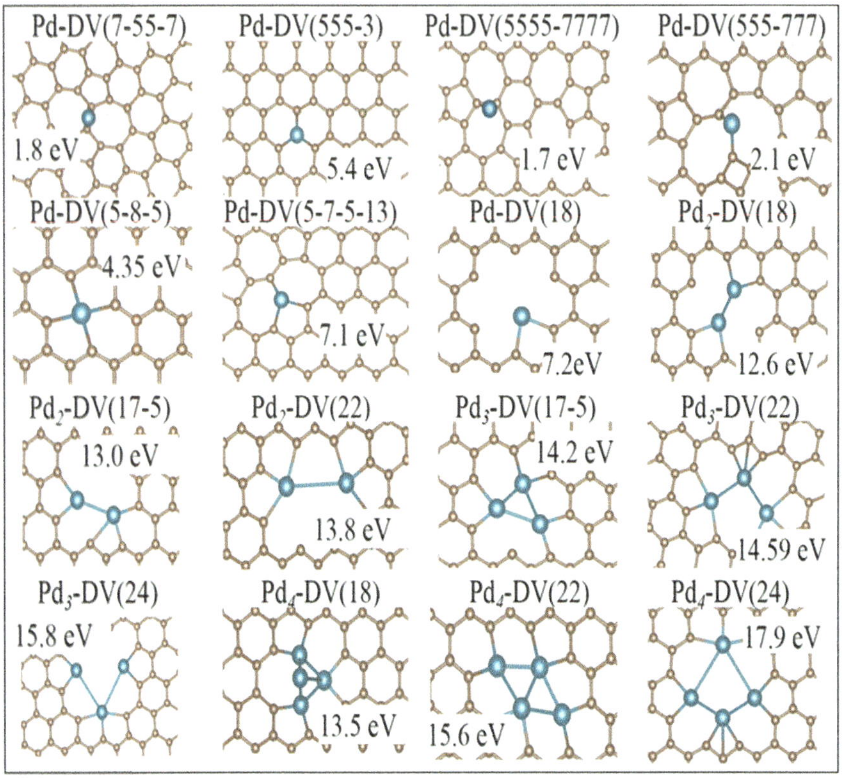

Fig. 9.5 Ground state structures of Pd_n clusters bound to defects in Graphene

binding palladium clusters to a larger void, the binding energy increased to 13.8 eV, 15.8 eV and 17.9 eV eV for 2, 3, 4 anchored Pd_n clusters respectively. As shown in Fig. 9.6a, there is almost a linear correlation between the number of anchored Pd sites and binding energy, and a further correlation to the size of the void. These trends indicate that the size of the void is important for strongly binding the Pd_n clusters, since large voids allow formation of multiple Pd-C bonds that enhance the binding. More importantly, the studies show that the binding within the cluster also increases upon deposition. Figure 9.6b compares the average binding energy of free and supported clusters and in all cases, the binding per atom is enhanced as the clusters are bound to the defects in graphene. There are two important outcomes. First, the enhanced intra-atomic binding reduces the leaching and secondly, the strong binding reduces the modility of the clusters on the surface which reduces the agglomeration of the clusters to grow into bigger sizes. The stability and the neglect towards agglomeration, while important for the recyclability and preservation of the catalyst are only a first step towards an improved catalyst. The key issue is if and how the binding to defects conspires the catalyst to increase its activity.

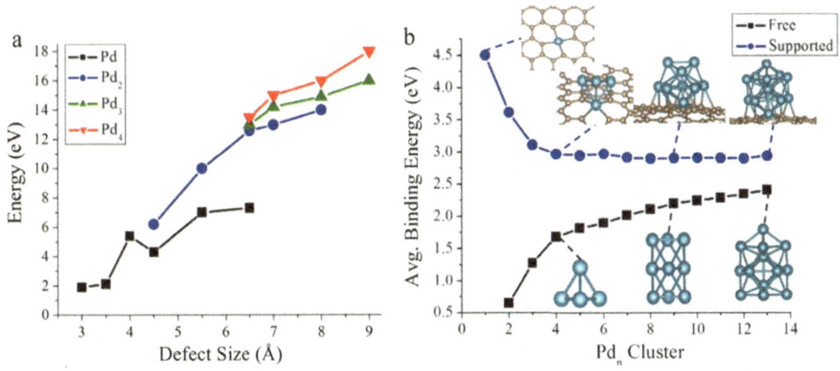

Fig. 9.6 Stability cluster covalently of Pd_n bond to defected graphene. **a** The binding energy of Pd_n, $n = 1$–4 on different sized graphene defects. **b** The average atomization energy of free Pd_n clusters and Pd_n on defected graphene

The Suzuki reaction follows a three-step pathway of oxidative addition, trans-metalation, and reductive elimination [65, 67, 68]. Figure 9.7 shows the model Suzuki reaction for different supported and free Pd clusters to gain a qualitative insight into the origin of the variations in catalytic activity. For the oxidative addition step, the Br-C bond of 4-bromobenzoic acid is cleaved by a Pd cluster/atom, indicating that this step requires a charge donation from the catalyst and support (Fig. 9.7, blue arrow). During the trans-metalation step, it was proposed that the charge flows

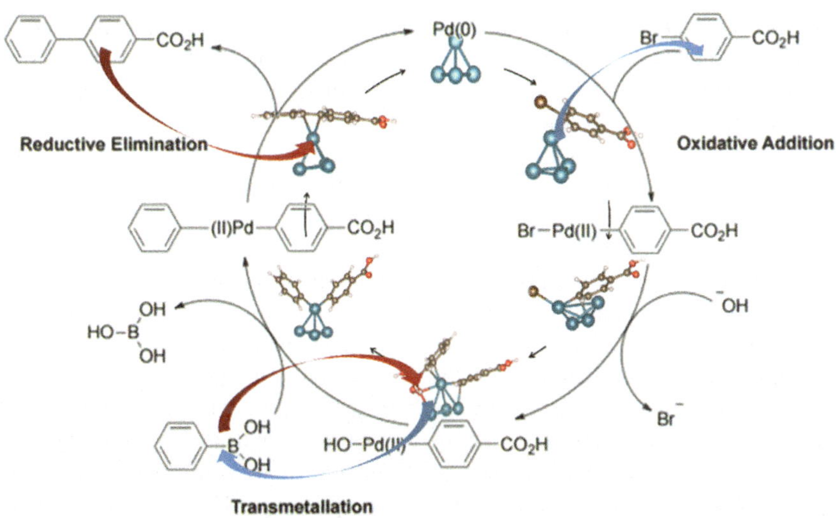

Fig. 9.7 Full reaction cycle of Pd catalyzed Suzuki reaction using Pd_4. Charge flow from catalyst to reagent is indicated by blue arrows, and charge flow from reagent to catalyst is indicated by red arrows

towards the phenyl boronic acid when breaking the carbon-boron bond, and then flows back when forming the palladium-boron bond. In the reductive elimination step, the carbon-carbon bond is formed and the product is released from the palladium surface. This process involves charge flowing back from the binary complex to the catalyst (Fig. 9.7, red arrow). Different steps of the catalytic cycle for the Suzuki cross-coupling reaction require the catalyst to serve alternately as an efficient charge donor or acceptor. The conducting nature of the graphene allows it to activate both the donation and acceptance of charge by exchanging charge with the cluster.

Also examined, was the individual steps and overall reaction activation energies for free Pd_n clusters (n = 1, 4, 13, and 14), providing insight into the microscopic process. The Pd atom [87, 88] and Pd_4 clusters were selected based on earlier studies [89] where it was found that the presence of Pd_1 and Pd_4 fragments in solution were critical for cross-coupling reactions. The results in Fig. 9.8a suggest that the activation energies are cluster size-dependant. It was found that Pd with a low coordination tends to have a lower activation energy than Pd with a higher coordination, although the support and ligands also play a significant role [90].

The activation energy of Pd_n clusters bound to a double atom vacancy sites in the graphene sheet were calculated to understand the effect of the Pd-support interaction, along with the activation energies for a Pd atom ligated by $P(Me)_3$ as a model of a homogeneous catalyst. For Pd_4, Pd_{13}, and Pd_{14}, the activation energies for clusters supported on reduced graphene were all lower than those for the corresponding free clusters, indicating strong Pd-support interaction that enhances Pd_n cluster reactivity in each step. Figure 9.8b shows the density of states of Pd_4 and Pd_4/G at the transition state for reductive elimination. The bonding and antibonding orbitals that control the formation of the C–C bond are significantly stabilized by the presence of the defected graphene. The stabilization is due to the positive charged state of the supported cluster as the support makes the catalyst a superior charge acceptor. The full Suzuki reaction pathway for Pd_4 and Pd_4/G is shown in Fig. 9.8c. Strikingly, the activation energy for all three steps are lower for the supported cluster than for the free cluster, illustrating the support activates both electron donating and electron withdrawing steps in the reaction cycle. These results strongly suggest that the supported Pd clusters are more effective catalysts because of lower activation energies resulting from the electronic interaction with the graphene support.

The charges on reagents and products were further analysed at each step for each cluster (Fig. 9.9) to identify the physical origin of the reduced activation energies. First let us consider, the trans-metalation step which involves both charge donation and back donation. As seen in Fig. 9.9a, the electron-rich ligand of the Pd/$P(Me)_3$ catalyst actively enhances charge donation, but the back flow of the charge is hindered which leads to a high activation energy. In contrast, graphene-supported Pd clusters, especially Pd_4/G, have lower activation energies for trans-metalation due to the conductive nature of graphene allowing it to accept charge. There is a clear inverse relationship between the net charge flow and the activation energies demonstrating that catalysts which effectively accept charge are more active for trans-metalation.

The relationship between charge flow and activation energy was also observed in the oxidative addition and reductive elimination steps (Fig. 9.9b, c). In oxidative

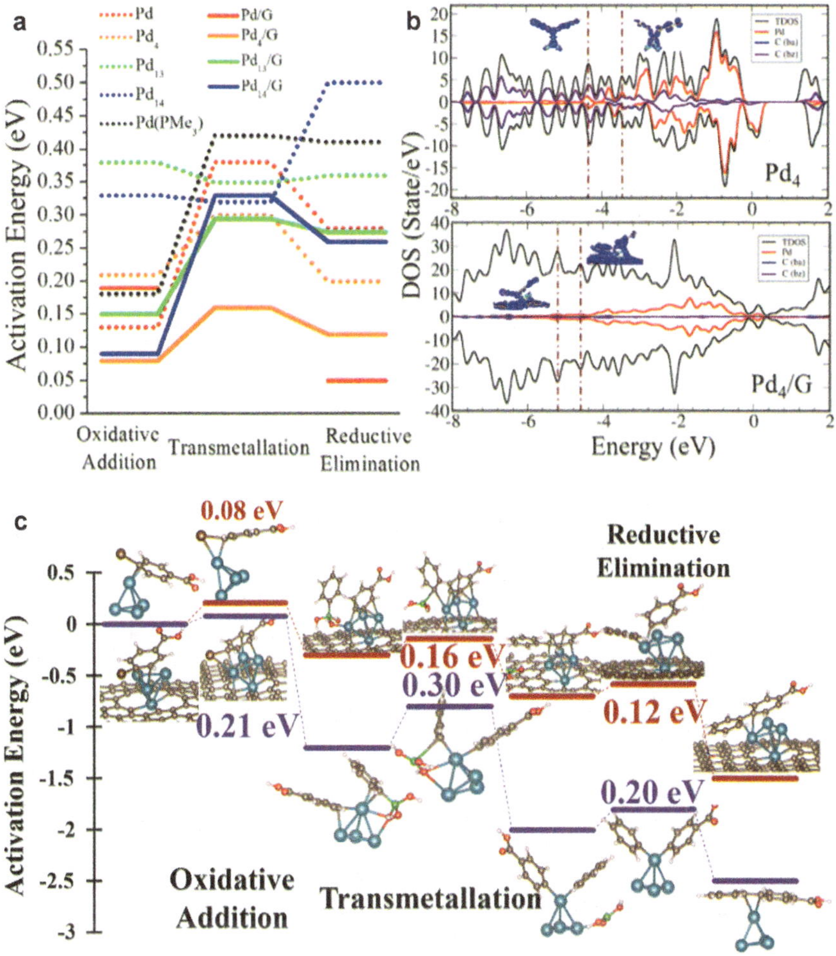

Fig. 9.8 **a** Activation energy of free and supported Pd_n clusters (n = 1, 4, 13, and 14, and $Pd(PMe_3)$) in the three Suzuki reaction steps. **b** Density of States (DOS) and bonding/antibonding orbitals at the transition state for reductive elimination for Pd_4 and Pd_4/G. (ba = benzoic acid, bz = benzene). (**c**) Full reaction pathways for Pd_4 and Pd_4/G

addition, charge donation by graphene lowers the activation energy. In Fig. 9.9b, clusters that accept charge from the reactant such as Pd_{13}, and Pd_{14} have high activation energies, while clusters that are better donors, such as Pd_4/G and Pd_{14}/G have lower activation energies. During the reductive elimination step, the lowest activation energy is observed for catalysts that readily accept charge. This is supported by the large activation energy for Pd/PMe_3, because the electron rich ligand causes the catalyst to act as a charge donor, while the better acceptors such as Pd/G and Pd_4/G have lower activation energies. To confirm graphene assists both charge donation and withdrawal, the adsorption energies of Br (acceptor) and P (donor) were calculated.

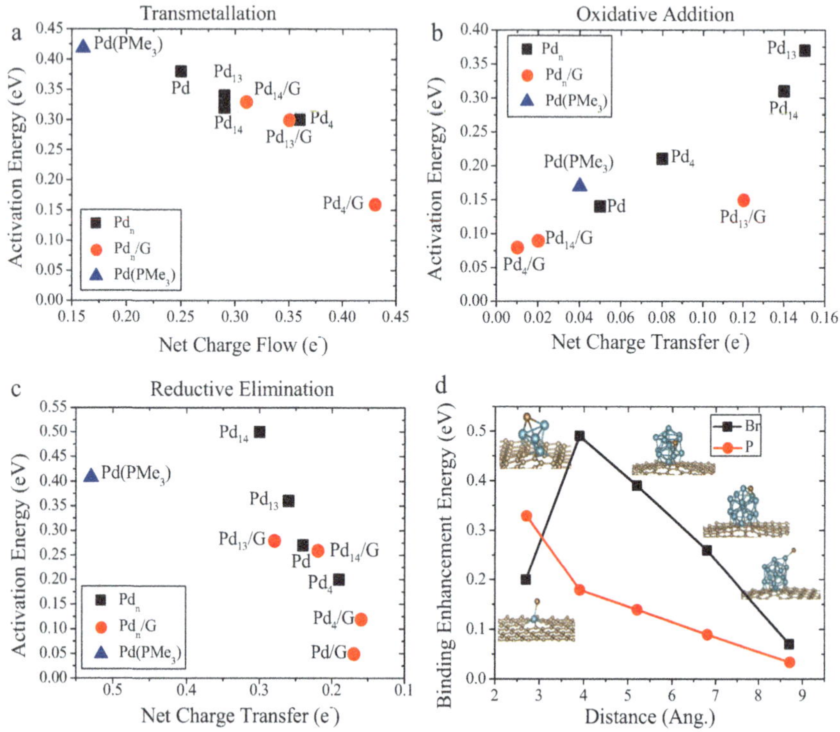

Fig. 9.9 Charge transfer and activation energy for free and supported Pd_n, and $Pd(PMe_3)$ during **a** trans-metalation, **b** oxidative addition and **c** reductive elimination. **d** Enhance Br and P binding energy correlated to the distance from the Pd active site to the lattice of graphene

The binding energy of both the Br and P are enhanced on the defected graphene as compared to the free clusters, confirming that the support enhances both charge donation and acceptance and that the graphene appears to function as a solid-state ligand.

9.5 Search for Alternate Cheaper Catalysts

To investigate the possibility of replacing Pd with a cheaper 3d transition metal, the activation energies for the free and supported Fe_{13}, Ni_{13}, and Cu_{13} clusters were calculated and compared with that for the free and supported Pd_{13} clusters in the full reaction cycle of the Suzuki reaction, as shown in Fig. 9.10. As mentioned above, the typical Suzuki reaction proceeds via three steps: oxidative addition, transmetallation, and reductive elimination. In the charge donating oxidative addition step, the activation energies of the free first-row transition metal clusters are significantly lower than that of the free Pd_{13} cluster. The Fe_{13} cluster exhibits extremely low

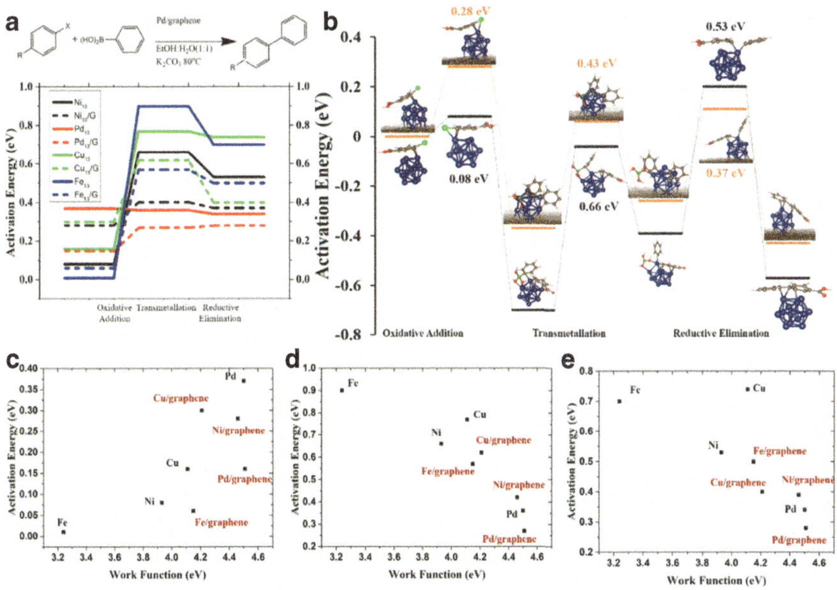

Fig. 9.10 **a** Activation energies of free and supported Fe, Cu, Ni and Pd clusters in full reaction cycle. **b** Reaction pathway of free and supported Ni_{13} cluster in Suzuki reaction. Activation Energies of free and supported Ni_{13}, Fe_{13}, Cu_{13} and Pd_{13} clusters as a function of work function of the corresponding metal cluster in the oxidative addition (**c**), transmetallation (**d**) and reductive elimination (**e**) steps

activation energy confirming that the Fe cluster is an effective charge donor. After deposition, the activation energy for the oxidative addition on the Ni_{13} cluster was found to raise significantly, while in contrast, supported Pd_{13} shows a lower activation energy than that of a free Pd_{13} cluster. For Fe_{13}, there is a small increase in the activation energy after deposition. It was noted that the activation energy for oxidative addition on both Fe clusters are extremely low, less than 0.1 eV. For Cu_{13}, the activation energy increases after being supported on graphene, most likely because the graphene lowers the d-band center of Cu, weaken the bindings, especially for the alkyl group that binds to the cluster during the oxidative addition step. To investigate if the changes are consistent with our hypothesis of charge donation/acceptance, the binding energy of a Br atom as a proxy to the charge donation was investigated. It was found that the change in activation energies of the free and supported Ni, Pd, and Cu clusters were highly correlated with the changes in Br adsorption energy, indicating a good measure of the charge donating ability of the cluster. These findings were consistent with the hypothesis that the charge donating capability of a cluster significantly affects its activity towards oxidative addition. Also, Fig. 9.10c–e displays the correlations between the work function and activation energy. For details we recommend to read the original articles [45, 81, 91].

For the charge withdrawing transmetallation and reductive elimination steps, the activation energies of the free Fe_{13}, Cu_{13}, and Ni_{13} clusters were found to be significantly higher than that of Pd_{13}. This suggests that Suzuki reactions catalyzed by the first-row transition metals will be hindered in these steps where clusters act as charge acceptors. After the deposition, the activation energies of supported Fe_{13}, Cu_{13} and Ni_{13} clusters were considerably reduced. For the supported Ni_{13} cluster, the activation energy was slightly higher than the free Pd_{13} cluster and had the lowest activation energy of the non-Palladium clusters. A similar trend was seen in the reductive elimination step that also requires the cluster to act as a charge acceptor. Among the studied systems, the Pd_{13} cluster on graphene has the lowest activation energy, the free Pd_{13} cluster has the second lowest, and the Ni_{13} on graphene has the lowest activation energy for the non-Palladium clusters. While the activation energies of Ni_{13}/graphene are higher than those of Pd_{13}/graphene, we next considered a different approach to further reduce the activation energy for Ni clusters.

Yuan et al. [81] also considered the possibility of reducing the activation energy of the supported catalysts by using bimetallic catalysts. $Ni_{12}Pd$, $Pd_{12}Ni$ and Ni_2Pd_2 clusters were tested in the Suzuki reaction and the results are shown in Fig. 9.11.

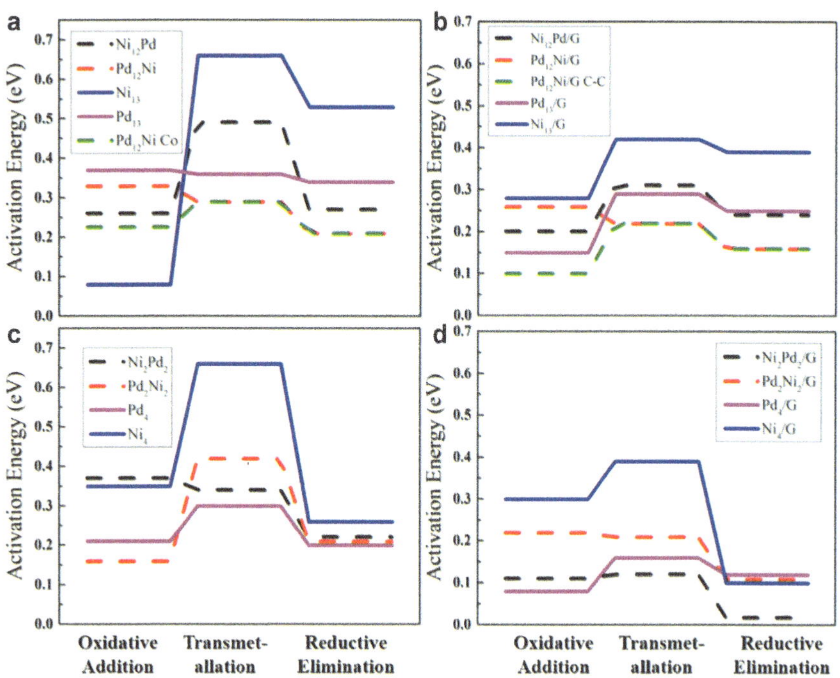

Fig. 9.11 The activation energy of **a** free $Pd_{12}Ni$ and $Ni_{12}Pd$ clusters, **b** supported $Pd_{12}Ni$ and $Ni_{12}Pd$ clusters, **c** free Pd_2Ni_2 and Ni_2Pd_2 clusters, and **d** supported Pd_2Ni_2 and Ni_2Pd_2 clusters. $Pd_{12}Ni$ C–C and $Pd_{12}Ni$/G C–C denote co-catalysis on the bimetallic cluster where the oxidative addition is performed on the Pd site and the transmetallation and reductive elimination steps occur on the Ni site

These results show that the electronic properties of the individual Pd and Ni active sites in the bimetallic cluster are significantly modified. As seen in Fig. 9.11b–d, the Ni sites of the bimetallic Pd/Ni clusters donate charge to the Pd sites, with the Pd sites becoming negatively charged, while the Ni sites become positively charged. This modifies the charge donating and withdrawing capability of the atoms, which may then translate to the modification of the activation energies.

In Fig. 9.11a–d, the activation energy for the free and supported 13-atom and 4-atom clusters are shown. For the free 13-atom clusters, Ni_{13} is most active for the oxidative addition step, and the $Pd_{12}Ni$ cluster (where the Ni atom is the active site) is most active in the transmetallation and reductive elimination steps. This supports the concept that the charge transfer from Ni to Pd results in a Ni site that is activated towards charge accepting steps such as transmetallation and reductive elimination. The activation energy then follows $Ni < Pd^{-}(Ni) < Ni^{+}(Pd) < Pd$ for the steps where the cluster acts as a charge donor. $Pd^{-}(Ni)$ denotes the Pd site on a bimetallic cluster where Ni has donated charge to Pd, and $Ni^{+}(Pd)$ denotes the Ni site on a bimetallic cluster where the Ni has donated charge to Pd. In contrast, for the steps where the cluster acts as a charge acceptor, the activation energy follows a trend as $Ni^{+}(Pd) < Pd < Pd^{-}(Ni) < Ni$ for transmetallation, and $Ni^{+}(Pd) < Pd^{-}(Ni) < Pd < Ni$ for reductive elimination. When different metal atoms are contained in the same cluster, it is possible for different steps in the reaction to be carried out on different active sites. Such a pathway may only occur when there are adjacent bimetallic surface atoms, this mechanism will not occur for core-shell bimetallic particles where the atoms are segregated into shells.

Depositing the bimetallic clusters on graphene results in a further decrease in the activation energy, as seen in Fig. 9.11b. After deposition, the lowest activation energy pathway follows a co-catalysis process on different Ni and Pd sites. For all three steps, the activation energies of the supported $Pd_{12}Ni$ cluster are all lower than that of the supported Pd_{13} cluster. Decreasing the cluster size to four atoms further lowers the activation energy of the catalysis. The activation energy of the free and supported Ni_2Pd_2 clusters are shown in Fig. 9.11c, d, and these clusters are found to be even more active than the 13-atom clusters. These results are consistent with the concept that the Ni becomes more positively charged in the bimetallic cluster, and hence it becomes a better charge acceptor, while Pd sites become more negatively charged and are more active in charge donating steps. By combining the graphene support that enhances charge transfer with the small size of the cluster and the bimetallic charge transfer, it was found that the supported Pd_2Ni_2 cluster is the most active cluster studied here towards the Suzuki cross-coupling reaction.

9.6 Summary

In this chapter, we have attempted to provide a conceptual framework for understanding the dramatic increase in catalytic activity that can be achieved by optimizing the electronic interactions between the metal cluster and the conducting support.

Various experimental investigations have revealed that the microwave irradiation induces defects into the graphene that can range from monovacancy to larger voids. These defects strongly bind the Pd clusters/nanoparticles that act as the catalyst. More important is the role of the support as the theoretical investigations demonstrate that the resulting metal-ligand interaction promotes electron flow between the metal cluster and the defected graphene. This is clearly due to the conductive nature of the support that facilitates any charge transfer to and from the reactants. However, the benefit of the support substrate is diminished if the metal is weakly immobilized. Preliminary theoretical results have found that graphene may also lower the activation energy of Ni_n clusters. These studies indicate that a bifunctional catalyst combining Ni and Pd where different reaction steps could be performed by different metallic sites may offer a more powerful catalyst. In addition to modifying the supporting cluster, one can envision enhancing the catalytic activity by making the support as a better donor/acceptor. This could be accomplished by doping the graphene with atoms or attaching ligands. These are interesting directions that need further work.

References

1. C.A. Busacca, D.R. Fandrick, J.J. Song, C.H. Senanayake, Adv. Synth. Catal. **353**, 1825–1864 (2011)
2. J. Magano, J.R. Dunetz, Chem. Rev. **111**, 2177–2250 (2011)
3. I. Shinkai, A.O. King, R.D. Larsen, Pure Appl. Chem. **66**, 1551 (1994)
4. I.P. Beletskaya, A.V. Cheprakov, in *The Mizoroki–Heck Reaction*, John Wiley & Sons, Ltd, 2009, pp. 51–132
5. A. Jutand, in *The Mizoroki–Heck Reaction*, John Wiley & Sons, Ltd, 2009, pp. 1–50
6. O. Bagheri, F. Sadegh, M. Moghadam, S. Tangestaninejad, V. Mirkhani, I. Mohammadpoor-Baltork, M. Safiri, Appl. Organomet. Chem. **28**, 337–346 (2014)
7. R.J.P. Corriu, J. Organomet. Chem. **653**, 20–22 (2002)
8. M.K. Das, J.A. Bobb, A.A. Ibrahim, A. Lin, K.M. AbouZeid, M.S. El-Shall, A.C.S. Appl, Mater. Interfaces **12**, 23844–23852 (2020)
9. C. Deraedt, D. Astruc, Acc. Chem. Res. **47**, 494–503 (2014)
10. J. Derosa, V.T. Tran, V.A. van der Puyl, K.M. Engle, Aldrichim. Acta **51**, 21–32 (2018)
11. E. Gioria, J. del Pozo, J.M. Martinez-Ilarduya, P. Espinet, Angew. Chem. Int. Ed. **55**, 13276–13280 (2016)
12. J.H. Hilmer, E. Champeil, M.K. Lakshman, Abstr. Pap. Am. Chem. Soc. **230**, U3264–U3264 (2005)
13. S. Jadhav, A. Jagdale, S. Kamble, A. Kumbhar, R. Salunkhe, RSC Adv. **6**, 3406–3420 (2016)
14. J.K. Kirsch, J.B. Johnson. Abstr. Pap. Am. Chem. Soc. **247** (2014)
15. A.H. Labulo, B.S. Martincigh, B. Omondi, V.O. Nyamori, J. Mater. Sci. **52**, CP14–9248 (2017)
16. R.C. Larock, D.R. Leach, Organometallics **1**, 74–81 (1982)
17. A. Lengar, C.O. Kappe, Org. Lett. **6**, 771–774 (2004)
18. A. Leyva-Perez, Dalton Trans. **46**, 15987–15990 (2017)
19. G. Li, R. Jin, Nanotechnology Reviews **2**, 529–545 (2013)
20. M.I. Lipschutz, T.D. Tilley, Angew. Chem. Int. Ed. **53**, 7290–7294 (2014)
21. M.-X. Liu, H.-P. Gong, Z.-J. Quan, X.-C. Wang, Synlett **29**, 330–335 (2018)
22. J.C. Lo, J. Gui, Y. Yabe, C.-M. Pan, P.S. Baran, Nature **516**, 343–348 (2014)
23. K. Mennecke, W. Solodenko, A. Kirschning, *Synthesis-Stuttgart*, 1589–1599 (2008)

24. J.A. Milligan, J.P. Phelan, S.O. Badir, G.A. Molander, Angew. Chem. Int. Ed. **58**, 6152–6163 (2019)
25. V. Misuk, A. Mai, K. Giannopoulos, D. Karl, J. Heinrich, D. Rauber, H. Loewe, J. Flow Chem. **5**, 43–47 (2015)
26. S. Moussa, A.R. Siamaki, B.F. Gupton, M.S. El-Shall, Acs Catal. **2**, 145–154 (2012)
27. E.I. Negishi, Acc. Chem. Res. **15**, 340–348 (1982)
28. M.H.G. Prechtl, J.D. Scholten, J. Dupont, Molecules **15**, 3441–3461 (2010)
29. H. Prokopcova, C.O. Kappe, Adv. Synth. Catal. **349**, 448–452 (2007)
30. J.A. Przyojski, K.P. Veggeberg, H.D. Arman, Z.J. Tonzetich, Acs Catal. **5**, 5938–5946 (2015)
31. K.S.M. Salih, Y. Baqi, Catalysts **10** (2020)
32. N. Shang, S. Gao, X. Zhou, C. Feng, Z. Wang, C. Wang, RSC Adv. **4**, 54487–54493 (2014)
33. S.E. Smith, A.R. Siamaki, B.F. Gupton, E.E. Carpenter, RSC Adv. **6**, 91541–91545 (2016)
34. A. Suzuki, Chem. Commun, 4759–4763 (2005)
35. P. Taladriz-Blanco, P. Herves, J. Perez-Juste, Top. Catal. **56**, 1154–1170 (2013)
36. M. Turks, C.J. Exner, C. Hamel, P. Vogel, *Synthesis-Stuttgart*, 1065–1074 (2009)
37. A. Vannucci, *Abstr.* Pap. Am. Chem. Soc. **249** (2015)
38. P. Vogel, C.M.R. Volla, R. Loska, S. R. Dubbaka, Abstr. Pap. Am. Chem. Soc. 237 (2009)
39. H. Wang, T. Song, X. Su, Z. Li, J. Wang, Acs Sustain. Chem. Eng. **8**, 1262–1267 (2020)
40. H.-X. Wang, H.-F. Wu, X.-L. Yang, N. Ma, L. Wan, Polyhedron **26**, 3857–3864 (2007)
41. M.-L. Xu, W. Huang, Tetrahedron Lett. **55**, 4230–4232 (2014)
42. L. Xue, Z. Lin, Chem. Soc. Rev. **39**, 1692–1705 (2010)
43. D.Y. Yang, Y.D. Wang, J.B. Yan, X. Huang, J. Chem. Res.-S, 192–193 (2002)
44. H. Yang, X. Yang, W. Tang, Tetrahedron **72**, 6143–6174 (2016)
45. Y. Yang, A.C. Reber, S.E. Gilliland, C.E. Castano, B.F. Gupton, S.N. Khanna, J. Catal. **360**, 20–26 (2018)
46. C.S. Yeung, V.M. Dong, Chem. Rev. **111**, 1215–1292 (2011)
47. K.S. Yoo, C.H. Yoon, K.W. Jung, J. Am. Chem. Soc. **128**, 16384–16393 (2006)
48. K.S. Yoo, C.H. Yoon, R.K. Mishra, Y.C. Jung, S.W. Yi, K.W. Jung, J. Am. Chem. Soc. **129**, 3771 (2007)
49. Y. Zhang, B. Feng, Chin. J. Org. Chem. **34**, 2406–2411 (2014)
50. C. Sambiagio, S.P. Marsden, A.J. Blacker, P.C. McGowan, Chem. Soc. Rev. **43**, 3525–3550 (2014)
51. F. Ullmann, Ber. Dtsch. Chem. Ges. **36**, 2382–2384 (1903)
52. F. Ullmann, P. Sponagel, Ber. Dtsch. Chem. Ges. **38**, 2211–2212 (1905)
53. I. Goldberg, Ber. Dtsch. Chem. Ges. **39**, 1691–1692 (1906)
54. R.F. Heck, J. Am. Chem. Soc. **91**, 6707–6714 (1969)
55. F.-S. Han, Chem. Soc. Rev. **42**, 5270–5298 (2013)
56. K. Sonogashira, Y. Tohda, N. Hagihara, Tetrahedron Lett. **16**, 4467–4470 (1975)
57. L. Cassar, J. Organomet. Chem. **93**, 253–257 (1975)
58. D. Haas, J.M. Hammann, R. Greiner, P. Knochel, Acs Catal. **6**, 1540–1552 (2016)
59. A.O. King, N. Okukado, E.-i. Negishi, J. Chem. Soc., Chem. Commun., 683–684 (1977)
60. C. Cordovilla, C. Bartolomé, J.M. Martínez-Ilarduya, P. Espinet, Acs Catal. **5**, 3040–3053 (2015)
61. J.K. Stille, Angew. Chem. Int. Ed. **25**, 508–524 (1986)
62. Y. Nakao, T. Hiyama, Chem. Soc. Rev. **40**, 4893–4901 (2011)
63. C.C.C. Johansson Seechurn, M.O. Kitching, T.J. Colacot, V. Snieckus, Angew. Chem. Int. Ed., **51**, 5062–5085 (2012)
64. Á. Molnár, Chem. Rev. **111**, 2251–2320 (2011)
65. M. Pagliaro, V. Pandarus, R. Ciriminna, F. Béland, P. Demma Carà, *ChemCatChem* **4**, 432–445 (2012)
66. S.R. Chemler, D. Trauner, S.J. Danishefsky, Angew. Chem. Int. Ed. **40**, 4544–4568 (2001)
67. I.W. Davies, L. Matty, D.L. Hughes, P.J. Reider, J. Am. Chem. Soc. **123**, 10139–10140 (2001)
68. S.P. Andrews, A.F. Stepan, H. Tanaka, S.V. Ley, M.D. Smith, Adv. Synth. Catal. **347**, 647–654 (2005)

69. Z. Niu, Q. Peng, Z. Zhuang, W. He, Y. Li, Chem. Eur. J. **18**, 9813–9817 (2012)
70. D. Astruc, F. Lu, J.R. Aranzaes, Angew. Chem. Int. Ed. **44**, 7852–7872 (2005)
71. J.M. Richardson, C.W. Jones, J. Catal. **251**, 80–93 (2007)
72. Y. Ji, S. Jain, R.J. Davis, J. Phys. Chem. B **109**, 17232–17238 (2005)
73. J.D. Webb, S. MacQuarrie, K. McEleney, C.M. Crudden, J. Catal. **252**, 97–109 (2007)
74. J.M. Richardson, C.W. Jones, Adv. Synth. Catal. **348**, 1207–1216 (2006)
75. J.G. de Vries, Dalton Trans., 421–429 (2006)
76. Y. Nishina, J. Miyata, R. Kawai, K. Gotoh, RSC Adv. **2**, 9380–9382 (2012)
77. Y. Yang, C.E. Castano, B.F. Gupton, A.C. Reber, S.N. Khanna, Nanoscale **8**, 19564–19572 (2016)
78. G.M. Scheuermann, L. Rumi, P. Steurer, W. Bannwarth, R. Mülhaupt, J. Am. Chem. Soc. **131**, 8262–8270 (2009)
79. G. Park, S. Lee, S.J. Son, S. Shin, Green Chem. **15**, 3468–3473 (2013)
80. K. Köhler, R.G. Heidenreich, S.S. Soomro, S.S. Pröckl, Adv. Synth. Catal. **350**, 2930–2936 (2008)
81. Y. Yang, PhD, Virginia Commonwealth University (2019)
82. A.R. Siamaki, A.E.R.S. Khder, V. Abdelsayed, M.S. El-Shall, B.F. Gupton, J. Catal. **279**, 1–11 (2011)
83. X. Hao, S. Barnes, J.R. Regalbuto, J. Catal. **279**, 48–65 (2011)
84. R. Kumar, J.-H. Oh, H.-J. Kim, J.-H. Jung, C.-H. Jung, W.G. Hong, H.-J. Kim, J.-Y. Park, I.-K. Oh, ACS Nano **9**, 7343–7351 (2015)
85. A. Bagri, C. Mattevi, M. Acik, Y.J. Chabal, M. Chhowalla, V.B. Shenoy, Nat. Chem. **2**, 581–587 (2010)
86. E. Gracia-Espino, G. Hu, A. Shchukarev, T. Wågberg, J. Am. Chem. Soc. **136**, 6626–6633 (2014)
87. A.V. Gaikwad, A. Holuigue, M.B. Thathagar, J.E. ten Elshof, G. Rothenberg, Chem. Eur. J. **13**, 6908–6913 (2007)
88. H. Yan, H. Cheng, H. Yi, Y. Lin, T. Yao, C. Wang, J. Li, S. Wei, J. Lu, J. Am. Chem. Soc. **137**, 10484–10487 (2015)
89. A. Leyva-Pérez, J. Oliver-Meseguer, P. Rubio-Marqués, A. Corma, Angew. Chem. Int. Ed. **52**, 11554–11559 (2013)
90. P.J. Ellis, I.J.S. Fairlamb, S.F.J. Hackett, K. Wilson, A.F. Lee, Angew. Chem. Int. Ed. **49**, 1820–1824 (2010)
91. Z. Yang, Y. Wang, M. Zhu, Z. Li, W. Chen, W. Wei, T. Yuan, Y. Qu, Q. Xu, C. Zhao, X. Wang, P. Li, Y. Li, Y. Wu, Y. Li, Acs Catal. **9**, 2158–2163 (2019)

Chapter 10
Metallo-Carbohedrenes and Their Reactivity

10.1 The Discovery of Metallo-Carbohedrenes

Since the discovery of C_{60} called *buckminsterfullerene* whose shape resembles that of geodesic domes, the so-called buckyballs have attracted extensive research interest and touted for many important applications [1–31]. Furthermore, extensive studies have been conducted upon the incorporation of metals such as lanthanum into the interior of C_{60} giving rise to promising applications such as in superconductive materials [32–35]. In addition, researchers have also found that a few carbon atoms in C_{60} cage can be replaced by other atoms without substantially destabilizing the entire fullerene cage [36–56].

Similar to fullerenes, Castleman and his coworkers discovered a new cluster species in 1992, named metallo-carbohedrenes or "Met-Cars" for short [57, 58]. It was more or less just a serendipity when they used the laser to probe various titanium reactions and noticed a very strong peak at 528 mass unit, as shown in Fig. 10.1 [59–61]. Repeated experiments were tried for the same reaction with a number of other hydrocarbon gases; also they substituted a different transition metal for titanium, and substituted a hydrocarbon composed of deuterium (the one-neutron isotope of hydrogen) as the raw material, as well as ^{13}C to use in the hydrocarbon raw material; finally they ascertained the discovery of this mystery material Ti_8C_{12} [57–60, 62, 63].

Following the discovery of the existence of stable Met-Cars clusters, the stoichiometry M_8C_{12} (M = Ti, V, Zr, Nb, and Hf) and their pentagonal dodecahedral structure of T_h symmetry attracted reasonable interest [57–60]. The eight equivalently bonded metal atoms were found to account for the unusual stability of this class of clusters. Subsequent investigations were also conducted by experimentalists and theoreticians with a reemphasis on the unusual stability of these Met-Cars, leading knowledge on the range of metals which form this interesting cage cluster species [64].

Z. Luo and S. N. Khanna, *Metal Clusters and Their Reactivity*,
https://doi.org/10.1007/978-981-15-9704-6_10

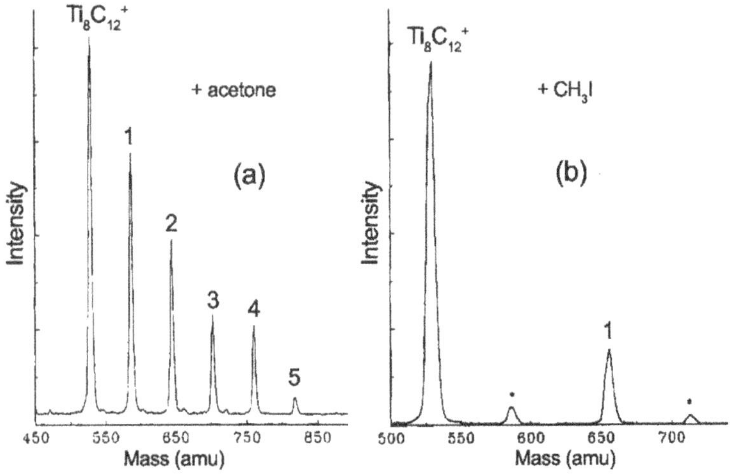

Fig. 10.1 **a** Mass spectra of products arising from reactions of $Ti_8C_{12}^+$ with acetone. The numerals indicate the number of acetone molecules associating onto $Ti_8C_{12}^+$. The apparent truncation at n = 4 is a consequence of the reactant pressure and not the nature of the bonding. **b** Mass spectra of products arising from reactions of $Ti_8C_{12}^+$ with methyl iodide. The numerals indicate the number of iodine atoms associating onto $Ti_8C_{12}^+$. The peaks marked by * are due to acetone impurity. Reproduced with permission from Ref. [68]. Copyright 1994 American Chemical Society

To elucidate the structure of the Met-Cars, researchers further examined how this new material reacted with the other chemicals. For example, they tried ammonia which bonds well to transition metals, and found the reaction product had exactly 8 ammonia molecules. As ammonia could only link up to titanium on the outside of the cluster, it was concluded that the structure of Ti_8C_{12} must have eight titanium atoms on the cluster surface; moreover, the atoms were in equivalent positions (indicating that every titanium atom bonded to three carbons, and every carbon bonded to two titanium atoms and another carbon). That severely constrained the possible cage structures like the well-known fullerenes and a water cluster where 20 water molecules on the outside, showing the species of pentagonal dodecahedrons. The structure of this mystery material Ti_8C_{12} was finally ascertained as a quasi-buckyball with 12 pentagonal faces where each face contains three carbon and two titanium atoms. It was then called metallo-carbohedrenes or Met-Cars for short, and has excited reasonable research interest [57–60, 62, 65–76]. It is interesting to mention that, although obvious similarity to fullerene C_{60}, it is not easy for the final piece of puzzle about the structure of Ti_8C_{12} to fall into place in January, 1992. It was said Castleman thought hard every day and night, even furiously scribbling in the dark campus while he was walking home from work [63].

> The basic lesson from all this is pretty simple, and not unique to our discovery: Try something outside of your formal plans. Be observant. Be tenacious—keep returning to puzzle things out. Always think about the potential significance of your observations
>
> .—A. W. Castleman, Jr.

10.2 Reactivity of Met-Cars

Since the discovery as a new cluster species referred to Metallo-Carbohedrenes in 1992 [57, 58], reasonable research interest has been stimulated to elucidate the physical and chemical properties of this new species. Researchers utilized metastable decay, collision-induced dissociation, and photo-dissociation techniques to probe the bonding and dissociation patterns of the Met-Cars. Interestingly, $Ti_8C_{12}^+$ was found to be inert toward oxygen etching reaction. Further, the reactions of $Ti_8C_{12}^+$ had been studied under thermal energy conditions and a high gas pressure at 0.7 Torr within a selected ion drift tube reactor (SIDT) [65].

Figure 10.1a shows a typical product distribution acquired from the reaction of $Ti_8C_{12}^+$ with acetone. It was noted that, based on a multiple-step attachment process, one to five acetone molecules bind to $Ti_8C_{12}^+$ leading to the adducts $Ti_8C_{12}^+(C_3H_6O)_{1-5}$. The multiple attachment processes of acetone onto $Ti_8C_{12}^+$ can be represented as,

$$Ti_8C_{12}^+ + xC_3H_6O \rightarrow Ti_8C_{12}^+(C_3H_6O)_x \tag{10.1}$$

where x was observed to be 1 to 5. In comparison, the reaction of $Ti_8C_{12}^+$ with methyl iodide (CH_3I) yields rare association products, as shown in the mass spectrum of Fig. 10.1b. The only apparent peak as numbered corresponds to the product $Ti_8C_{12}I^+$. There were no changes when varying the pressures of MeI from 0.2 to 0.8 mTorr. This reaction process indicates the dissociation of CH_3I, which differs from an association reaction as Eq. 10.1 but resembles the reactivities of pure Al_n^- clusters and $Al_nI_x^-$ species toward CH_3I in the flow tube reactor [77, 78]. The selective reactivity of the $Ti_8C_{12}^+$ is still elusive and open to further exploration.

However, the reactivity of Met-Cars with methanol is different from the reactivity towards acetone and methyl iodide. It was worth mentioning that $Ti_8C_{12}^+$ takes up eight methanol molecules, as shown in Fig. 10.2a, b. In the case of a very low partial pressure of methanol being presented, the reaction terminates at eight methanol attachments. Similarly, this multiple attachment processes of methanol onto $Ti_8C_{12}^+$ can be represented as,

$$Ti_8C_{12}^+ + xCH_3OH \rightarrow Ti_8C_{12}^+(CH_3OH)_x \tag{10.2}$$

where x was observed to be 1 to 8. However, at a much higher pressure of methanol, all the species $Ti_8C_{12}^+(CH_3OH)_{0-7}$ disappeared while $Ti_8C_{12}^+(CH_3OH)_8$ interestingly survives and dominates the entire mass spectrum, as shown in Fig. 10.2b. This observation demonstrated reasonable activity of $Ti_8C_{12}^+$ and the unique stability of $Ti_8C_{12}^+(CH_3OH)_8$ where the eight sites of Ti atoms were taken by attaching methanol molecules one by one. As the oxygen atom of a methanol molecule has a lone pair electron, its electronegativity renders the chemisorption of methanol on $Ti_8C_{12}^+$ through an O–Ti bond. No products of $Ti_8C_{12}^+(CH_3OH)_{x>8}$ were observed,

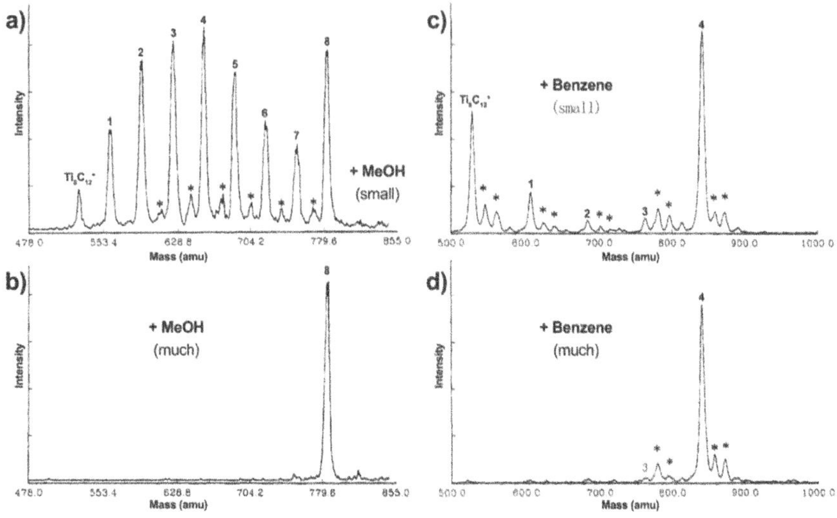

Fig. 10.2 Mass spectra of products arising from reactions of $Ti_8C_{12}^+$ with methanol obtained at a very low partial pressure of methanol (**a**), and a much higher pressure of methanol (**b**). The number stands for the number of methanol molecules associating onto $Ti_8C_{12}^+$. Note that association reactions terminate at the eighth step. Mass spectra of products arising from reactions of $Ti_8C_{12}^+$ with benzene: **c** obtained at a very low partial pressure of benzene; **d** obtained at a much higher partial pressure of benzene. The number stands for the number of benzenes associating onto $Ti_8C_{12}^+$. Note that association reactions terminate at the fourth step. Reproduced with permission from Ref. [65]. Copyright 1993 American Chemical Society

indicating the distinctive stability of $Ti_8C_{12}^+(CH_3OH)_8$ with fully occupied Ti atom sites.

Reaction of $Ti_8C_{12}^+$ with benzene was found to produce $Ti_8C_{12}^+(C_6H_6)_x$ with a limitation of x up to 4 indicating the association reactions terminate at the fourth step for $Ti_8C_{12}^+$ encountering with benzene molecules. The $Ti_8C_{12}^+(C_6H_6)_4$ was found to dominate the observed products and appear as the only one product surviving in the presence of large quantity of benzene. In view of the unique geometry of $Ti_8C_{12}^+$ by possessing 12-faces and each face having three carbon atoms and two Ti atoms, it is reasonable for the benzene ring to be flat-adsorption on the four nonadjacent faces ascribed to the delocalized π-electrons of benzene. The reactivity of $Ti_8C_{12}^+$ with four benzene molecules was expressed as,

$$Ti_8C_{12}^+ + 4C_6H_6 \rightarrow Ti_8C_{12}^+(C_6H_6)_4 \quad (10.3)$$

Note that all these reactions involved only an association mechanism but no evidence for destructive eliminations to the Met-Cars due to their decent stability.

Some further investigations have further eliminated the association reactivity of $Ti_8C_{12}^+$ with the polar molecule systems. In particular, the association reactions were

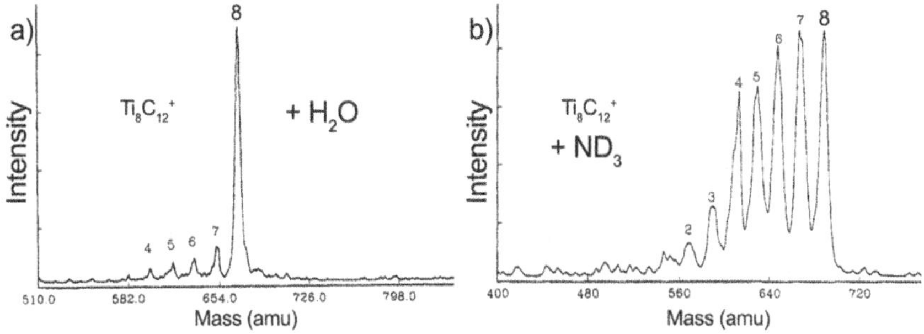

Fig. 10.3 Mass spectra of products arising from reactions of $Ti_8C_{12}^+$ with **a** H_2O and **b** ND_3. The number indicates the number of H_2O (or ND_3) associating onto $Ti_8C_{12}^+$. Note that association reactions terminate at the eighth step. Reproduced with permission from Ref. [65]

also found to be terminated at the eighth step for water and ammonia system, as seen in Fig. 10.3. They can be represented as

$$Ti_8C_{12}^+ + 8H_2O \rightarrow Ti_8C_{12}^+(H_2O)_8 \tag{10.4}$$

$$Ti_8C_{12}^+ + 8ND_3 \rightarrow Ti_8C_{12}^+(ND_3)_8 \tag{10.5}$$

Further researches in this area have found that the Met-Cars are not limited to $Ti_8C_{12}^+$. For example, researchers have also found $Nb_8C_{12}^+$ sharing the equivalent physics and chemistry as $Ti_8C_{12}^+$. Figure 10.4 shows two photoionization mass spectra measured for vanadium/carbon and niobium/carbon cluster distributions at 215 nm. $V_8C_{12}^+$ and $Nb_8C_{12}^+$ appear in these mass spectra with sharp peak in resolution because both V and Nb metals have only one naturally occurring isotope, indicating that the Met-Car clusters have reasonable abundances in the molecular beam. Nevertheless, there might also be additional clusters with equal or greater

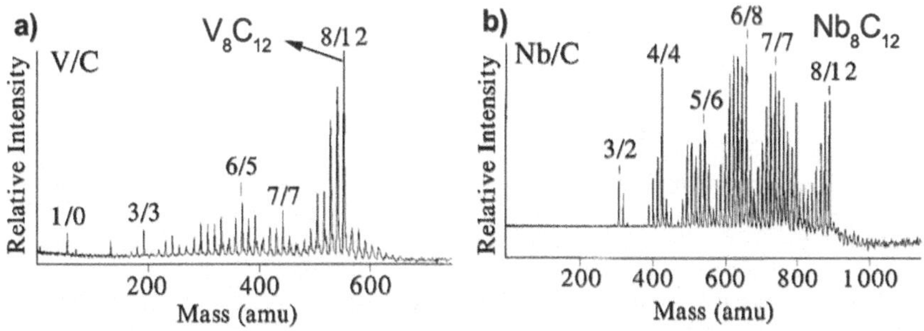

Fig. 10.4 Photoionization mass spectra for $V_8C_{12}^+$ and $Nb_8C_{12}^+$ clusters at 215 nm. Laser power dependence studies show that these spectra are the result of multiphoton absorption of at least two photons. Reproduced with permission from Ref. [79]. Copyright 1996 American Chemical Society

density but not being detected assuming they have higher ionization potentials. Two factors determine the intensity of neutral clusters through the photoionization spectroscopy: (i) the density of the respective clusters in the molecular beam, and (ii) the ionization cross sections of the clusters. It is therefore not appropriate to conclude a formation mechanism for the Met-Car clusters and their stabilities just from the mass abundance.

Figure 10.5a, b present the reactivities of $Nb_8C_{12}^+$ with acetone (C_3H_6O) and methyl iodide (CH_3I), where the products $Nb_8C_{12}^+(C_3H_6O)_x$ (x = 1–5) and $Nb_8C_{12}I^+$ were observed showing similar reactivity of $Nb_8C_{12}^+$ as that of $Ti_8C_{12}^+$. The reaction

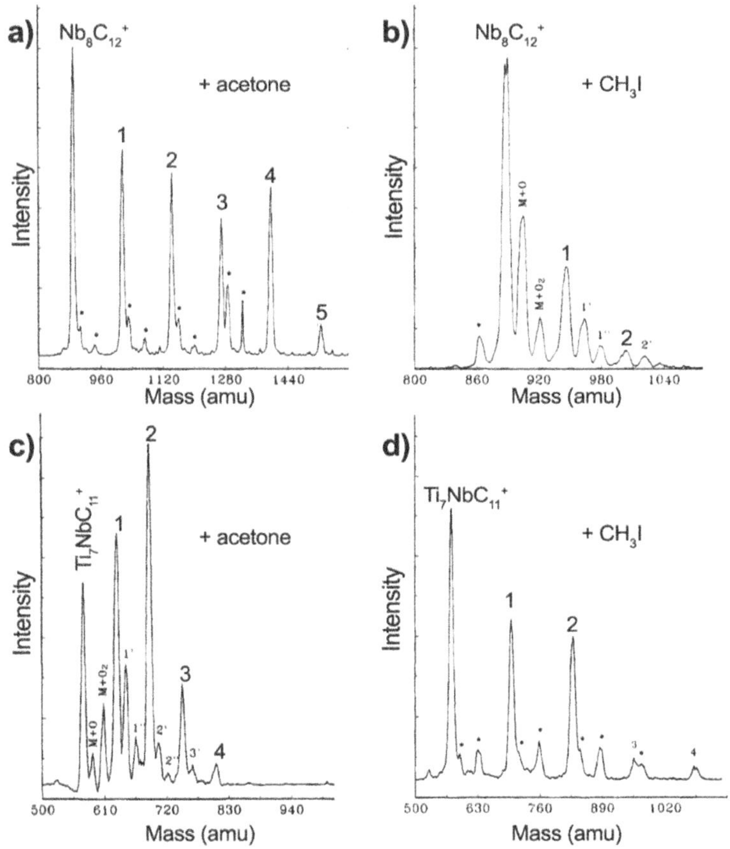

Fig. 10.5 Mass spectra of products arising from reactions of $Nb_8C_{12}^+$ with acetone (**a**) and methyl iodide (**b**). The numerals indicate the number of acetone molecules or iodine atoms associating onto $Nb_8C_{12}^+$ respectively. **c** Mass spectra of products arising from reactions of $Ti_7NbC_{12}^+$ with acetone: the numerals indicate the number of associations of acetone molecules onto $Ti_7NbC_{12}^+$; $Ti_7NbC_{12}O^+$ (primed) and $Ti_7NbC_{12}O_2^+$ (double primed). **d** Mass spectra of products arising from reactions of $Ti_7NbC_{12}^+$ with methyl iodide, where the numerals indicate the number of iodine atoms associating onto $Ti_7NbC_{12}^+$. All peaks marked by * are due to impurity. Reproduced with permission from Ref. [68]. Copyright 1994 American Chemical Society

studies in relatively high gas pressures not only demonstrated the similar chemistry for various Met-Cars, but also indicated reasonable stability of these novel species [72].

Further investigations were also extended to Ti–Nb alloy Met-Cars, such as $Ti_7NbC_{12}^+$. However, it was found that the reactivity of $Ti_7NbC_{12}^+$ takes on difference from $Ti_8C_{12}^+$. As an Nb atom possesses five valence electrons (one more electron than that of a titanium atom), that is, the cluster $Ti_7NbC_{12}^+$ has a total of 80 electrons which is equal to that of the neutral Ti_8C_{12}. Note that, although in the sense of all electrons are paired, 80 electrons do not lead to a closed shell according to the Jellium model [78, 80–86]. As results, $Ti_7NbC_{12}^+$ was found to be very reactive toward acetone, as shown in Fig. 10.5c where the observed products indicated that $Ti_7NbC_{12}^+$ react and take oxygen atoms off acetone to form $Ti_7NbC_{12}O(C_3H_6O)_x^+$ species. When an acetone molecule approaches the positively charged Met-Car cluster, the ion-dipole interaction tends to shift the positive charge at the oxygen's point of encountering due to the lone-pair electrons (electronegativity) of the oxygen atom. This effect was also noted in the reactions of C_{60}^{2+} with polar molecules [87]. So the abstraction of oxygen most likely occurs at the dipole-induced charge centers of $Ti_7NbC_{12}^+$ cluster. Also, the $Ti_7NbC_{12}^+$ was found to react with methyl iodide giving rise to the products of bonding one to four iodine atoms ($Ti_7NbC_{12}I_4^+$) from methyl iodide, as shown in Fig. 10.5d.

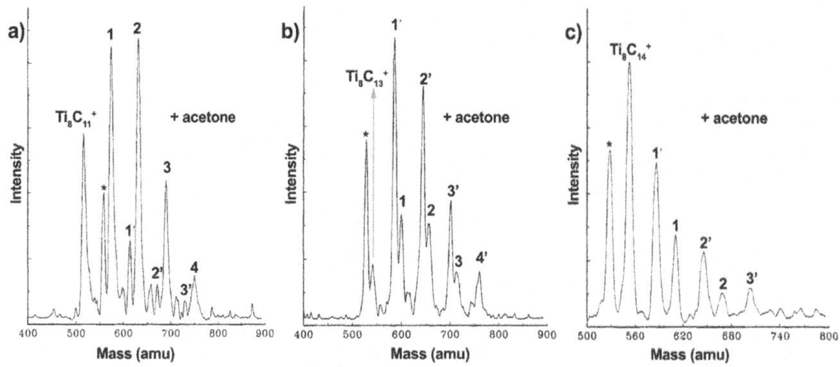

Fig. 10.6 **a** Product distribution arising from multistep reactions of $Ti_8C_{11}^+$ with 0.44 mTorr of acetone. The peak labeled by 1 represents addition of an acetone molecule to $Ti_8C_{11}^+$ via an association reaction. The peak marked by * is the chemical reaction product $Ti_8C_{11}^+$-($COCH_3$). Sequential association products onto the peaks marked 1 and * are labeled n and n', respectively. **b** Product distribution arising from multistep reactions of $Ti_8C_{13}^+$ with 0.47 mTorr of acetone. The peak labeled by 1 corresponds to an association product onto $Ti_8C_{13}^+$, while an * denotes the chemical reaction product $Ti_8C_{12}^+$. Other peaks marked by n and n' correspond to association products onto $Ti_8C_{13}^+$ and its product respectively. **c** Product distribution arising from multistep reactions of $Ti_8C_{14}^+$ with 0.44 mTorr of acetone. The peaks marked 1 and 2 represent acetone molecules associating onto $Ti_8C_{14}^+$. The peak marked by * corresponds to the chemical reaction product $Ti_8C_{12}^+$. The remaining peaks labeled by n' are association products coordinating onto the $Ti_8C_{12}^+$ product. Reproduced with permission from Ref. [70]. Copyright 1995 American Chemical Society

Researchers also compared the stability and reactivity of $Ti_8C_{11}^+$, $Ti_8C_{13}^+$, and $Ti_8C_{14}^+$ [70]. Figure 10.6 presents the product distribution arising from multistep reactions of $Ti_8C_{11}^+$, $Ti_8C_{13}^+$, and $Ti_8C_{14}^+$ with acetone. Unlike the $Ti_8C_{12}^+$, the contiguous dissimilation of other titanium-carbon clusters exhibited some degree of chemical reactivity toward acetone, mainly seen as association reactions. In addition to the reacting via association channels, the carbon-poor cluster $Ti_8C_{11}^+$ was found to break the chemical bonds of acetone; however, the carbon-rich clusters $Ti_8C_{13}^+$ and $Ti_8C_{14}^+$ could undergo a loss of carbon leading to the normal Met-Cars $Ti_8C_{12}^+$ [70].

Considering the cluster reactivity is generally dependent on the structural stability, a large or small HOMO–LUMO gap, and whether or not a closed-shell electronic configuration, recently Berkdemir et al. [64] investigated the structural and electronic properties of the M-substituted metallocarbohedrynes based on Ti_8C_{12} with a C_{3v} symmetry using the spin-polarized density functional theory (DFT) calculations with the plane-wave basis set and the generalized gradient approximation for the exchange-correlation functional. They examined Be, Mg, Ca, Sr, and Ba in place of Ti as these metals consist of two valence electrons less than Ti, as well as Sc and Y which have one less valence electron than Ti. As showing in Table 10.1, the HOMO–LUMO gaps of the M-substituted Ti_8C_{12} metallocarbohedrynes are in the range of

Table 10.1 The binding energy per atom (E_b) and the HOMO–LUMO gap (Δ_{HL}) of the M-substituted metallocarbohedrynes as well as the Ti_8C_{12} metallocarbohedryne, and their dicationic isomers

Structure	E_b/eV per atom	Δ_{HL}/eV
Ti_8C_{12}	−6.477	0.146
$Ti_8C_{12}^{2+}$	−6.083	1.735
Ti_7BeC_{12}(1)	−6.359	0.748
Ti_7BeC_{12}(2)	−6.300	0.855
Ti_7MgC_{12}(1)	−6.206	0.715
Ti_7MgC_{12}(2)	−6.231	0.966
Ti_7CaC_{12}(1)	−6.278	0.747
Ti_7CaC_{12}(2)	−6.311	0.979
Ti_7SrC_{12}(1)	−6.256	0.729
Ti_7SrC_{12}(2)	−6.285	0.882
Ti_7BaC_{12}(1)	−6.300	0.787
Ti_7BaC_{12}(2)	−6.321	0.864
$Ti_6Sc_2C_{12}$(1)	−6.403	0.982
$Ti_6Sc_2C_{12}$(2)	−6.421	0.975
$Ti_6Sc_2C_{12}$(3)	−6.425	1.223
$Ti_6Sc_2C_{12}$(4)	−6.431	1.294
$Ti_6Y_2C_{12}$(1)	−6.393	0.865
$Ti_6Y_2C_{12}$(2)	−6.410	0.892
$Ti_6Y_2C_{12}$(3)	−6.411	1.151
$Ti_6Y_2C_{12}$(4)	−6.400	1.005

0.715–0.979 eV for the case of Be, Mg, Ca, Sr, and Ba; while a range of 0.865–1.294 eV calculated for the systems involving Sc and Y. These results provided a reference to understand the stability and reactivity of the metallocarbohedrynes and their derivatives [2].

10.3 Deposition of Met-Cars

Castleman and coworkers also demonstrated the ability to produce Met-Cars in the solid state [61], but the difficulties in finding a suitable solvent for isolating Met-Cars is a challenge. An alternative method involving the deposition of mass-gated species was employed through soft-landing of the gas-phase clusters [88]. Such experiments were conducted to produce Met-Cars from the direct laser vaporization of metal/graphite composite pellets, and then the mass-selected Met-Cars were deposited onto special carbon-covered TEM-sample grids, as sketched in Fig. 10.7. Initial morphology characterization of the soft-landed Zr_8C_{12} Met-Cars has been performed with high-resolution transmission electron microscope (HRTEM) [88].

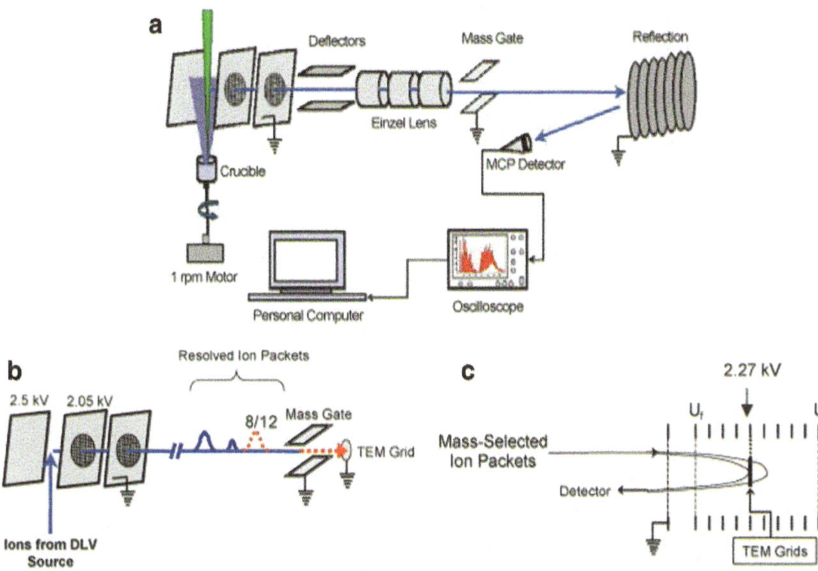

Fig. 10.7 Schematic of the newly constructed time-of-flight mass spectrometer. **a** The overall configuration, **b** hard-landing and **c** soft-landing deposition arrangements. Note that during the soft-landing deposition of the mass gated packets, the species are deposited with a small distribution of energies, arising from the spatial distribution in the extraction region of the time-of-flight. Reproduced with permission from Ref. [88]. Copyright 2003 Elsevier

References

1. H.W. Kroto, J.R. Heath, S.C. O'Brien, R.F. Curl, R.E. Smalley, Nature **318**, 162–163 (1985)
2. P.M. Allemand, K.C. Khemani, A. Koch, F. Wudl, K. Holczer, S. Donovan, G. Gruner, J.D. Thompson, Science **253**, 301–303 (1991)
3. F. Diederich, R.L. Whetten, C. Thilgen, R. Ettl, I. Chao, M.M. Alvarez, Science **254**, 1768–1770 (1991)
4. D.M. Poirier, T.R. Ohno, G.H. Kroll, Y. Chen, P.J. Benning, J.H. Weaver, L.P.F. Chibante, R.E. Smalley, Science **253**, 646–648 (1991)
5. T. Guo, M.D. Diener, Y. Chai, M.J. Alford, R.E. Haufler, S.M. McClure, T. Ohno, J.H. Weaver, G.E. Scuseria, R.E. Smalley, Science **257**, 1661–1664 (1992)
6. K.H. Johnson, D.P. Clougherty, M.E. McHenry, Science **255**, 1490 (1992)
7. C.M. Varma, J. Zaanen, Science **255**, 1490–1491 (1992)
8. C.S. Yannoni, M. Hoinkis, M.S. Devries, D.S. Bethune, J.R. Salem, M.S. Crowder, R.D. Johnson, Science **256**, 1191–1192 (1992)
9. O. Zhou, G.B.M. Vaughan, Q. Zhu, J.E. Fischer, P.A. Heiney, N. Coustel, J.P. McCauley, A.B. Smith, Science **255**, 833–835 (1992)
10. M.H. Ge, K. Sattler, Science **260**, 515–518 (1993)
11. S.W. McElvany, J.H. Callahan, M.M. Ross, L.D. Lamb, D.R. Huffman, Science **260**, 1632–1634 (1993)
12. S.W. McElvany, M.M. Ross, N.S. Goroff, F. Diederich, Science **259**, 1594–1596 (1993)
13. D.T. Colbert, J. Zhang, S.M. McClure, P. Nikolaev, Z. Chen, J.H. Hafner, D.W. Owens, P.G. Kotula, C.B. Carter, J.H. Weaver, A.G. Rinzler, R.E. Smalley, Science **266**, 1218–1222 (1994)
14. J.E. Fischer, Science **264**, 1548–1549 (1994)
15. A. Lappas, K. Prassides, K. Vavekis, D. Arcon, R. Blinc, P. Cevc, A. Amato, R. Feyerherm, F.N. Gygax, A. Schenck, Science **267**, 1799–1802 (1995)
16. K. Livingston, Science **268**, 1637 (1995)
17. F. Diederich, C. Thilgen, Science **271**, 317–323 (1996)
18. B.R. Eggen, M.I. Heggie, G. Jungnickel, C.D. Latham, R. Jones, P.R. Briddon, Science **272**, 87–89 (1996)
19. J.W. Mintmire, Science **272**, 45–46 (1996)
20. J. Liu, A.G. Rinzler, H.J. Dai, J.H. Hafner, R.K. Bradley, P.J. Boul, A. Lu, T. Iverson, K. Shelimov, C.B. Huffman, F. Rodriguez-Macias, Y.S. Shon, T.R. Lee, D.T. Colbert, R.E. Smalley, Science **280**, 1253–1256 (1998)
21. R. F. Service, Science **280**, 2044 (1998)
22. L. Marques, M. Mezouar, J.L. Hodeau, M. Nunez-Regueiro, N.R. Serebryanaya, V.A. Ivdenko, V.D. Blank, G.A. Dubitsky, Science **283**, 1720–1723 (1999)
23. C.A. Reed, K.C. Kim, R.D. Bolskar, L.J. Mueller, Science **289**, 101–104 (2000)
24. S.Q. Zhou, C. Burger, B. Chu, M. Sawamura, N. Nagahama, M. Toganoh, U.E. Hackler, H. Isobe, E. Nakamura, Science **291**, 1944–1947 (2001)
25. J.F. Bai, A.V. Virovets, M. Scheer, Science **300**, 781–783 (2003)
26. M.J. Moses, J.C. Fettinger, B.W. Eichhorn, Science **300**, 778–780 (2003)
27. H. Sakurai, T. Daiko, T. Hirao, Science **301**, 1878 (2003)
28. S.Y. Xie, F. Gao, X. Lu, R.B. Huang, C.R. Wang, X. Zhang, M.L. Liu, S.L. Deng, L.S. Zheng, Science **304**, 699 (2004)
29. K. Komatsu, M. Murata, Y. Murata, Science **307**, 238–240 (2005)
30. P.A. Troshin, A.G. Avent, A.D. Darwish, N. Martsinovich, A.K. Abdul-Sada, J.M. Street, R. Taylor, Science **309**, 278–281 (2005)
31. K. Kurotobi, Y. Murata, Science **333**, 613–616 (2011)
32. L. Bao, P. Peng, X. Lu, Acc. Chem. Res. **51**, 810–815 (2018)
33. R.D. Johnson, M.S. Devries, J. Salem, D.S. Bethune, C.S. Yannoni, Nature **355**, 239–240 (1992)
34. W. Andreoni, A. Curioni, Proceedings of the Symposium on Recent Advances in the Chemistry and Physics of Fullerenes and Related Materials **96**, 569–577 (1996)

35. Q.Y. Kong, Y.F. Shen, J. Xu, L. Zhao, J. Zhuang, S.X. Qian, Y.F. Li, X. Li, Y.H. Lin, R.F. Cai, Chem. Phys. Lett. **341**, 447–454 (2001)
36. Y. Miyamoto, N. Hamada, A. Oshiyama, S. Saito, Phys. Rev. B **46**, 1749–1753 (1992)
37. F. Chen, D. Singh, S.A. Jansen, J. Phys. Chem. **97**, 10958–10963 (1993)
38. S.H. Wang, F. Chen, Y.C. Fann, M. Kashani, M. Malaty, S.A. Jansen, J. Phys. Chem. **99**, 6801–6807 (1995)
39. H.J. Muhr, R. Nesper, B. Schnyder, R. Kotz, Chem. Phys. Lett. **249**, 399–405 (1996)
40. C.G. Ding, J.L. Yang, X.Y. Cui, C.T. Chan, J. Chem. Phys. **111**, 8481–8485 (1999)
41. T. Nakamura, K. Ishikawa, K. Yamamoto, T. Ohana, S. Fujiwara, Y. Koga, Phys. Chem. Chem. Phys. **1**, 2631–2633 (1999)
42. C.G. Ding, Y.L. Yang, R.S. Han, K.L. Wang, Phys. Rev. A **64**, 043201 (2001)
43. T. Nakamura, K. Ishikawa, A. Goto, M. Ishihara, T. Ohana, Y. Koga, Diamond Relat. Mater. **10**, 1228–1230 (2001)
44. X. Yang, G.C. Wang, Z.F. Shang, Y.M. Pan, Z.S. Cai, X.Z. Zhao, Phys. Chem. Chem. Phys. **4**, 2546–2553 (2002)
45. Z.Y. Yang, X.F. Xu, G.C. Wang, Z.F. Shang, Z.S. Cai, Y.M. Pan, X. Z. Zhao, J. Mol. Struct.-Theochem **618**, 191–200 (2002)
46. D. Gonzalez- Rodriguez, T. Torres, D.M. Guldi, J. Rivera, M.A. Herranz, L. Echegoyen, J. Am. Chem. Soc. **126**, 6301–6313 (2004)
47. H.Y. Rao, Z.G. Le, Chin. Chem. Lett. **15**, 246–249 (2004)
48. P. Cigler, M. Kozisek, P. Rezacova, J. Brynda, Z. Otwinowski, J. Pokorna, J. Plesek, B. Gruner, L. Doleckova-Maresova, M. Masa, J. Sedlacek, J. Bodem, H.G. Krausslich, V. Kral, J. Konvalinka, Proc. Natl. Acad. Sci. U. S. A. **102**, 15394–15399 (2005)
49. T.M. Simeon, I. Yanov, J. Leszczynski, Int. J. Quantum Chem. **105**, 429–436 (2005)
50. C.K. Acharya, C.H. Turner, J. Phys. Chem. B **110**, 17706–17710 (2006)
51. L. Viani, M.C. dos Santos, Solid State Commun. **138**, 498–501 (2006)
52. J.Q. Hou, H.S. Kang, J. Phys. Chem. A **111**, 1111–1116 (2007)
53. I. Garg, H. Sharma, K. Dharamvir, V.K. Jindal, J. Comput. Theor. Nanosci. **8**, 642–655 (2011)
54. M. Anafcheh, R. Ghafouri, Struct. Chem. **23**, 1921–1929 (2012)
55. M. Anafcheh, R. Ghafouri, F. Naderi, Physica E-Low-Dimensional Systems & Nanostructures **44**, 1992–1998 (2012)
56. T. Manning, R.R. Gokal, C. Konda, R. Hetzel, D. Lovingood, D. Phillips, Fullerenes, Nanotubes, Carbon Nanostruct. **21**, 1–11 (2013)
57. B.C. Guo, S. Wei, J. Purnell, S. Buzza, A.W. Castleman Jr., Science **256**, 515–516 (1992)
58. B.C. Guo, K.P. Kerns, A.W. Castleman Jr., Science **255**, 1411–1413 (1992)
59. S. Wei, B.C. Guo, J. Purnell, S. Buzza, A.W. Castleman Jr., J. Phys. Chem. **96**, 4166–4168 (1992)
60. S. Wei, B.C. Guo, J. Purnell, S. Buzza, A.W. Castleman Jr., Science **256**, 818–820 (1992)
61. S.F. Cartier, Z.Y. Chen, G.J. Walder, C.R. Sleppy, A.W. Castleman Jr., Science **260**, 195–196 (1993)
62. A.W. Castleman Jr., B.C. Guo, S. Wei, Z.Y. Chen, Plasma Phys. Controlled Fusion **34**, 2047–2051 (1992)
63. A.W. Castleman, Jr., The discovery of Met-Cars, R&D Innovator (1993)
64. C. Berkdemir, A.W. Castleman Jr., J.O. Sofo, Phys. Chem. Chem. Phys. **14**, 9642–9653 (2012)
65. B.C. Guo, K.P. Kerns, A.W. Castleman Jr., J. Am. Chem. Soc. **115**, 7415–7418 (1993)
66. S. Wei, B.C. Guo, J. Purnell, S.A. Buzza, A.W. Castleman Jr., J. Phys. Chem. **97**, 9559–9561 (1993)
67. H.T. Deng, B.C. Guo, K.P. Kerns, A.W. Castleman Jr., Int. J. Mass Spectrom. Ion Processes **138**, 275–281 (1994)
68. H.T. Deng, B.C. Guo, K.P. Kerns, A.W. Castleman, J. Phys. Chem. **98**, 13373–13378 (1994)
69. J. Purnell, S. Wei, A.W. Castleman Jr., Chem. Phys. Lett. **229**, 105–110 (1994)
70. K.P. Kerns, B.C. Guo, H.T. Deng, A.W. Castleman Jr., J. Am. Chem. Soc. **117**, 4026–4029 (1995)
71. H.T. Deng, K.P. Kerns, A.W. Castleman Jr., J. Am. Chem. Soc. **118**, 446–450 (1996)

72. H. Sakurai, A.W. Castleman Jr., J. Phys. Chem. A **101**, 7695–7698 (1997)
73. H. Sakurai, A.W. Castleman Jr., J. Phys. Chem. A **102**, 10486–10492 (1998)
74. H. Sakurai, A.W. Castleman Jr., J. Chem. Phys. **111**, 1462–1466 (1999)
75. H. Sakurai, S.E. Kooi, A.W. Castleman Jr., J. Cluster Sci. **10**, 493–507 (1999)
76. B.D. Leskiw, K.L. Knappenberger, A.W. Castleman Jr., J. Chem. Phys. **117**, 8321–8326 (2002)
77. D.E. Bergeron, P.J. Roach, A.W. Castleman Jr., N.O. Jones, J.U. Reveles, S.N. Khanna, J. Am. Chem. Soc. **127**, 16048–16053 (2005)
78. D.E. Bergeron, P.J. Roach, A.W. Castleman Jr., N. Jones, S.N. Khanna, Science **307**, 231–235 (2005)
79. L.R. Brock, M.A. Duncan, J. Phys. Chem. **100**, 5654–5659 (1996)
80. W. Ekardt, Phys. Rev. B **29**, 1558–1564 (1984)
81. M. Brack, Rev. Mod. Phys. **65**, 677–732 (1993)
82. W. Ekardt, *Metal clusters* (Wiley, Berlin, 1999)
83. B.D. Leskiw, A.W. Castleman Jr., Chem. Phys. Lett. **316**, 31–36 (2000)
84. A.W. Castleman Jr., A.C. Harms, R.E. Leuchtner, Z. Phys, D-Atoms Mol. Clusters **19**, 343–346 (1991)
85. P.J. Roach, W.H. Woodward, A.C. Reber, S.N. Khanna, A.W. Castleman Jr., Phys. Rev. B **81**, 195404 (2010)
86. R.E. Leuchtner, A.C. Harms, A.W. Castleman Jr., J. Chem. Phys. **94**, 1093–1101 (1991)
87. S. Petrie, G. Javahery, D.K. Bohme, J. Am. Chem. Soc. **115**, 1445–1450 (1993)
88. L. Gao, M.E. Lyn, D.E. Bergeron, A.W. Castleman, Int. J. Mass Spectrom. **229**, 11–17 (2003)

Chapter 11
Cluster Dissociation, Intracluster Reactivity and Effect of the Ligands

The physical properties of small metal clusters, such as ionization potentials, electron affinities, and dissociation energies, are largely different from those of the bulk materials. Theoretical calculations have suggested that small clusters have close-packed structures but that their stabilities change dramatically with the cluster sizes. Cluster size distributions based on mass spectrometry and fragmentation patterns provide information on particularly stable clusters; however, it is difficult to measure the dissociation energies as a function of cluster sizes. Early studies of cluster dissociation retrospect to 1980s, such as the one by Gingerirch [1] using conventional Knudsen cell techniques. Following that, cluster thermal dissociation [2–6], photodissociation (or laser-induced dissociation) [7–18] and collision-induced dissociation (CID) [19–38] have been extensively and meticulously studied, enabling the measurement of cluster dissociation energies. From analysis of internal cluster structure and fragmentation dynamics relating to total energy and electron population, it is discerned that the regimes of system evolution are associated with a few processes, including energy transfer/accumulation (by photo excitation or thermal treatment or enough collisions), contingent isentropic expansion, dynamical fragment formation with energy thresholds, de-excitation and thermalization. The spontaneous structural relaxation of metal clusters, driven by thermal/collision or photoexcitation [33], could not only induce direct fragmentation/dissociation but also give rise to intracluster redox reactions with size-dependent branching ratios, especially for transition-metal involved clusters which possess different valence states (or oxidation states) and varied charge distribution of the atoms in the cluster [34].

Z. Luo and S. N. Khanna, *Metal Clusters and Their Reactivity*,
https://doi.org/10.1007/978-981-15-9704-6_11

11.1 Cluster Dissociation

11.1.1 Collision-Induced Dissociation

Collision-induced dissociation (CID) has been extensively investigated [19–22, 35], and through CID the cluster dissociation energies can be determined by modelling the measured cross sections and hence checking out the collision energy threshold for dissociation. For example, utilizing the CID method, Armentrout et al. [20] measured the cross sections for Nb_n^+ (n = 2–6) clusters with Xe gas, as shown in Fig. 11.1.

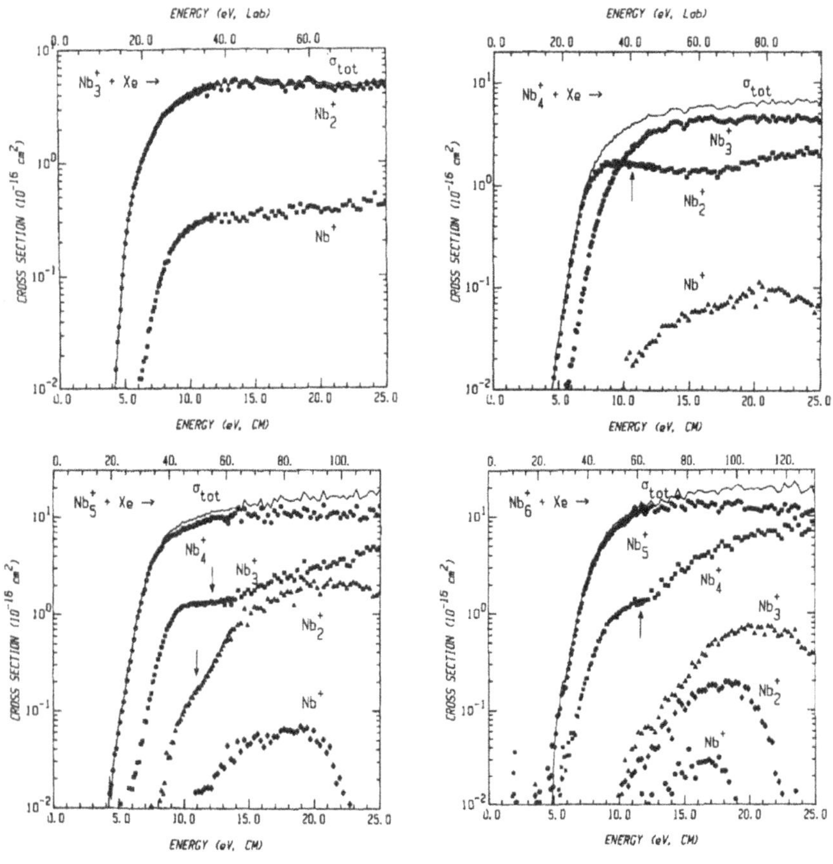

Fig. 11.1 Collision-induced dissociation of niobium cluster ions, Nb_n^+ + Xe, n = 2–6, parts a–e. The CID cross sections, measured in Å^2, are plotted as functions of collision energy in the center-of-mass (lower x-axis) and laboratory (upper x-axis) frames. Total cross sections for dissociation are shown as solid lines. Arrows indicate thresholds for higher energy processes that produce fragment ions, discussed in the text. Reproduced with permission from Ref. [20]. Copyright 1989 American Chemistry Society

It was found that Nb_n^+ readily dissociate to all possible ionic fragments, with the lowest energy dissociation pathway as "$Nb_n^+ \rightarrow Nb_{n-1}^+ + Nb$", and the largest neutral fragments at collision energies of less than 10 eV. Evidence was also presented for the loss of multiple Nb atoms from the Nb_n^+ cluster at collision energies >10 eV. The varied CID product channels provided information of thermodynamics, qualitative cross section energy dependences and the relative energy thresholds, enabling to fully understand the mechanisms of formation and dissociation for such metal clusters.

Anderson [21] studied mass-selected aluminum cluster cations Al_n^+ (n = 2–7) by xenon over an energy range of 0–10 eV. In order to circumvent this problem that clusters could be formed with broad distributions of internal and translational energy, they developed a technique to partially thermalize the cluster ions by forcing them to pass through a radio-frequency trap filled with a buffer gas [21, 36]. The use of a cooling trap helps improved the clusters to reach complete thermalization. The cooling trap was constructed from a stack of six 1.95 mm thick stainless-steel plates spaced 1.5 mm apart; and each plate had a maze-like slot 6.8 mm wide milled through it. The stack of slotted plates was capped on top and bottom by additional solid plates, which enables the assembly to form a labyrinthine channel of rectangular cross section (c.a., 6.8 mm wide, 22 mm high). Alternate slotted plates were connected to opposite phases of an RF tank circuit (3.5 MHz, 210V_{rms}), and the capping plates were connected to positive biased voltage with respect to the DC potential of the slotted plates [21]. The RF voltage created an effective potential preventing low energy ions from escaping through the sides of the channel, while the DC voltages preventing an escape from the top and bottom. In operation, this trap was filled through gas inlets in the top plate with ~3 mTorr of helium. Cluster ions entered the cooling trap with a broad distribution of translational energy, and lose energy in ~10 collisions with the helium when going through the entrance channel. A small DC potential prevents the ions from escaping back to the source, allowing them only exiting by diffusing through the gas-filled maze hence becoming thermalized in the process [21]. The effectiveness of the cooling process was also applied to measure the residence time distributions for different sized clusters simply by pulsing the cluster ions into the cooling trap [21].

Accordingly, the cross sections for each product channel were calculated using the formula as

$$\sigma(E)_i = n \cdot L \cdot \frac{(S + B)_i(E) - B_i(E)}{I_0(E)} \tag{11.1}$$

in which E refers to the collision energy, n is the target gas number density, L is the effective scattering cell length, $(S + B)_i(E)$ is the intensity of product ion i with the scattering cell full, $I_0(E)$ is the incident reagent ion intensity while $B_i(E)$ is the product ion intensity together with buffer gas flowing into the vacuum chamber. Assuming the total scattering was kept to be relatively small enough (c.a., ~2%) to avoid perturbations of the collision energy by non-reactive collisions, the error in this approximate formula is neglectable [21].

Integral cross sections for CID of Al_{2-7}^+ by xenon were measured for all product ions at collision energies ranging from 0.0 to 10.0 eV. As results, a similar trend existed for Al_{2-6}^+ on the collision energy dependence of the total CID cross sections. However, Al_7^+ displays a higher threshold and a slower rise with increasing collision energy, which indicates its better stability as a 20-electron cluster species. Another common trend is that, for all reagent clusters, the two lowest energy channels both involve the loss of a single atom hence produce Al^+ or Al_{n-1}^+ as ionic products, expressed as [21],

$$Al_n^+ + Xe \rightarrow Al_{n-1} + Al^+ + Xe \tag{11.2}$$

$$Al_n^+ + Xe \rightarrow Al_{n-1}^+ + Al + Xe \tag{11.3}$$

The threshold CID method has also been applied to study the fragmentation patterns and to measure the dissociation energies of small anionic copper clusters (Cu_n^-, n = 2–8) and their monocarbonyls (Cu_nCO^-, n = 3–7). Similar to the above case for Al_n^-, the main reaction channels for the bare clusters Cu_n^- (n = 2–8) were also found to be the loss of an atom and loss of a dimer,

$$Cu_n^- + Xe \rightarrow Cu_{n-1}^- + Xe + Cu \tag{11.4}$$

$$Cu_n^- + Xe \rightarrow Cu_{n-2}^- + Xe + Cu_2 \tag{11.5}$$

Furthermore, it is interesting to mention that all the clusters Cu_n^- (n = 2–8) were found to take a reaction channel as Eq. 11.4, but among them only Cu_{3-5}^- and Cu_7^- support another reaction channel as Eq. 11.5. As Cu_7^- is an 8-electron stable cluster which is consistent with closed shells in the jellium model, and Cu_7^- has the highest dissociation energy, so it is reasonable for Cu_8^- to take a dominant reaction pathway as Eq. 11.4 leading to the species Cu_7^- while give rare opportunity for it to undertake a reaction channel as Eq. 11.5. Moreover, the dissociation energies for the loss of a Cu atom from bare copper cluster anions also show even-odd alternation, indicating the different stability of the odd and even electron copper cluster systems. Copper cluster monocarbonyls (Cu_nCO^-) were also found to undertake the similar CID reactions, but the main reaction channel is loss of CO. Similar to Cu_7^-, the species Cu_5CO^- also has eight valence electrons and displays the highest carbonyl desorption energy [37].

While CID is inherently a low-resolution method for the measurements of dissociation energies, it also provides a tool to measure good quality physical data on systems which are not complicated by spectroscopic methods, such as approximate ionization potentials. From the product branching ratios and cross section magnitudes based on the CID investigations, qualitative structural information of the correlated clusters could also be derived [21]. It is still notable that, high energy collisions of metal clusters could be also associated with fragmentation induced by thermal effect (known as thermal dissociation) [2, 4–6, 38] and even shock waves [39].

11.1.2 Photodissociation

Another aspect is about photodissociation or photofragmentation, which is usually attained by intense laser radiation (i.e., the aforementioned LID) [7–10] or just photo-induced thermal desorption [3, 40]. Photodissociation/photodepletion spectroscopy is regarded as one of the most powerful techniques applicable for cluster ions. Early photodissociation investigation was also utilized to obtain spectral information of gas-phase ions [41]; while actually both spectroscopic and thermodynamic information can be obtained from the photodissociation studies [42–44].

Figure 11.2 shows a typical experimental setup in Terasaki group [45] for photodissociation studies based on a tandem TOF mass spectrometer. The metal cluster ions were generated by the LaVa source using a Nd: YAG laser. An Ar gas was mixed with a buffer He gas to produce M_NAr^+ clusters. The ions produced were extracted by a pulsed electric field into the TOF spectrometer, and the mass-selected cluster ions were then irradiated with a tunable pulsed laser for photodissociation. Fragment ions were mass-analyzed by the secondary TOF equipped with a reflectron, and recorded as the varying wavelength of the dissociation laser.

On such as instrument, the photodissociation of Ag_n^+ and Mn_N^+ clusters were studied [45]. Considering that the electronic energy on photoexcitation readily converts to vibrational energies of internal modes of a cluster, such process corresponds to unimolecular dissociation in a statistical manner, that is, bond dissociation energies are the decisive parameters. The dissociation yield as a function of the photon energy enable to work out an optical absorption spectrum, known as action spectroscopy. Figure 11.3A shows such spectra of the partial photodissociation cross sections of Mn_3^+. Manganese cluster ions are one of the suited species

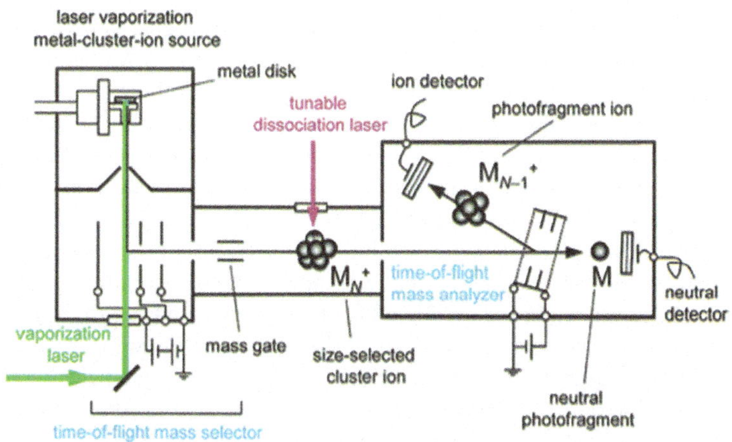

Fig. 11.2 Experimental setup for photodissociation spectroscopy of cluster ions based on a photofragment-detection scheme by a tandem TOF mass spectrometer. Reproduced from Ref. [45]. Copyright 2000 American Chemistry Society

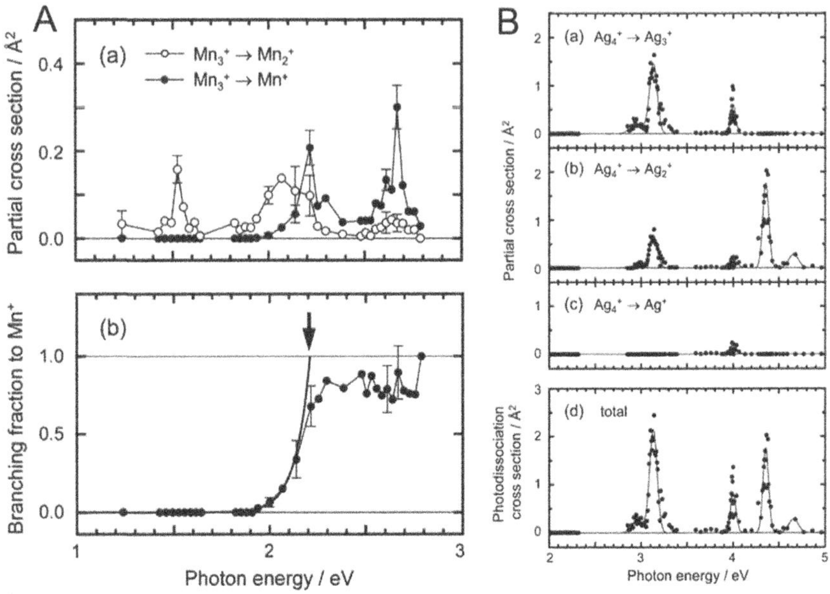

Fig. 11.3 **A** Spectra of the partial photodissociation cross sections of Mn_3^+: open (solid) circles are for the dissociation channel to Mn_2^+(Mn^+). **b** Branching fraction of the Mn^+ channel as a function of the photon energy. The thick solid curve fits the onset behavior, which is explained by the internal-energy distribution of primary Mn_3^+. The arrow indicates the threshold energy of the Mn^+ formation from Mn_3^+. **B** Photodissociation action spectra of Ag_4^+: **a–c** partial cross sections for the processes producing Ag_3^+, Ag_2^+, and Ag^+, respectively; **d** total cross section. The solid curves represent fitting to Gaussian profiles. Reproduced from Ref. [45]. Copyright 2000 American Chemistry Society

for the photodissociation studies as their bond dissociation energies are relatively weak thus readily dissociation with the removal of one or two Mn atoms even in the radiation of visible light. As the photon energy varied from 1.2 to 2.8 eV, Mn_2^+ was observed as the dominant product below 2.0 eV, whereas Mn^+ gradually takes over above it at the further increased photon energy. Such monotonic change in the branching fraction showed that the dissociation of Mn_3^+ proceeds with sequential loss of manganese atoms, with dependence on the excess energy. It is notable that Mn_2^+ and Mn_3^+ bear ferromagnetic coupling between local spins, in contrast to anti-ferromagnetic bulk manganese. In contrast, the photodissociation of Ag_4^+ found a different way, as shown in Fig. 11.3B. This finding may be related to the relatively strong Ag-Ag bond energy and emissivity of silver clusters in an argon matrix, where the clusters are prevented from dissociation and stay in the excited electronic state until they fluoresce (~1 ns). This difference could also be associated with the temperature of the clusters studied, as thermal motions and structural relaxation play an important role in the evolution of metallic properties as well as size dependence.

Among others, the obtained photodissociation spectrum of VFe^+ took on two absorption maxima at 260–340 nm as well as a dissociation threshold at 380 nm, suggesting a bond energy $D^o(V^+–Fe) = 75 \pm 5$ kcal/mol, which is in good agreement with that determined by ion-molecule bracketing [46]. Similar photodissociation studies on $Mn_2{}^+$ revealing the bonding energy of Mn^+–Mn have also been reported [7, 47]. Comparing with a CID method as mentioned above, the dissociation thresholds obtained using the photodissociation approaches are sharper and generally more accurate if it is discriminated between one-photon and multiphoton processes.

Early studies showed the mass spectrometry analysis of preselected $Fe_6{}^+$ ions in the absence and presence of a dissociation light field. In contrast, in the presence of ~4 mJ/cm² per pulse of 2.33 eV radiation in the dissociation zone, multiple photo-product ions Fe^+_{2-5} were observed with well-resolved peak shape characteristic. From these observations, it was concluded that fragmentation of the $Fe_6{}^+$ ion occurs faster than the time scale of the mass interrogation. Such behavior is typical of the iron cluster cation photodissociation events, and has been found to be applicable to other metal cluster systems, such as the cations composed of nickel and niobium ranging in size from two to ten atoms [8]. Also addressed, was the fluence dependence of the different product channels for the $Fe_6{}^+$ irradiated with 2.33 eV light. The calculated solid curves through the experimental fractional populations of the $Fe_5{}^+$, $Fe_4{}^+$, and $Fe_3{}^+$ products were generated using a single adjustable parameter, the absorption cross section, in the context of a simple stepwise absorption/fragmentation scheme:

$$Fe_6^+ \xrightarrow{h\nu} Fe_5^+ + Fe \xrightarrow{h\nu} Fe_4^+ + 2Fe \xrightarrow{h\nu} Fe_3^+ + 3Fe \xrightarrow{h\nu} \cdots \quad (11.6)$$

As results, the only one-photon dissociation process of the parent ion at this energy was found to be the ejection of a neutral Fe atom. If the kinetic scheme is simply based on Eq. 11.6, all products will be the outcome of dissociation involving the loss of one Fe atom per photon absorbed. However, as the time scale for fragmentation was faster than assessable in the apparatus, the identity of the absorber was not determined. That is, there exist alternative but indistinguishable dissociation reactions which yield the same kinetic prediction but may involve absorption of multiple photons at the loss of one Fe atom, expressed as the following schemes:

$$\begin{array}{ccccccccc}
Fe_6^+ & \xrightarrow{h\nu} & (Fe_6^+)^* & \xrightarrow{h\nu} & (Fe_6^+)^{**} & \xrightarrow{h\nu} & (Fe_6^+)^{***} & \xrightarrow{h\nu} & \cdots \\
\downarrow & & \downarrow & & \downarrow & & & & \\
Fe_5^+ + Fe & & (Fe_5^+)^* + Fe & & (Fe_5^+)^{**} + Fe & & & & \\
 & & \downarrow & & \downarrow & & & & \\
 & & Fe_4^+ + Fe & & (Fe_4^+)^* + Fe & & & & \\
 & & & & \downarrow & & & & \\
 & & & & Fe_3^+ + Fe & & & &
\end{array} \quad (11.7)$$

Therefore, it is difficult to discriminate between one-photon and multiphoton processes, leaving a challenge to ensure correct conclusions when bracket cluster dissociation energies using the photodissociation measurements. In order to eliminate such uncertainty, ion-molecule reactions (e.g., with ascertained bond activation energies) and collision-induced dissociation can be utilized simultaneously as a reference to validate the energy values.

In comparison, photodissociation of diatomic metal cluster systems are relatively simple dissociation process "$M_2^+ \rightarrow M^+ + M^{+}$". A typical study on this was reported by *Hettich* and *Freiser* in 1987 [10]. With a focus on heteronuclear metal dimer ions MFe^+ (M = Sc, Ti, V, Cr, Fe, Co, Ni, Cu, Nb, and Ta) and utilizing Fourier transform mass spectrometry, they studied the photodissociation of these cluster systems, which enabled a method of probing the fundamental bonding nature between two bare transition-metal atoms. Both M^+ and Fe^+ were observed as photoproducts and the one having a relatively lower ionization potential dominates the products respectively. The photodissociation spectra of MFe^+ obtained by monitoring the fragmentation of MFe^+ as a function of wavelength are displayed in Fig. 11.4. Among these MFe^+ species, broad absorptions in the ultraviolet and visible spectral regions were observed, revealing cross sections from 0.06 Å^2 (for VFe^+) to 0.62 Å^2 (for $CrFe^+$). Bond energies obtained by noting the photoappearance onsets are in a range between 48 kcal/mol (for $ScFe^+$) and 75 kcal/mol (for VFe^+), which is in good agreement with the values obtained by ion-molecule bracketing techniques. Moreover, ionization potentials for these MFe^+ can also be calculated comparing the ionic dimers with their neutral metal dimer counterparts, as was found to be in the range of 5.4 eV (for VFe) to 7.4 eV (for TaFe) [10].

11.1.3 Coulomb Explosion

Coulomb explosion is generally a process in which a molecule moving with high velocity strikes a solid and the electrons that bond the molecule are torn off rapidly in violent collisions with the electrons of the solid; as a result, the molecule is suddenly transformed into a cluster of charged atomic constituents that then separate under the influence of their mutual Coulomb repulsion. Coulomb explosions are most studied using a particle accelerator which is normally employed in nuclear physics. It could also be done under a narrow powerful laser beam, where a small amount of solid explodes into plasma of ionized atomic particles; with their low masses, outer valence electrons responsible for chemical bonding are easily stripped from atoms, leaving them positively charged. Coulomb explosion has become a mechanism for coupling electronic excitation energy from intense electromagnetic fields into atomic motion. Given a mutually repulsive state between atoms whose chemical bonds are broken, the material explodes into a small plasma cloud of energetic ions with higher velocities than that seen in thermal emission.

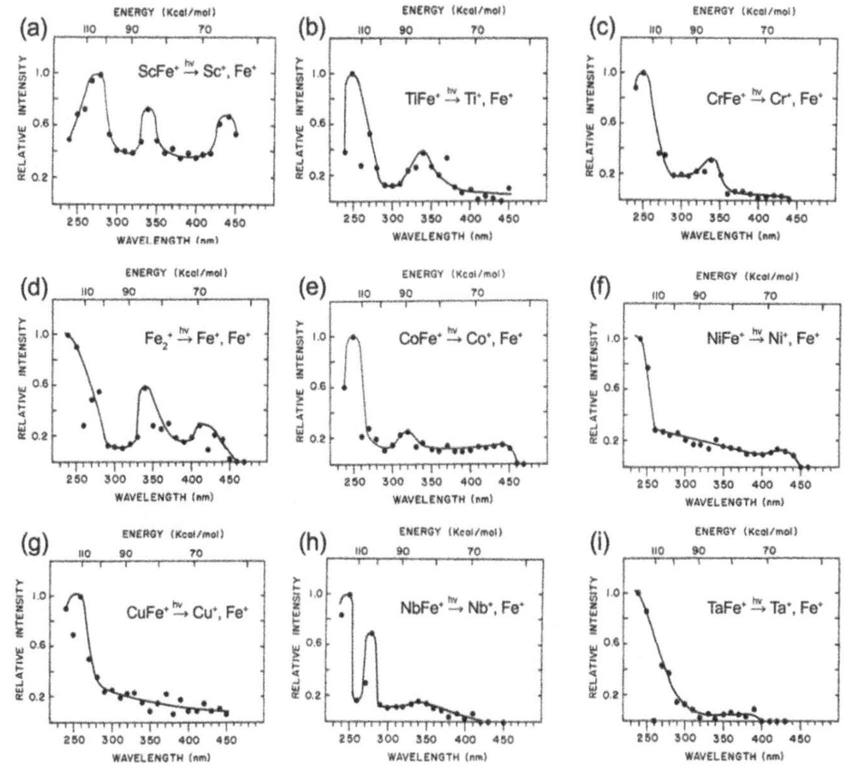

Fig. 11.4 The photodissociation spectra of $ScFe^+$ (**a**), $TiFe^+$ (**b**), $CrFe^+$ (**c**), $Fe_2{}^+$ (**d**), $CoFe^+$ (**e**), $NiFe^+$ (**f**), $CuFe^+$ (**g**), $NbFe^+$ (**h**), and $TaFe^+$ (**i**) obtained by monitoring the appearance of Fe^+ and M^+ (M = Sc, Ti, Cr, Fe, Co, Ni, Cu, Nb, and Ta) as a function of wavelength. Reproduced with permission from Ref. [10]. Copyright 1987 American Chemical Society

Coulomb explosion experiments generally serve two main purposes: (1) to yield valuable information on the interactions of fast ions with solids; (2) to determine the stereochemical structures of molecular-ion projectiles. For example, Coulomb explosion and photodissociation channel of $H_2{}^+$ and $D_2{}^+$ have been investigated with 790 nm, sub-100 fs laser pulses by employing a high-resolution photofragment imaging technique [48]. At intensities close to the threshold for Coulomb explosion, they observed a peak structure in the Coulomb explosion kinetic energy spectra for both $H_2{}^+$ and $D_2{}^+$, which was attributed to the different dissociation energies on vibrationally excited states of the molecules. Preservation of vibrational structure during the Coulomb explosion suggests ionization at a critical internuclear distance. When using pulses with durations of 200–500 fs, three Coulomb explosion kinetic energy groups were observed with different angular distributions in both $H_2{}^+$ and $D_2{}^+$.

The valence electron cloud of metal clusters provides a finite fermion system with remarkable properties as electronic shell effects and strong optical absorption in a narrow frequency band [49–51]. The surface-plasmon resonance of metal clusters depends sensitively on the geometry of the cluster and thus provides an ideal handle for analyzing and for controlling cluster dynamics [52, 53]. The dynamical scenarios become more involved when clusters stay in contact with a substrate, either embedded inside or deposited on a surface. Although it is the aim of the present contribution to explore, little has been done in the regime of highly non-linear dynamics on metal clusters induced by intense laser fields [54–59]. Recently Fehrer et al. [60] investigated the dynamical evolution of a Na_8 cluster embedded in Ar matrices of various sizes, Na_8Ar_N (N = 30 to 1048). This system was excited by an intense short laser pulse leading to high ionization stages, and the subsequent highly non-linear motion of the cluster under Ar environment was analyzed in terms of trajectories, shapes, and energy flow. The most prominent effects were found on the temporary stabilization of high charge states for several ps, and sudden stopping of the Coulomb explosion of embedded Na_8 clusters [60]. Döppner et al. [51] reported the charging of free silver clusters in strong laser fields by dual-pulse excitation over a broad cluster size range. Depending on the laser intensity and the cluster size, an optical delay between 0.45-ps and 13.5-ps was found necessary to drive the plasmon mode of the cluster into resonance with the laser field and allow for an effective charging of the system. Note that the optimum time delay to reach maximum charging changes with the cluster size and inversely with the laser intensity.

11.2 Intracluster Reactivity

Several investigations on metal clusters and metal cluster complex shed light on the intracluster reactivity [34, 61–88]. For example, Fox et al. [34] reported the intracluster reactivity on hydrated vanadium cations $V^+(H_2O)_n$ due to absorption of black body radiation, as shown in Fig. 11.5. At a pressure of about 4×10^{-10} mbar in the cell region, the fragmentation is induced by the black body background radiation. As shown in Fig. 11.5a, the initial cluster distribution which also contained $V(OH)_2^+(H_2O)_m$ was observed after accumulating the ions for 2 s in the cell; however, after a reaction delay of 2 s, the clusters shifted to lower masses, and a distinct shift in favor of the $V(OH)_2^+(H_2O)_m$ clusters was observed. The $V^+(H_2O)_m$ completely disappeared after 10 s, while the $V(OH)_2^+(H_2O)_m$ further evaporate water until the final product $V(OH)_2^+(H_2O)_3$ was observed. This trend was also observed for small hydrated vanadium cations (Fig. 11.5b), as well as mass-selected $V^+(H_2O)_{10}$ (Fig. 11.5c). In brief, besides the loss of water ligands, the $V^+(H_2O)_n$ clusters were found to display two different intracluster redox reactions with size-dependent branching ratios, resulting in $V(OH)^+(H_2O)_n$ or $V(OH)_2^+(H_2O)_n$ clusters together with a concurrent release of atomic hydrogen. These behaviors reflect the properties of the transition metals to form stable compounds in a variety of oxidation states [34].

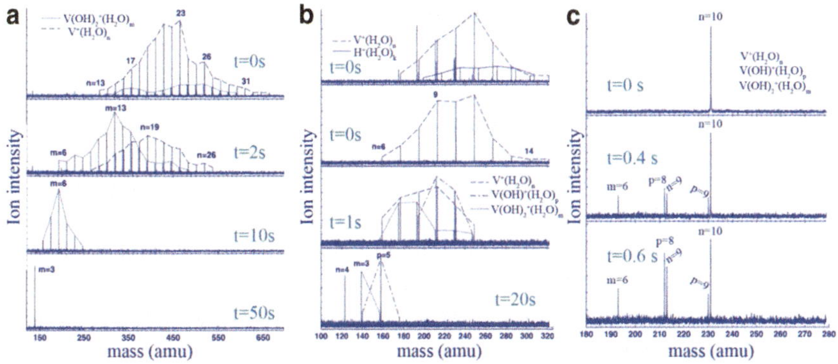

Fig. 11.5 **a** Mass spectra showing the fragmentation of $V^+(H_2O)_n$ after variable reaction delays. **b** Mass spectrum showing a distribution of small hydrated vanadium cations, $V^+(H_2O)_n$, n = 7–18. Almost every $V^+(H_2O)_n$ peak is accompanied by hydroxide-containing cluster species, $V(OH)^+(H_2O)_{n-1}$ and $V(OH)_2^+(H_2O)_{n-2}$. In addition, protonated water clusters, $H^+(H_2O)_m$, have been produced in the cluster source. After storing these clusters for 1 s in the FT-ICR cell, hydrated vanadium hydroxide species, $V(OH)^+(H_2O)_p$ and $V(OH)_2^+(H_2O)_m$, have been formed due to absorption of black body radiation. The different cluster species further evaporate water. After 20 s, the most dominant products are $V(OH)^+(H_2O)_5$, $V(OH)_2^+(H_2O)_3$, and $V^+(H_2O)_4$. **c** Mass spectra of the black body radiation induced unimolecular reactions of size-selected $V^+(H_2O)_{10}$ with a delay of 0 s, 0.4 s, 0.6 s after trapping. The main fraction of the clusters undergoes an intracluster reaction to form $V(OH)^+(H_2O)_{8,9}$. Smaller fractions fragment to $V^+(H_2O)_9$ or undergo an intracluster reaction to form $V(OH)_2^+(H_2O)_6$. Reproduced with permission from Ref. [34]. Copyright 2002 Royal Society of Chemistry

Intracluster ion-molecule reactions have also been studied for several other systems, such as Ti^+ with C_2H_5OH and CF_3CH_2OH clusters where the influence of fluorine substituents on chemical reactivity was discussed [62], also intramolecular hydrogen bonding interactions could enhance cluster stability and its reactivity [89]. A recent investigation on intracluster $V^+(CH_3COOR)_n$ (R=CH_3, C_2H_5, and C_2D_5) complexes in gas phase was presented by *Paul* et al. [90] The mass spectral examination indicated that the presence of a major sequence of cluster series as $V^+(CH_3COO)(CH_3COOR)_n$, indicating the insertion of V^+ into the C–O bond of CH_3COOR within the heteroclusters, followed by alkyl radical elimination. On the other hand, the products of $V^+(OR)$ and $V^+(CH_3)(OR)$ were interpreted to arise from the insertion of V^+ into the C(O)–O bond of the ester group, followed by CH_3CO and CO elimination. In addition, the VO^+ ion is present throughout the mass spectra, indicating that insertion of V^+ into the C=O bond of CH_3COOR also occurs. Note that such selective reactivity may also involve competition mechanism, including the metal-atom insertion and fragmentation [62, 90].

11.3 Effect of Ligands on Reactivity

In addition to the intracluster interaction and reaction, there are a few previously published studies that have demonstrated the effect of the ligand on the reactivities of the metal complexes [91–96]. Meanwhile, ongoing efforts are devoted to explore the reactivity and catalysis of ligand-protected metal clusters of the ligand always bring forth a significant influence in the practical experiments. It is believed that, to minimize the reactivity (maximize the stability) of a ligand-protected metal cluster, the metallic core should correspond to a spheroidal geometric structure and an even distribution of surface charges; meanwhile, it is preferred for the ligands to be located in a balanced position on opposite sides of the metallic core [97]. An example of the ligand effect on the reactivity of typical aluminum clusters and their iodides (c.a., $Al_{13}I_x^-$, $Al_{14}I_y^-$) was illustrated in unexpected ways, as shown in Fig. 11.6. The icosahedral core $Al_{13}I_x^-$ were found to do not react with methanol even in the presence of one unbalanced iodine ligand. This is because Al_{13}^- has a closed electronic shell with a high-lying LUMO which enacts an energy penalty when it binds to the lone pair electrons on the methanol molecule, giving rise to an exceptionally poor Lewis acid.

It is notable that, although the ground state $Al_{13}I_2^-$ is unreactive, a complementary active site could be induced if there are iodine atoms on adjacent aluminum atoms, resulting in likely reaction with methanol. In comparison, the $Al_{14}I_y^-$ clusters having an adatom-decorated core embody more clearly the effect of ligands on the metal cluster reactivity, as shown in Fig. 11.6b. DFT calculations found that, clusters with iodine bound to the adatom site are reactive with methanol, regardless of the site of the iodine ligands. Since the ligands can induce such complementary active pairs, these

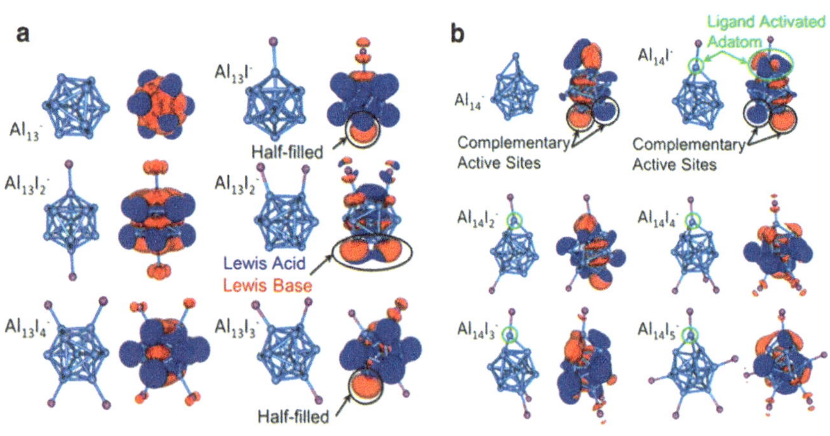

Fig. 11.6 **a** Geometric structures and frontier orbitals of Al_{13}^- and $Al_{13}I_x^-$, where a higher energy isomer $Al_{13}I_2^-$, with two neighboring iodine atoms on the same side of Al_{13}^- icosahedron, shows Lewis acid/base sites. **b** Geometric structures and frontier orbitals of Al_{14}^- and $Al_{14}I_x^-$, showing ligand-induced active sites

results explain the tendency of ligand-protected clusters toward compact metallic cores with closed geometric shells.

On the other hand, the gas-phase reactivity of metal clusters repeatedly suggested that, species having a non-uniform charge distribution enable to generate complementary active sites (i.e., both Lewis acid site and Lewis base site that could accept or donate electrons), which promotes the metal cluster reactions with polar molecules [98–106]. On this point, the chemical stability of a small metal cluster (although essentially determined by geometric and electronic structure) could be maximized when i) the cluster has a closed electronic shell that corresponds to a large HOMO-LUMO gap (c.a., >1 eV, along with likely large atomic binding energies, and/or electron removal/addition energies); and (ii) the charge density is evenly distributed on the cluster thus prohibiting the presence of active sites.

On this point, a joint experimental and theoretical study of aluminum iodides reacting with methanol showed that, the addition of odd/even number of iodine ligands to aluminum superatoms may activate or passivate the cluster (Fig. 11.7, Left). For example, $Al_{14}I_3^-$ was found to bear a closed electronic shell, and this cluster was not activated with respect to methanol as the three iodine adatoms has an at balanced positions on the metallic core. No surprise, $Al_{13}I_2^-$ and $Al_{13}I_4^-$ have an icosahedral 13-atom core and balanced ligands, and they were found to be passivated by the iodine ligands hence inert in methanol. In contrast, the iodine atoms of $Al_9I_3^-$ are located on unbalanced positions, enabling to induce active sites on the cluster. In brief, even if a cluster is protected by ligands, it still could be reactive especially when the geometric considerations are not met. Nevertheless, it is worth to point out that, one may need to treat with the ligand protection on balanced position in an objective and sensible light. For example, the unique gyro-like structure of Al_{13}^+ and cluster-π interaction induce uneven distribution of charges on the 13-atoms

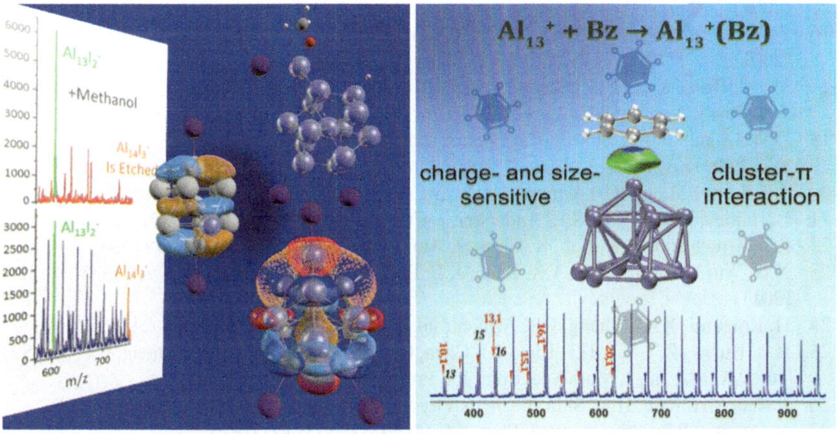

Fig. 11.7 (**Left**) The reactivity of $Al_nI_m^-$ clusters with MeOH showing the iodine passivated aluminum clusters. (**Right**) Cluster–π interactions cause altered reactivity of $Al_n^{\pm,0}$ clusters with benzene, with enhanced stability of Al_{13}^+Bz

cluster, enabling a strong electrostatic attraction and orbital interactions in Al_{13}^{+}Bz cluster with reasonable stability [107]. Also, proper ligand accommodation could causes anti-centrosymmetric structure of a ligand-protected metal clusters such as $[Au_{13}Cu_4(PPh_3)_4(SPy)_8]^+$ [108].

References

1. K.A. Gingerich, Faraday Symp. Chem. Soc. **14**, 109–125 (1980)
2. R.C. Dunbar, T.B. McMahon, D. Tholmann, D.S. Tonner, D.R. Salahub, D.Q. Wei, J. Am. Chem. Soc. **117**, 12819–12825 (1995)
3. K. Koyama, S. Kudoh, K. Miyajima, F. Mafune, Chem. Phys. Lett. **625**, 104–109 (2015)
4. V.M. Akulin, C. Brechignac, A. Sarfati, Phys. Rev. Lett. **75**, 220–223 (1995)
5. K.P. Remya, T. Udayabhaskararao, T. Pradeep, J. Phys. Chem. C **116**, 26019–26026 (2012)
6. T.T. Tsong, Surf. Sci. **50**, 621–625 (1975)
7. M.F. Jarrold, A.J. Illies, M.T. Bowers, J. Am. Chem. Soc. **107**, 7339–7344 (1985)
8. P.J. Brucat, L.S. Zheng, C.L. Pettiette, S. Yang, R.E. Smalley, J. Chem. Phys. **84**, 3078–3088 (1986)
9. M.E. Geusic, T.J. McIlrath, M.F. Jarrold, L.A. Bloomfield, R.R. Freeman, W.L. Brown, J. Chem. Phys. **84**, 2421–2422 (1986)
10. R.L. Hettich, B.S. Freiser, J. Am. Chem. Soc. **109**, 3537–3542 (1987)
11. S.H. Yoon, J.H. Moon, K.M. Choi, M.S. Kim, Rapid Commun. Mass Spectrom. **20**, 2201–2208 (2006)
12. F. Aguirre, J. Husband, C.J. Thompson, K.L. Stringer, R.B. Metz, J. Chem. Phys. **119**, 10194 (2003)
13. Y.-C. Zhao, Z.-G. Zhang, J.-Y. Yuan, H.-G. Xu, W.-J. Zheng, Chin. J. Chem. Phys. **22**, 655–662 (2009)
14. R.-Z. Li, H.-G. Xu, G.-J. Cao, Y.-C. Zhao, W.-J. Zheng, Chin. J. Chem. Phys. **24**, 572–579 (2011)
15. J.H. Moon, Y.S. Shin, H.J. Cha, M.S. Kim, Rapid Commun. Mass Spectrom. **21**, 359–368 (2007)
16. Z.W. Fu, L.M. Russon, M.D. Morse, P.B. Armentrout, Int. J. Mass Spectrom. **204**, 143–157 (2001)
17. M. McDonald, B.H. McGuyer, F. Apfelbeck, C.H. Lee, I. Majewska, R. Moszynski, T. Zelevinsky, Nature **534**, 122–126 (2016)
18. D.W. Chandler, P.L. Houston, J. Chem. Phys. **87**, 1445–1447 (1987)
19. S.K. Loh, D.A. Hales, L. Lian, P.B. Armentrout, J. Chem. Phys. **90**, 5466–5485 (1989)
20. S.K. Loh, L. Lian, P.B. Armentrout, J. Am. Chem. Soc. **111**, 3167–3176 (1989)
21. L. Hanley, S.A. Ruatta, S.L. Anderson, J. Chem. Phys. **87**, 260–268 (1987)
22. P.B. Armentrout, S.K. Loh, K. Ervin, J. Am. Chem. Soc. **106**, 1161–1163 (1984)
23. X.N. Wu, J.B. Ma, B. Xu, Y.X. Zhao, X.L. Ding, S.G. He, J. Phys. Chem. A **115**, 5238–5246 (2011)
24. H. Koizumi, X.-G. Zhang, P.B. Armentrout, J. Phys. Chem. A **105**, 2444–2452 (2001)
25. R.B. Huang, Z.Y. Liu, H.F. Liu, L.H. Chen, Q. Zhang, C.R. Wang, L.S. Zheng, F.Y. Liu, S.Q. Yu, X.X. Ma, Int. J. Mass Spectrom. Ion Processes **151**, 55–62 (1995)
26. J. Ree, Y.H. Kim, H.K. Shin, J. Chem. Phys. **124**, 074307 (2006)
27. N. Aristov, P.B. Armentrout, J. Phys. Chem. **90**, 5135–5140 (1986)
28. E.M. Moustafa, I. Ritacco, E. Sicilia, N. Russo, T. Shoeib, Phys. Chem. Chem. Phys. **17**, 12673–12682 (2015)
29. Z.Y. Liu, Z.C. Tang, R.B. Huang, Q. Zhang, L.S. Zheng, J. Phys. Chem. A **101**, 4019–4025 (1997)

30. Z. Parsons, C. Leavitt, T. Duong, G.S. Groenewold, G.L. Gresham, M.J. Van Stipdonk, J. Phys. Chem. A **110**, 11627–11635 (2006)
31. T. Su, M.T. Bowers, J. Am. Chem. Soc. **95**, 1370–1373 (1973)
32. A.K. Lam, R.A. O'Hair, Rapid Commun. Mass Spectrom. **24**, 1779–1790 (2010)
33. L. Capron, W.Y. Feng, C. Lifshitz, B.L. Tjelta, P.B. Armentrout, J. Phys. Chem. **100**, 16571–16576 (1996)
34. B.S. Fox, I. Balteanu, O.P. Balaj, H. Liu, M.K. Beyer, V.E. Bondybey, Phys. Chem. Chem. Phys. **4**, 2224–2228 (2002)
35. B. Xu, Y.-X. Zhao, X.-L. Ding, Q.-Y. Liu, S.-G. He, J. Phys. Chem. A **117**, 2961–2970 (2013)
36. L. Hanley, S.L. Anderson, Chem. Phys. Lett. **122**, 410–414 (1985)
37. V.A. Spasov, T.H. Lee, K.M. Ervin, J. Chem. Phys. **112**, 1713–1720 (2000)
38. H. Vach, A. Demartino, M. Benslimane, M. Chatelet, F. Pradere, J. Chem. Phys. **100**, 8526–8536 (1994)
39. V.N. Kondratyev, in *Similarities and Differences between Atomic Nuclei and Clusters: Toward a Unified Development of Cluster Science*, eds. Y. Abe, I. Arai, S. M. Lee and K. Yabana, Amer Inst Physics, Melville, pp. 447–450 (1998)
40. G. Luttgens, N. Pontius, P.S. Bechthold, M. Neeb, W. Eberhardt, Phys. Rev. Lett. **88**, 076102 (2002)
41. R.C. Dunbar, in *Gas Phase Ion Chemistry*, ed. M. T. Bowers, Academic Press, New York, vol. 3 (1984), ch. 20
42. R.L. Hettich, B.S. Freiser, J. Am. Chem. Soc. **108**, 2537–2540 (1986)
43. C.J. Cassady, B.S. Freiser, J. Am. Chem. Soc. **106**, 6176–6179 (1984)
44. R.L. Hettich, T.C. Jackson, E.M. Stanko, B.S. Freiser, J. Am. Chem. Soc. **108**, 5086 (1986)
45. A. Terasaki, J. Phys. Chem. A **111**, 7671–7689 (2007)
46. R.L. Hettich, B.S. Freiser, J. Am. Chem. Soc. **107**, 6222–6226 (1985)
47. K. Ervin, S.K. Loh, N. Aristov, P.B. Armentrout, J. Phys. Chem. **87**, 3593–3596 (1983)
48. D. Pavicic, LMU München (2004)
49. M. Brack, Rev. Mod. Phys. **65**, 677–732 (1993)
50. W.A. De Heer, Rev. Mod. Phys. **65**, 611–676 (1993)
51. T. Doppner, T. Fennel, P. Radcliffe, J. Tiggesbaumker, K.H. Meiwes-Broer, Eur. Phys. J. D **36**, 165–171 (2005)
52. P.G. Reinhard, E. Suraud, *Introduction to cluster dynamics* (Wiley, New York, 2003)
53. F. Calvayrac, P.G. Reinhard, E. Suraud, C.A. Ullrich, Physics Reports-Review Section of Physics Letters **337**, 493–578 (2000)
54. R.W. Farley, P.J. Ziemann, R.G. Keesee, A.W. Castleman, Jr., Z. Phys. D: At., Mol. Clusters, **25**, 267–273 (1993)
55. D.A. Card, D.E. Folmer, S. Sato, S.A. Buzza, A.W. Castleman Jr., J. Phys. Chem. A **101**, 3417–3423 (1997)
56. D.E. Folmer, L. Poth, E.S. Wisniewski, A.W. Castleman Jr., Chem. Phys. Lett. **287**, 1–7 (1998)
57. D.E. Folmer, E.S. Wisniewski, S.M. Hurley, A.W. Castleman Jr., Proc. Natl. Acad. Sci. U. S. A. **96**, 12980–12986 (1999)
58. A.W. Castleman, Jr., in *Physics and Chemistry of Clusters*, eds. E. E. B. Campbell and M. Larsson, vol. 117, pp. 143–168 (2001)
59. L. Poth, E.S. Wisniewski, A.W. Castleman Jr., Am. Sci. **90**, 342–349 (2002)
60. F. Fehrer, P.M. Dinh, M. Baer, P.G. Reinhard, E. Suraud, Eur. Phys. J. D **45**, 447–458 (2007)
61. J.F. Garvey, W.R. Peifer, M.T. Coolbaugh, Acc. Chem. Res. **24**, 48–54 (1991)
62. Y.-M. Koo, T.K. Kim, D.W. Jung, K.-W. Jung, J. Phys. Chem. A **110**, 13724–13730 (2006)
63. D.E. Clemmer, Y.M. Chen, N. Aristov, P.B. Armentrout, J. Phys. Chem. **98**, 7538–7544 (1994)
64. A.C. Harms, S.N. Khanna, A.B. Chen, A.W. Castleman Jr., J. Chem. Phys. **100**, 3540–3544 (1994)
65. D. Tholmann, D.S. Tonner, T.B. McMahon, J. Phys. Chem. **98**, 2002–2004 (1994)
66. C. Berg, T. Schindler, G. Niednerschatteburg, V.E. Bondybey, J. Chem. Phys. **102**, 4870–4884 (1995)

67. M. Sanekata, F. Misaizu, K. Fuke, S. Iwata, K. Hashimoto, J. Am. Chem. Soc. **117**, 747–754 (1995)
68. H. Watanabe, S. Iwata, K. Hashimoto, F. Misaizu, K. Fuke, J. Am. Chem. Soc. **117**, 755–763 (1995)
69. M. Beyer, C. Berg, H.W. Gorlitzer, T. Schindler, U. Achatz, G. Albert, G. NiednerSchatteburg, V.E. Bondybey, J. Am. Chem. Soc. **118**, 7386–7389 (1996)
70. M. Sanekata, F. Misaizu, K. Fuke, J. Chem. Phys. **104**, 9768–9778 (1996)
71. T. Schindler, C. Berg, G. NiednerSchatteburg, V.E. Bondybey, Chem. Phys. Lett. **250**, 301–308 (1996)
72. C. Berg, U. Achatz, M. Beyer, S. Joos, G. Albert, T. Schindler, G. Niedner-Schatteburg, V.E. Bondybey, Int. J. Mass Spectrom. **167**, 723–734 (1997)
73. U. Achatz, S. Joos, C. Berg, T. Schindler, M. Beyer, G. Albert, G. Niedner-Schatteburg, V.E. Bondybey, J. Am. Chem. Soc. **120**, 1876–1882 (1998)
74. C. Berg, M. Beyer, U. Achatz, S. Joos, G. Niedner-Schatteburg, V.E. Bondybey, Chem. Phys. **239**, 379–392 (1998)
75. U. Buck, C. Steinbach, J. Phys. Chem. A **102**, 7333–7336 (1998)
76. R.C. Dunbar, T.B. McMahon, Science **279**, 194–197 (1998)
77. M. Beyer, U. Achatz, C. Berg, S. Joos, G. Niedner-Schatteburg, V.E. Bondybey, J. Phys. Chem. A **103**, 671–678 (1999)
78. A. Irigoras, J.E. Fowler, J.M. Ugalde, J. Am. Chem. Soc. **121**, 574–580 (1999)
79. S.E. Rodriguez-Cruz, R.A. Jockusch, E.R. Williams, J. Am. Chem. Soc. **121**, 8898–8906 (1999)
80. D.C. Sperry, A.J. Midey, J.I. Lee, J. Qian, J.M. Farrar, J. Chem. Phys. **111**, 8469–8480 (1999)
81. B.S. Fox, M.K. Beyer, U. Achatz, S. Joos, G. Niedner-Schatteburg, V.E. Bondybey, J. Phys. Chem. A **104**, 1147–1151 (2000)
82. C.J. Mundy, J. Hutter, M. Parrinello, J. Am. Chem. Soc. **122**, 4837–4838 (2000)
83. M. Sena, J.M. Riveros, Chem. Eur. J. **6**, 785–793 (2000)
84. U. Achatz, B.S. Fox, M.K. Beyer, V.E. Bondybey, J. Am. Chem. Soc. **123**, 6151–6156 (2001)
85. B.S. Fox, M.K. Beyer, V.E. Bondybey, J. Phys. Chem. A **105**, 6386–6392 (2001)
86. J.J. Gilligan, A.W. Castleman Jr., J. Phys. Chem. A **105**, 5601–5605 (2001)
87. F. Mercuri, C.J. Mundy, M. Parrinello, J. Phys. Chem. A **105**, 8423–8427 (2001)
88. B.S. Fox, O.P. Balaj, I. Balteanu, M.K. Beyer, V.E. Bondybey, J. Am. Chem. Soc. **124**, 172–173 (2002)
89. M. Bhadra, J.Y.C. Lee, R.E. Cowley, S. Kim, M.A. Siegler, E.I. Solomon, K.D. Karlin, J. Am. Chem. Soc. **140**, 9042–9045 (2018)
90. D. Paul, K. Hong, T.K. Kim, J.-S. Oh, K.-W. Jung, Int. J. Mass Spectrom. **315**, 15–21 (2012)
91. H. Shao, Y. Wang, C.W. Bielawski, P. Liu, Acs Catal. **10**, 3820–3827 (2020)
92. Dv. Sokolski, Y.A. Dorfman, Rakitska. Tl, *Doklady Akademii Nauk Sssr*, **201**, 123 (1971)
93. B.C. Guo, A.W. Castleman, Int. J. Mass Spectrom. Ion Processes **113**, R1–R5 (1992)
94. V.N. Setkina, N.K. Baranets, A.G. Ginzburg, V. Zdanovic, M.N. Nefedova, D.N. Kursanov, J. Organomet. Chem. **61**, 287–300 (1973)
95. M. Anbar, D. Meyerste, *Transactions of the Faraday Society*, **65**, 1812 (1969)
96. D. Meyerstein, W.A. Mulac, *Transactions of the Faraday Society* **65**, 1818 (1969)
97. H. Yang, Y. Wang, J. Lei, L. Shi, X. Wu, V. Maekinen, S. Lin, Z. Tang, J. He, H. Haekkinen, L. Zheng, N. Zheng, J. Am. Chem. Soc. **135**, 9568–9571 (2013)
98. K. Shimamura, F. Shimojo, R.K. Kalia, A. Nakano, K.-I. Nomura, P. Vashishta, Nano Lett. **14**, 4090–4096 (2014)
99. P.J. Roach, W.H. Woodward, A.W. Castleman Jr., A.C. Reber, S.N. Khanna, Science **323**, 492–495 (2009)
100. A.C. Reber, S.N. Khanna, P.J. Roach, W.H. Woodward, A.W. Castleman Jr., J. Phys. Chem. A **114**, 6071–6081 (2010)
101. A.C. Reber, P.J. Roach, W.H. Woodward, S.N. Khanna, A.W. Castleman Jr., J. Phys. Chem. A **116**, 8085–8091 (2012)

102. M.B. Abreu, C. Powell, A.C. Reber, S.N. Khanna, J. Am. Chem. Soc. **134**, 20507–20512 (2012)
103. W.H. Woodward, A.C. Reber, J.C. Smith, S.N. Khanna, A.W. Castleman Jr., J. Phys. Chem. C **117**, 7445–7450 (2013)
104. Z. Luo, J.C. Smith, C. Berkdemir, A.W. Castleman Jr., Chem. Phys. Lett. **590**, 63–68 (2013)
105. D.M. Cox, D.J. Trevor, R.L. Whetten, A. Kaldor, J. Phys. Chem. **92**, 421–429 (1988)
106. F. Shimojo, S. Ohmura, R.K. Kalia, A. Nakano, P. Vashishta, Phys. Rev. Lett. **104**, 126102 (2010)
107. M. Yang, H. Zhang, Y. Jia, B. Yin, Z. Luo, J. Phys. Chem. A **124**, 4087–4094 (2020)
108. R. Anumula, A.C. Reber, P. An, C. Cui, M. Guo, H. Wu, Z. Luo, S.N. Khanna, Nanoscale **12**, 14801–14807 (2020)

Chapter 12
Charge Transfer and the Harpoon Mechanism

12.1 Introduction

An important concept in the interaction of metals with nonmetal atoms/molecules is that of charge transfer (CT). The charge-transfer reactivity is extensively investigated and mostly seen for an ion/neutral species reaction in which the total charge on the reactant ion is transferred initially to the reactant neutral species so that the reactant ion becomes a neutral entity [1–38] Some of the possible reactions of ions M^{2+}, M^+ and M^- with a neutral species Y are categorized in terms of the above definitions as follows [39]:

$$M^{2+} + Y \rightarrow M^+ + Y^+ \text{(Partial charge transfer)} \tag{12.1}$$

$$M^+ + Y \rightarrow M^{2+} + Y + e^- \text{(Charge stripping)} \tag{12.2}$$

$$M^- + Y \rightarrow M^+ + Y + 2e^- \text{(Charge stripping \& charge inversion)} \tag{12.3}$$

These are all ion/neutral species reactions and charge permutation reactions. Besides, transient CT states frequently show up when two gas-phase species collide with each other; the CT process may lead to chemical reactions when there is a suitable difference in the electron affinity and ionization potential of the colliding species. Such reactivity is examined in terms of harpoon-type reactions between alkali-metal atoms and halogen molecules, where the harpoon mechanism is demonstrated to allow for a largely increased reactive cross section due to the valence electron transfer from the alkali metal atom (M) to the halogen molecule (X_2),

$$M + X_2 \rightarrow \left(M^+ + X_2^- \rightarrow\right) MX + X \text{ (Harpoon mechanism)} \tag{12.4}$$

The Sect. 12.3 is partly reproduced from *Chem. Phys. Lett.*, 2013, 590, 63–68.

Z. Luo and S. N. Khanna, *Metal Clusters and Their Reactivity*,
https://doi.org/10.1007/978-981-15-9704-6_12

Here X generally refers to halogen, while M aims at alkali metals. Researches on this interesting topic retrospect to 1986 when Dudley R. Herschbach, Yuan T. Lee, and John C. Polanyi received the Nobel Prize in Chemistry for elucidating the collision dynamics of elementary chemical reactions [40]. Their research has been of great importance for the development of a new field of research in chemistry (i.e., reaction dynamics) and has provided a much more detailed understanding of how chemical reactions occur, including the discovery of harpoon mechanism. In a very recent study, Castleman considered the validity of the harpoon mechanism in gas-phase cluster reactivity of coinage metal clusters [41]. Interesting reactions of copper and silver cluster anions toward chlorine were observed and the harpoon mechanism is identified for "$[Cu_8]^-/[Ag_8]^- + Cl_2$". This finding revealed the harpoon mechanism in cluster reactivity which remains a subject of increasing interest and activity as a bridge in probing atoms and macroscopic matter.

12.2 Charge-Transfer Reactions of Clusters

Charge transfer reactions between ionic and neutral species are basic processes in physics and chemistry. In a simple case of ion–atom collisions, the development of technology has largely improved the experimental approaches to control beams of atoms and atomic ions with much ease and over an increasing range of impact energies [42]. On the other hand, the theoretical demonstration for the elementary processes has reached a high level of accuracy and enables predictions in the cases where charge transfer is difficult to be investigated by experimental method. Further progress has also been made for molecule/cluster targets with the advances in the controllable production of molecular beams and clusters [43–45]. In particular, charge transfer involving metal clusters has attracted reasonable research interest in the late decades [46–51].

As an electron jump is often the first step of a chemical reaction, the direct observation of an electron transfer from an atom/cluster to a molecule/particle is of primary interest in surface science and catalysis. A fundamental question is for which cluster size the colliding partner interacts either with the whole cluster or only a part of it; and from which cluster size does the interaction resemble an atom-surface interaction? Insights into the evolution of charge transfer properties between clusters and molecules/atoms as a function of particle size is very important to address these questions. By collisional neutralization of mass selected cluster ions [52–55], researchers have been able to estimate cluster ionization potentials through experimental data [56], which concerns the availability of electron transfer to form the correlative neutral clusters [50].

Early work by Bréchignac et al. [46] reported a study of charge exchange between mass-selected sodium cluster ions Na_n^+ and Cs atoms,

$$Na_n^+ + Cs \rightarrow Na_n + Cs^+, \ (1 \leq n \leq 21) \tag{12.5}$$

In this case, they measured the cross section for this process to lie between 40 to 10 Å^2. The reionization of the neutral products Na_n obtained from "$Na_n^+ + Cs$" collisions showed that most of them do not undergo any fragmentation [46], suggesting that charge transfer is an efficient way to produce the mass selected neutral species.

Figure 12.1b displays a typical spectrum for Na_9^+ after redispersion but without reionization, where the peaks correspond to surviving parent ion (Na_9^+), fragment

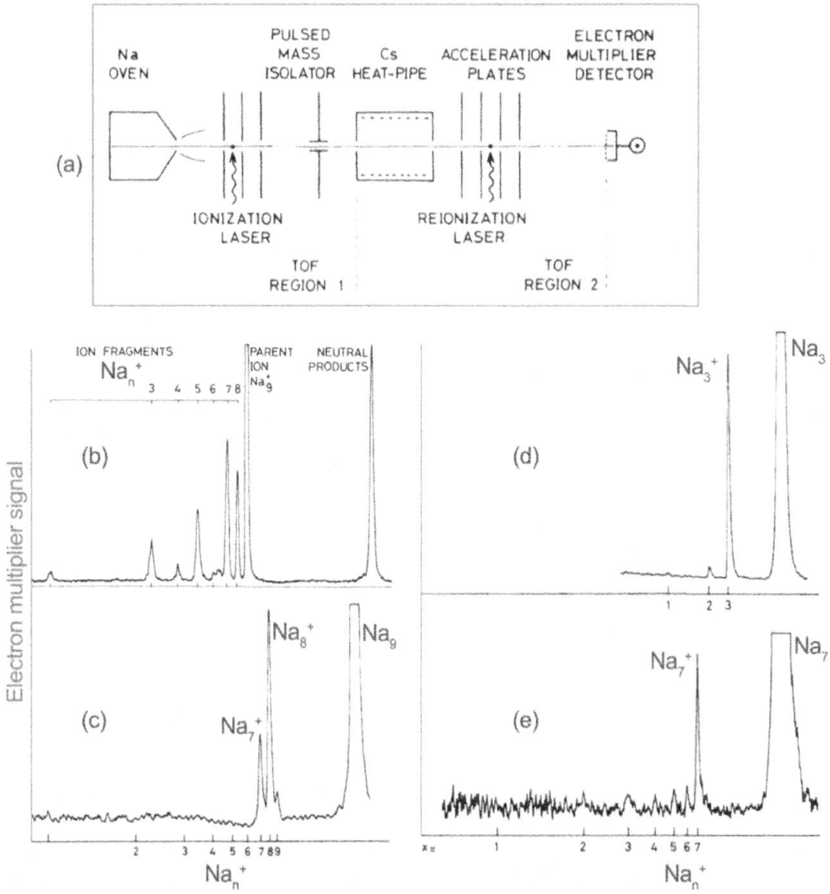

Fig. 12.1 **a** Schematic diagram showing the essential elements of the apparatus in studying the charge transfer between Na_n and Cs. The first two acceleration plates serve as the angle-limiting apertures. **b** Typical mass spectrum of Na_9 showing the essential features of the collision and dissociation processes. Note that the neutral-product peak is detected without reionization and the fragment ions are dispersed in the acceleration region after the heat pipe. **c–e** The reionization mass spectra of the neutral charge-transfer products from Na_9^+, Na_3^+ and Na_7^+ parent ions. The large peaks on the right-hand side correspond to the neutral products and the sharp peaks correspond to the narrow distribution of the ionized neutrals. Reproduced with permission from Ref. [46]. Copyright 1988 American Physical Society

ion arising from collision-induced dissociation (Na^+_{1-8}), and fast neutrals arising principally from the charge-transfer process. In contrast, Fig. 12.1c shows a series of reionization spectra for Na_9, Na_3, and Na_7, respectively. The spectra of Na_3 and Na_7 showed mostly the parent, evidently due to charge exchange. In contrast, for Na_9 the reionization spectrum displays $Na_8{}^+$ and $Na_7{}^+$ dominating the mass spectrum, indicating that an evaporation of a monomer is associated with the charge transfer, expressed as [46]

$$Na^+_n + Cs \rightarrow Na_{n-p} + Na_p + Cs^+, \ (1 \le n \le 21, \ p < n) \quad (12.6)$$

Higher masses of Na^+_n (e.g., Na_{21}) were also noted to exhibit evaporation of a single atom in the re-ionization spectrum. Similar experiment on "K^+_n + Cs" was also observed. Measurements on Na^+_n and K^+_n charge transfer cross-sections with cesium atoms have been interpreted using the Rapp and Francis formalism [47].

Extensive investigations regarding the charge transfer and fragmentation in collisions of metal clusters have been undertaken in several other groups [50, 57–95]. The charge transfer and fragmentation in collisions of alkali metals attracted major interest, either theoretically through a microscopic framework called nonadiabatic quantum molecular dynamics, or experimentally through mass spectrometry. Figure 12.2 displays such a study on Li^+_n clusters. Alike to the above, clusters were produced as an evaporative ensemble containing some internal energy; thus, they partially undergo uni-molecular dissociation during their propagation in the TOF. Under such experimental conditions, the dissociation is dominated by evaporation of a neutral monomer, hence a pathway as Eq. 12.7.

$$Li^{2+}_{31} \rightarrow Li^{2+}_{30} + Li \quad (12.7)$$

Apparently, the dissociation ratio $Li_{30}{}^{2+}/Li_{31}{}^{2+}$ depends on the two time windows of the experiment: the residence time in the accelerating region (t_1) and the propagation time in the drift tube of the TOF (t_2), as shown in Fig. 12.2A. Products of the dissociation process propagate in the first drift tube with the center-of-mass velocity of the parent; and then they are spatially resolved into individual mass packets in the second drift tube with dependence on the V_R. When the collision cell is activated (i.e., the Cs pressure is larger than zero), the structure displayed by the retarding field images the combined effects of dissociation and charge transfer. By comparing the spectra obtained cell-on and cell-off, the signals that are exclusively due to CT can be identified.

Figure 12.2B presents three mass spectra for $Li_{31}{}^{2+}$ corresponding to the V_R values at 1250 and 2500 V, compared with the case of no collisions and no electrostatic analysis (i.e., $V_R = 0$ V) respectively. Charge transfer results in a neutral Li_{31} cluster which still gives chance to evaporate one and two Li atoms leading to smaller neutral species Li_{30} and Li_{29} (seen as cationic $Li_{30}{}^+$ and $Li_{29}{}^+$ after the reionization). Based on the experimental observations, the following channels are demonstrated: (i) CT occurring for both the parent and its dissociation product,

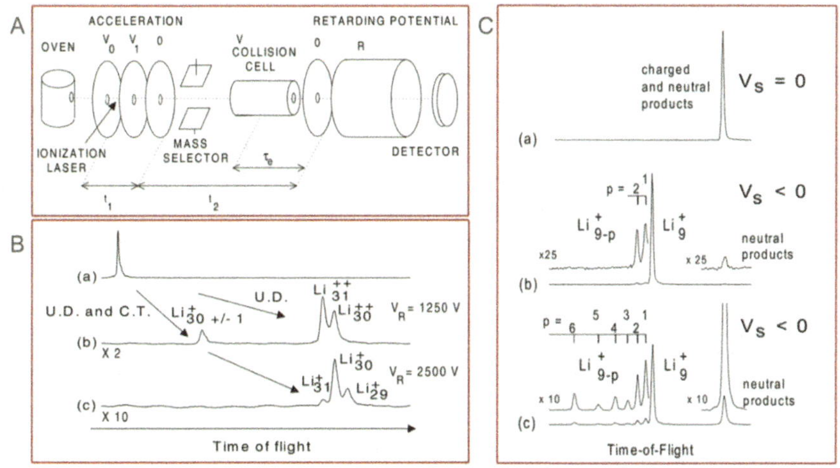

Fig. 12.2 **A** Experimental setup, where the $Li_{31}{}^{2+}$ clusters were produced and mass selected, and then interacted with a vapor. The collision products were charge and mass analyzed at t_2 by TOF spectrometry. **B** TOF mass spectrum for $Li_{31}{}^{2+}$ in case of no collisions and no electrostatic analysis (**a**); and then TOF charge and mass analysis of the collision products at t_2 for $V_R = 1250$ V (b) and 2500 V (c) respectively. **(C)** Time-of-flight spectra obtained after "$Li_9{}^+$ + Cs" collisions at an energy of 5000 eV. Trace (**a**): no electro-static dispersion, $V_s = 0$. Trace (b): $\sigma_{C:T}\, n_A l \leq 1$, no collisions. The charged fragments coming from dissociation are mass dispersed ($V_s < 0$). Trace (c): single collision regime, and dispersion of the charged collision fragments ($V_s < 0$). The light charged fragments come from C.I.D. Reproduced with permission from Ref. [50]. Copyright 2000 Springer Nature

$$Li_{31}^{2+} + Cs \rightarrow Li_{31}^{2+} + Cs^+;\ Li_{30}^{2+} + Cs \rightarrow Li_{30}^{2+} + Cs^+ \tag{12.8}$$

(ii) evaporation of excited singly charged products,

$$Li_{31}^{+} \rightarrow Li_{30}^{+} + Li \rightarrow Li_{29}^{+} + Li_2;\ Li_{30}^{+} \rightarrow Li_{29}^{+} + Li \tag{12.9}$$

Similar experimental results on "$Li_9{}^+$ + Cs" have also been examined at a laboratory energy of 5000 eV, as displayed in Fig. 12.2C [50]. These procedures allowed to measure the signal of neutral clusters produced from mass selected M_n^+ cluster parents, arising from three different physical processes:

(a) $M_n^+ \rightarrow M_{n-1}^+ + M;\ M_n^+ \rightarrow M_{n-2}^+ + M_2$, Unimolecular Decay (U.D.);

(b) $M_n^+ + A \rightarrow M_{n-q1-2q2}^+ + q_1 M + q_2 M_2 + A$, Collision Induced Dissociation (C.I.D.);

(c) $M_n^+ + A \rightarrow M_n + A^+$, Cluster neutralization by Charge Transfer (C.T.). (12.10)

Assuming the CT cross sections for medium-size singly charged metal clusters barely depend on cluster size, one can deduce the absolute value of the CT cross section from Beer's law [50]. For example, Bréchignac et al. [50]. measured the cross-sections for collisional charge transfer between singly charged free clusters M_n (M = Li, Na; n = 1...50) and atomic targets A (cesium, potassium) as a function of collisional relative velocity in an energy range of 1–10 keV. For each cluster size, the experimental values of the charge transfer cross-section $\sigma(\upsilon)$ were fitted with a universal parametric curve with two independent parameters, i.e., the maximum cross-section (σ_m) and the corresponding velocity (υ_m). For small size clusters (n ≤ 15), the $\sigma(\upsilon)$ characteristic parameters for the "M_n^+ + A" showed strong variations with the number of atoms in the cluster, as shown in Fig. 12.3. It was demonstrated

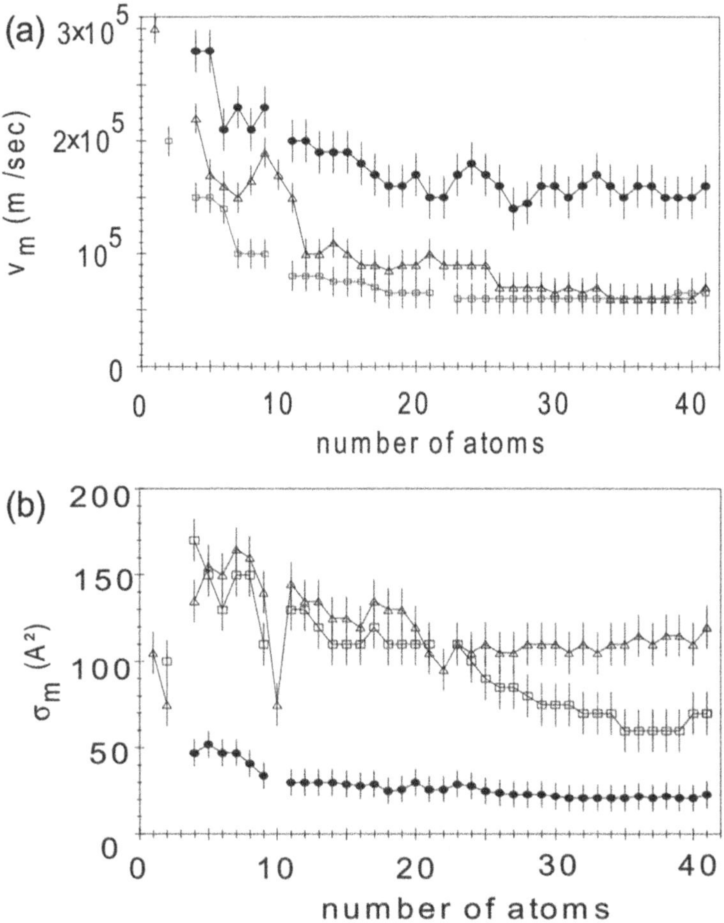

Fig. 12.3 **a** Representation of the velocity corresponding to the maximum charge transfer cross-section (υ_m), obtained by a fit of the data, as a function of cluster atom number for Na_n^+ + Cs (□), Li_9^+ + Cs (Δ) and Li_9^+ + K (•) collisional systems. **b** Representation of the absolute C.T. cross-section at the maximum of the velocity profile (σ_m), as a function of cluster atom number for Na_n^+ + Cs (□), Li_9^+ + Cs (Δ) and Li_9^+ + K (•) collisional systems. Reproduced with permission from Ref. [50]. Copyright 2000 Springer Nature

that the charge transfer patterns observed for various collisional systems present similarities, which appear more sensitive to cluster quantum size effects than to collision energy defects. The σ_m and υ_m parameters showed differences in both their size evolution, and their absolute values varied in terms of projectile and target electronic structures [50, 68].

It is notable that integral and exclusive charge-transfer cross sections can be understood only if all types of fragmentation processes including statistical decay are considered. Moreover, the influence of the cluster structure (isomers, temperature, size) on measured and measurable cross sections is associated with the different charge transfer channels, as well as fragmentation aforementioned in Eq. 12.10 [96].

12.3 The Harpoon Mechanism[1]

The harpoon mechanism is properly described in the reactivity of halogens with alkali metal atoms. One current question is whether the harpoon mechanism can account for the reactive behavior of microscopic charged systems. In order to examine this question, a here-to-fore issue is to study the operative mechanism for the reactivity of coinage metal clusters and chlorine in the gas phase. Gas-phase collision theory provides a first principles approach that accounts for the reaction rate, ν, based on collisions between two species [97]:

$$\nu\,\alpha\,\sigma\sqrt{\frac{8\mathrm{k_B}}{\pi\mu}}\cdot N_A\exp\left(-\frac{\mathrm{E_s}}{\mathrm{RT}}\right)\cdot[\mathrm{A}][\mathrm{B}] \tag{12.11}$$

This equation displays an Arrhenius-like form, where σ is the collisional cross section; $\mathrm{k_B}$ is Boltzmann's constant; T is the temperature; μ is the reduced mass $\mu = (\mathrm{m_A} + \mathrm{m_B})/(\mathrm{m_A m_B})$; $\mathrm{N_A}$ is Avogadro's constant; [A] and [B] are the concentrations of the two species; and the exponential portion refers to a factor associated with the activation energy (Ea) which is the minimum kinetic energy needed for a successful reaction, where R is the universal gas constant. As a simpler expression, Eq. 12.11 can also be written as $\mu = k[\mathrm{A}][\mathrm{B}]$ indicating that the rate of reaction is proportional to the reactant concentrations. Note that the experimental value ν is generally smaller than that calculated from the kinetic theory because not only must the molecules collide with enough kinetic energy but they also must come together in a specific relative orientation to activate the reaction. Therefore, a steric factor, P, should be included with a range "$0 \leq P \leq 1$" where the two limits indicate that either none or all the relative orientations lead to a reaction. Therefore the new rate constant should follow the form [97]:

[1]This section is partly reproduced from Chem. *Phys. Lett.*, 2013, 590, 63–68.

$$k = \sigma^* \sqrt{\frac{8k_B T}{\pi \mu}} \cdot N_A \exp\left(-\frac{E_a}{RT}\right) \quad (12.12)$$

where $\sigma^* = p\sigma$ comparing with Eq. 12.11. For many cases, very low P values are observed, indicating that the reactions have stringent orientation requirements. However, some reactions are found to have a $P > 1$. A well-known example is the reaction of K atoms with Br_2 where the K atom plucks a Br atom out of the Br_2 molecule [98, 99]. In this reaction, an electron leaps from the metal atom (i.e., a harpoon) to the halogen, which results in a coulomb attraction between the metal and halogen which extends the cross section for their reactive encounter [99–102], resulting in a $p = \frac{\sigma^*}{\sigma} > 1$. In other words, the distance r^* at which the reaction can successfully occur is larger than the distance (r) needed for reactants in a non-reactive collision. This surprising conclusion has coined the harpoon mechanism [63, 103]. In addition to the reactivity of alkali metals with halogens, the harpoon mechanism helped explain the reactivity of lanthanide cations with fluorocarbons [104, 105], as well as hydrogen with electronically excited alkali-metal atoms (and alkaline earth metal atoms) which has attracted significant research interest in view of the advantages provided by nonadiabatic processes in chemical reactions [69, 106–108].

Clusters are known as a bridge between atomic and macroscopic matter. Understanding cluster reactivity can help develop tunable materials with possible catalytic or energetic qualities [109–112]. In view of the atomic electron configurations of Cu:[Ar]$3d^{10}4s^1$ and Ag:[Kr]$4d^{10}5s^1$, it is likely that $[Cu_8]^-$ and $[Ag_8]^-$ will behave similarly to an alkali-metal atom, hence their reactivity towards chlorine may follow the harpoon model with products of "$[Cu_8Cl]^-$ +Cl" and "$[Ag_8Cl]^-$ + Cl".

The reaction apparatus leading to these findings was based on an instrument that has been previously described [112]. The magnetron sputtering source (MagS-source) [113] ensures a better cluster yield and a tunable distribution of the silver and copper cluster anions. A DC power supply (TDK-Lambda Americas Inc., GENESYS™ 750 W/1500 W) was used to provide the high voltages needed for the magnetron sputtering. The silver and copper disks (99.99% pure, Φ50-mm, and 6-mm thickness) were obtained from Kurt J. Lesker Company. Ultrahigh purity Argon (Praxair, Inc., purity > 99.99%) was used as the working gas for the sputtering, while high purity helium (Praxair, Inc., purity > 99.995%) was presented at the rear of the magnetron chamber to carry the clusters through an adjustable iris (nozzle) into a laminar flow tube where they encountered the reactant gas Cl_2. The cluster species were extracted into a differentially pumped ion guide vacuum system and analyzed by a quadrupole mass spectrometer (Extrel CMS).

All calculations were carried out using density functional theory (DFT) at the B3LYP/LanL2DZ level of theory [114, 115] as implemented in Gaussian 03 package [116]. The models were configured by GaussView software (Version 4.1) with optimized geometry. The ground states of all the $[Cu_n]^-$/$[Ag_n]^-$ clusters display a spin multiplicity of singlet for even-electron clusters and doublet for odd-electron systems.

Figure 12.4a presents a mass spectrum of copper cluster anions after exposure to chlorine gas, where the dominant products are assigned to $[Cu_nCl_{n+1}]^-$ (n = 1–6) species which have been demonstrated as a starting point in the formation of ionic crystals [117]. Besides these $[Cu_nCl_{n+1}]^-$ products, it is important to note that $[Cu_8Cl]^-$ appears as a distinctive peak among the reaction products, in fact the only peak that belongs to the $[Cu_nCl]^-$ series. Figure 12.4b shows the reactivity of silver cluster anions with chlorine, where three classes of reaction products are observed: (i) $[Ag_nCl_{n+1}]^-$, (ii) $[Ag_nCl_2]^-$ and (iii) $[Ag_nCl]^-$. The $[Ag_nCl_{n+1}]^-$ species were observed to only in the small mass range ($n \leq 4$), seen as $[AgCl_2]^-$, $[Ag_2Cl_3]^-$, $[Ag_3Cl_4]^-$, and $[Ag_4Cl_5]^-$, which were inferred to react based on the same mechanism as in the formation of the $[Cu_nCl_{n+1}]^-$ series. When a $[Ag_n]^-$ cluster collides with a Cl_2 molecule, the first-step reaction is expected to follow one of the following channels:

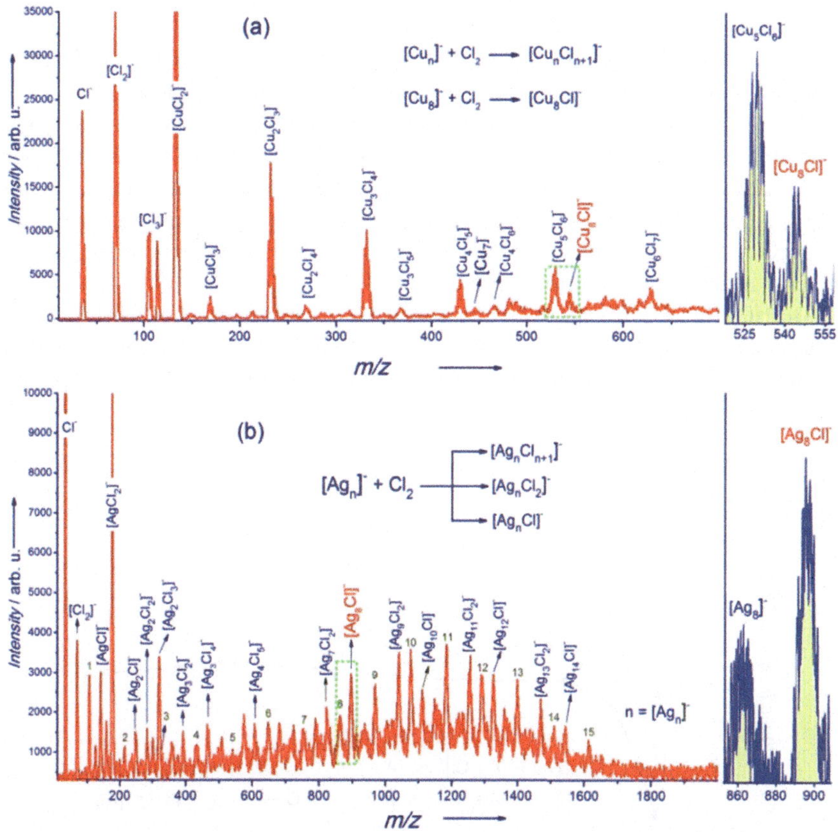

Fig. 12.4 The reaction of $[Cu_n]^-$ and $[Ag_n]^-$ clusters with Cl_2. **a** A representative mass spectrum showing the reaction product distribution of $[Cu_n]^-$ with Cl_2 (5.2-sccm flow rate); **b** mass distribution of $[Ag_n]^-$ after reacting with Cl_2 (1.2-sccm flow rate). The peaks of $[Cu_8Cl]^-$ and $[Ag_8Cl]^-$ are enlarged on the right of a and b, respectively

$$[Ag_n]^- + Cl_2 \rightarrow Ag_{n-1} + [AgCl_2]^- \rightarrow \quad (12.13)$$

$$[Ag_n]^- + Cl_2 \rightarrow [Ag_nCl_2]^- \rightarrow \quad (12.14)$$

$$[Ag_n]^- + Cl_2 \rightarrow AgCl + [Ag_{n-1}Cl]^- \rightarrow \quad (12.15)$$

where Eq. 12.13 is responsible for the dominant product $[AgCl_2]^-$, while Eqs. 12.14 and 12.15 show the likely pathways in forming $[Ag_nCl_2]^-$ and $[Ag_nCl]^-$, respectively; also, an additional arrow in each equation indicates possible successive reactions. Note that the $[Ag_nCl]^-$ and $[Ag_nCl_2]^-$ species appearing in the larger mass range ($8 \leq n \leq 14$) display an odd-even alternation (Fig. 12.4b). This agrees with previous theoretical findings that the calculated incremental binding energies, spin excitation energies, and HOMO-LUMO gaps of $[Ag_n]^-$ clusters all display an even/odd oscillation, which corresponds to their even/odd selective reactivity [118]. As seen in Fig. 12.4b, the intensity ratio of $[Ag_8Cl]^-$ to $[Ag_8]^-$ is larger than that of the other observed $[Ag_nCl]^-$ clusters and their correlated $[Ag_n]^-$ product clusters when $8 \leq n \leq 14$. In order to better demonstrate this observation, Fig. 12.5 displays the logarithmic intensity ratio between the $[Ag_nCl]^-$ ($n = 8, 10, 12, 14$) clusters and their $[Ag_n]^-$ product counterparts with respect to chlorine flow rates. These curves do not follow a linear or exponential function; in particular, the logarithmic intensity ratio of $[Ag_8Cl]^-$ vs. $[Ag_8]^-$ shows an obvious difference from the others (the values are larger than zero when chlorine is present) [41].

The HOMO-LUMO gaps of the $[Ag_n]^-$ and $[Cu_n]^-$ clusters are shown in Fig. 12.6a, c. In general, HOMO-LUMO gaps are associated with the ability of electron gain/loss and help predict cluster reactivity; however, the HOMO-LUMO gaps of $[Ag_8]^-$ and

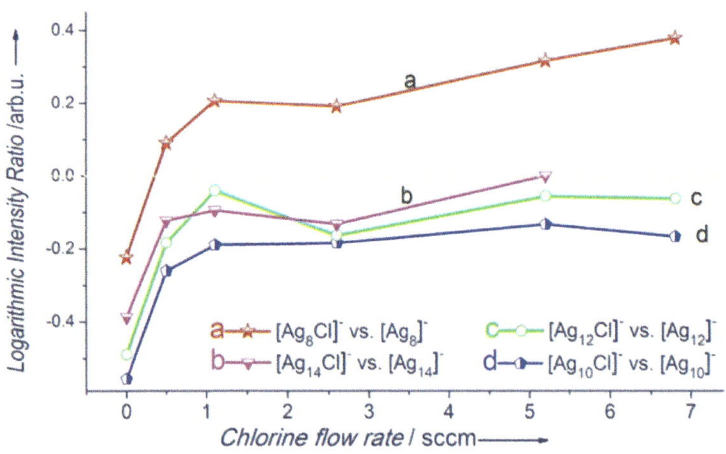

Fig. 12.5 The logarithmic intensity ratio of the product peaks $[Ag_nCl]^-$ versus $[Ag_n]^-$ at different flow rates of Cl_2. The points represent the data and the lines are drawn only for a connection

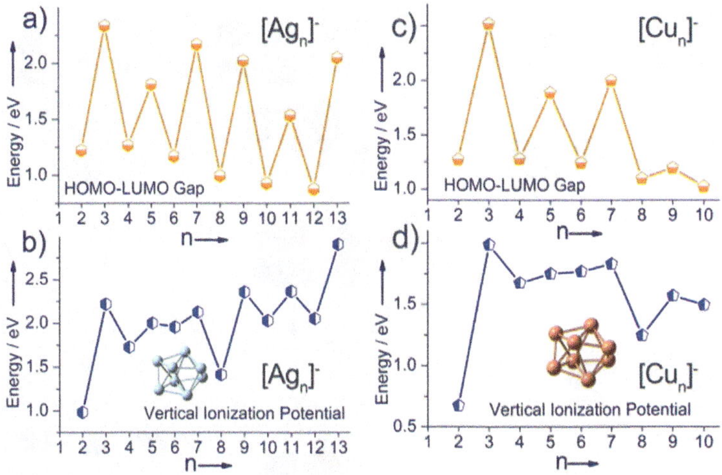

Fig. 12.6 HOMO-LUMO gaps and vertical ionization potentials (VIP) of silver (**a/b**) and copper (**c/d**) clusters. The VIP of $[Cu_n]^-$ ($2 \leq n \leq 10$) and $[Ag_n]^-$ ($2 \leq n \leq 13$) were calculated using density functional theory (DFT) at the B3LYP/LanL2DZ level of theory as implemented in Gaussian. The inset images showing the optimized structures of $[Ag_8]^-$ and $[Cu_8]^-$

$[Cu_8]^-$ are not distinct in any way. For both Ag and Cu cluster anions, the vertical ionization potentials (VIPs) were defined as VIP = E_{total} ($[Ag_n]^-$)−E_{total} ($[Ag_n]^0$), (E_{total} is the total energy of the optimized species including zero-point vibrational energies), and found that both $[Ag_8]^-$ and $[Cu_8]^-$ exhibit small VIP values compared to their adjacent clusters (Fig. 12.6b, d). Small VIP values of $[Cu_8]^-$ and $[Ag_8]^-$ allow them to readily donate electrons to chlorine. Note that the clusters $[Ag_2]^-$ and $[Cu_2]^-$ also display small VIP values as the 3 valence electrons imply a 2-electron closed shell plus an additional unpaired electron. In comparison, the $[Cu_8]^-$ and $[Ag_8]^-$ clusters exhibit 9 valence electrons and the delocalized nearly-free-electron-gas (NFEG) orbitals are best described as $|1S^2 \mid 1P^2 \mid 1P^4 \mid 2S^1 \,|$.

Figure 12.7 compares the calculated orbitals of the anionic $[Ag_8]^-$ and $[Cu_8]^-$ clusters with that of a potassium atom, indicating the similarities between the clusters and the atom in orbital shape and electron occupation. It is very likely that the reactivity of chlorine with $[Cu_8]^-$ and $[Ag_8]^-$ clusters is akin to that of alkali metals with halogens through a harpoon mechanism as mentioned above. This reaction begins with an electron transfer (from $[Ag_8]^-/[Cu_8]^-$ to Cl_2) which creates a force with the electron acting as a harpoon, bringing the reagents together until the newly formed chlorine molecule ($[Cl_2]^-$) rapidly dissociates, ejecting a Cl atom and forming the product $[Ag_8Cl]^-/[Cu_8Cl]^-$, expressed as,

$$[Ag_8]^-(/[Cu_8]^-) + Cl_2 \rightarrow [Ag_8Cl]^-(/[Cu_8Cl]^-) + Cl \tag{12.16}$$

To estimate the value of P for the harpoon-like reaction in Eq. 12.15, a simple model for the harpoon mechanism has been established by calculating the distance at

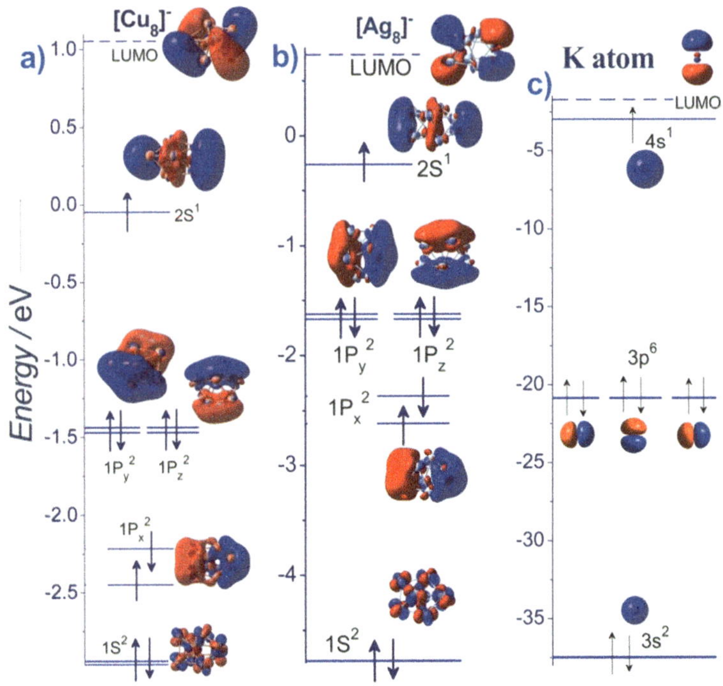

Fig. 12.7 The calculated orbital pattern of the anionic clusters $[Cu_8]^-$ (**a**) and $[Ag_8]^-$ (**b**), compared with that of a K atom (**c**)

which it becomes energetically favorable for the electron to leap from the metal atom (cluster) to the halogen molecule. There are three contributions to the energy involved in Eq. 12.15, the ionization energy of $[Cu_8]^-$/$[Ag_8]^-$, the electron affinity (EA) of a halogen molecule, and the Columbic interaction potential between $[Cu_8]^-$/$[Ag_8]^-$ and Cl_2. The energy difference between the ionization potential of an alkali-metal atom and the electron affinity (EA) of a halogen molecule is matched by the columbic attraction between the two [99, 103]. The only difference between the present cluster systems and the traditional alkali-halide reaction is that, when $[Cu_8]^-$ or $[Ag_8]^-$ donates an electron to Cl_2, it becomes a neutral cluster (i.e., $[Cu_8]^0$ or $[Ag_8]^0$) interacting with a charged molecule (i.e., $[Cl_2]^-$). As a charged particle/molecule "a" can induce a dipole moment in the neutral molecule/cluster "b", a simple energy expression can be formed for the harpoon model between $[Cu_8]^-$/$[Ag_8]^-$ and Cl_2.

$$\mathrm{VIP} - \mathrm{EA} - \varphi_{\mathrm{ab}}^{(\mathrm{C,ind}\,\mu)} = 0 \tag{12.17}$$

Here VIP refers to the vertical ionization potential of the electron donor ($[Cu_8]^-$/$[Ag_8]^-$); EA is the electron affinity of the electron acceptor (Cl_2); while $\varphi_{ab}^{(C,\mathrm{ind}\,\mu)}$ represents the potential energy between the charge and the induced dipole moment which, according to the basic principles of electrodynamics [119], given as:

$$\varphi_{\text{ab}}^{(C,\text{ind}\mu)} = -\frac{C_a^2 \cdot ab}{4\pi \varepsilon_0 r^4} \tag{12.18}$$

where C_a is the charge of particle/molecule "a", while α_b is the polarizability of the neutral molecule/cluster "b"; and r refers to the distance between the two components. Consideration of the interaction potential in Eq. 12.17 neglects the centrifugal barrier and translational energy dependence which is involved in the Langevin-Gioumousis-Stevenson reaction cross section for ion-dipole interactions, and it was not considered because it does not significantly affect the results.

Based on Eqs. 12.17 and 12.18, considering the values 2.38 eV (EA of Cl_2) [120], 1.42 eV (VIP of $[Ag_8]^-$) and 337.25 au (α of $[Ag_8]^{0}$; au, atomic unit) from the calculation results, one attains $r_1^* = 5.22\text{Å}$ and a cross section $\sigma_1^* = \pi r_1^{*2} = 85.56\text{Å}^2$ for the system "$[Ag_8]^- + Cl_2$". Consequently, the steric factor is estimated as $P_1 = \frac{\sigma_1^*}{\sigma_2} = 1.09$ in this reaction. Similarly, using the calculated values: VIP ($[Cu_8]^-$) = 1.25 eV, and $\alpha\left(Cu_8^0\right) = 265.60\text{au}$, we have also evaluated the reactive cross section for "$[Cu_8]^- + Cl_2$", where $r_2^* = 4.72$ Å $\sigma_2^* = 125.90$ Å^2 and $P_2 = 1.05$. (All the subscripts 1 or 2, refer to the system for "$[Ag_8]^- + Cl_2$" or "$[Cu_8]^- + Cl_2$", respectively). It is noteworthy that the calculated values of both P_1 and P_2 are greater than 1, strongly supporting that the harpoon mechanism is consistent with the cluster reactivity of $[Cu_8]^-$/$[Ag_8]^-$ towards chlorine [121, 122].

DFT calculations were performed to examine the interaction potential between a $[Cu_8]^-$/$[Ag_8]^-$ cluster and an approaching Cl_2 molecule to provide further insight into the harpoon-type cluster reaction. Figure 12.8a, b displays the charge distribution of $[Ag_8]^-$ and $[Cu_8]^-$. The calculations show that such reactions most-likely undertake a one-dimensional pathway with the molecular long axis of Cl_2 perpendicular to a surface of the HOMO profile of $[Ag_8]^-$/$[Cu_8]^-$, as shown in Fig. 12.8c, d. Dozens of models were considered with different distances (i.e., r values) to estimate the interaction energy within the binary systems $[Ag_8{\bullet\bullet}Cl_2]^-$ and $[Cu_8{\bullet\bullet}Cl_2]^-$. The calculated relative energies (plotted in Fig. 12.8e, f) fit well with a typical energy profile of molecular interactions based on the van der Waals equation [123]:

$$\mathrm{U(r)} = -\frac{\mathrm{a}}{\mathrm{r^m}} + \mathrm{b} \cdot \exp(-\mathrm{c} \cdot \mathrm{r}) + \mathrm{d} \tag{12.19}$$

where U is the potential energy; r is the intermolecular distance; a, b, c, d, and m are fitting parameters. This equation can be easily separated into two parts: "$\mathrm{U(r)_{att}} = -\frac{\mathrm{a}}{\mathrm{r^m}}$" and "$\mathrm{U(r)_{rep}} = \mathrm{b} \cdot \mathrm{e}^{-\mathrm{c \cdot r}} + \mathrm{d}$" with the corresponding curves of attractive and repulsive potentials involved in bringing the two entities together. The reagents approach and first feel feeble forces of repulsive character until they cross at r^* where the electron transfers from the HOMO of the $[Ag_8]^-$/$[Cu_8]^-$ to the anti-bonding LUMO orbital on the Br_2 which results in the steep descending attractive potential [124, 125]. The crossing of the two potential curves indicates the distance r^* at which the harpoon-type reaction occurs with an increased cross-section (πr^{*2}) [63].

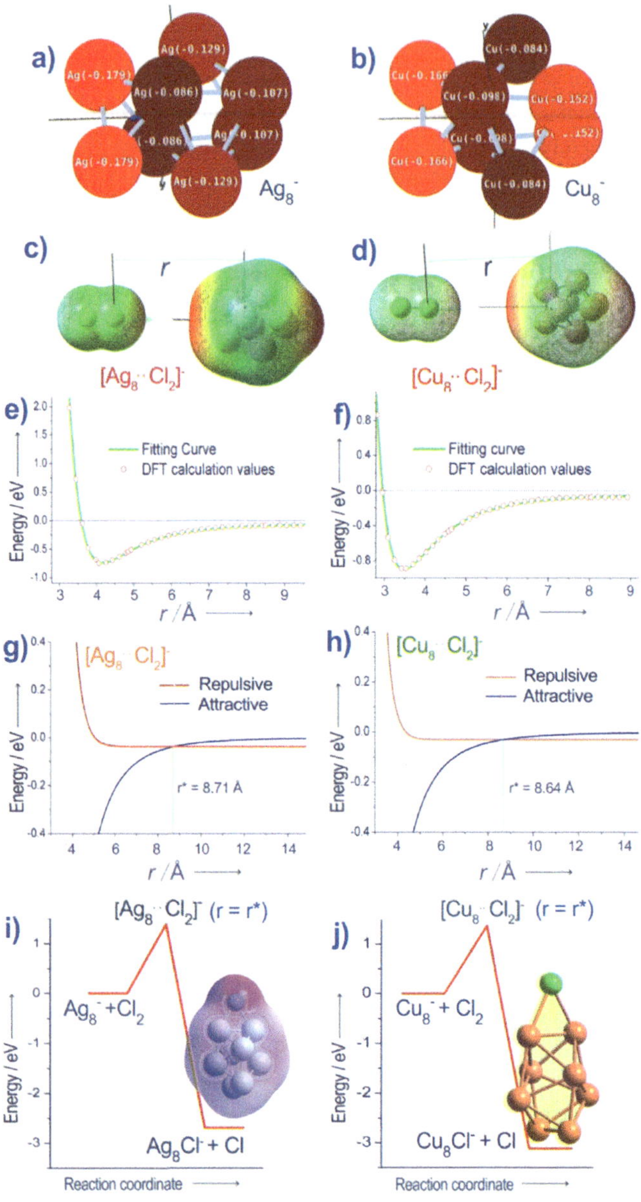

Fig. 12.8 Geometry and energy of $[Ag_8]^-/[Cu_8]^-$ approaching and reacting with a chlorine molecule. **a, b** The calculated charge distribution of $[Ag_8]^-/[Cu_8]^-$; **c, d** the most-likely orientation for a successful reactive encounter by the harpoon mechanism, mapped with HOMO surfaces; **e, f** the calculated relative energy of $[Ag_8{\cdot\cdot}Cl_2]^-$ and $[Cu_8{\cdot\cdot}Cl_2]^-$ at various intermolecular distances (r), corrected with zero-point-vibrational energy; **g, h** Repulsive and attractive potential energy diagram to illustrate the harpoon model for "$[Ag_8]^- + Cl_2$" and "$[Cu_8]^- + Cl_2$". I, **j** Simplified reaction coordinate diagrams of "$[Ag_8]^- + Cl_2$" and "$[Cu_8]^- + Cl_2$, showing the starting reactants, the final products, and the electron transfer transition state

For the $[Ag_8..Cl_2]^-$ and $[Cu_8..Cl_2]^-$ systems, a solution to Eq. 12.19 was calculated by utilizing Mathematica software, and subsequently, two curves representing the attractive potential and the repulsive potential were plotted based on this solution, as shown in Fig. 12.8g, h. For "$[Ag_8]^-$ + Cl_2" (Fig. 12.8g), the intersection of the two curves is at $r_1^{*'} = 8.71\acute{Å}$. Based on intersection, the reactive cross section was calculated as $\sigma_1^{*'} = \pi\left(r_1^{*'}\right)^2 = 238.33\acute{Å}^2$ and $P_1' = \frac{\pi\left(r_1^{*'}\right)^2}{\sigma_1} = 3.16$. (The accent signs refer to values calculated using the second DFT method). Similarly, the calculation results show that the crossing of the two potential curves for the system of "$[Cu_8]^- + Cl_2$" (Fig. 12.8h) occurs at $r_1^{*'} = 8.64\acute{Å}$, which indicates a cross section of $\sigma_2^{*'} = \pi\left(r_2^{*}\right)^2 = 234.52\acute{Å}$ and a $P_2' = 3.68$.

The optimized ground-state structure of $[Cu_8]^-$ bears the same symmetry (C_{2v}) as $[Ag_8]^-$ and $[Cu_8Cl]^-$ exhibits a similar structure to $[Ag_8Cl]^-$ [126, 127]. Considering $\Delta E = E(M_8) + E(Cl_2) - E(M_8Cl) - E(Cl)$, they calculated the energy difference and found that the reactivities of $[Ag_8]^-$ and $[Cu_8]^-$ toward Cl_2 are exothermic, as shown in Fig. 12.8i, j ($\Delta E_{Ag} = 2.69$ eV and $\Delta E_{Cu} = 3.11$ eV respectively from the calculation). This conforms to the general principles when alkali metals react with a halogen to produce alkali-halides [128]. While the ground states ($^1\sum^+$) of alkali halide molecules exhibit the characteristics of a closed valence shell, the alkali-like halide cluster anions reveal an electronic structure with an electron count similar to a closed valence shell as well. Figure 12.9 illustrates the simplified reaction diagrams of '$[Ag_8]^- + Cl_2$' and '$[Cu_8]^- + Cl_2$', showing the starting reactants, the final products, and the electron-transfer transition state. Insights into the reactivity of chlorine with the alkali-metal-like clusters will be of interest to other researchers working on obtaining a better understanding of the reaction mechanisms of such superatoms [109, 129], as well as the possible applications of copper and silver halides via soft-landing [130] or cluster assembly.

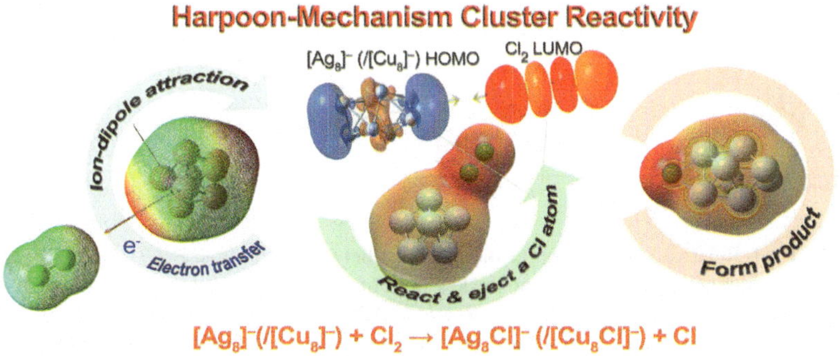

Fig. 12.9 Simplified reaction diagrams of '$[Ag_8]^-$ + Cl_2' and '$[Cu_8]^-$ + Cl_2', showing the starting reactants, the final products, and the electron-transfer transition state

12.4 Dependence on the Ionization Energy

Previous studies have revealed that the reactivity metal clusters, especially the charge-transfer reactions, often exhibit close dependence on the vertical/adiabatic electron-detachment energy (VDEs/ADEs), or simply named as vertical/adiabatic ionization energy (VIE/AIE) [131]. Fig. 12.10a, b shows an illustration of the ADE-dependence of Ag_n^- and Au_n^- clusters in reacting with O_2 [132, 133]. For the reaction of Ag_n^- with O_2 [133], some stable clusters whose VDEs are above the threshold of 3.0 eV (Fig. 12.10b) were observed to be inert as the ADE is too high to transfer electrons to the π*-antibonding orbital of O_2. Similar ADE-dependence was also observed in the reactions of Ag_n^- clusters with NO [133, 134], pertaining to similar electron transfer mechanisms. It is worth mentioning that, the large silver clusters (>1 nm) or nanocrystals could be controlled by its global electronic properties (instead of a bunch of discrete electronic states), allowing the influence of cluster structure to be small or even negligible. There is a similar case for gold clusters, as seen in Fig. 12.10c, d. Besides the odd-even alternation, it is interesting to note an ADE threshold at ~3.5 eV which separates the Au_n^- to be two parts, active or inert toward O_2. The ADE-dependence was also found in the reactions with CH_3I [135–138] and other metal clusters also undergo similar mechanisms [132, 133, 139, 140].

Recently the charge-transfer reactions of anionic copper clusters Cu_n^- ($n = 7$–37) with NO were also studied, as given in Fig. 12.11a. Interestingly, it was found that

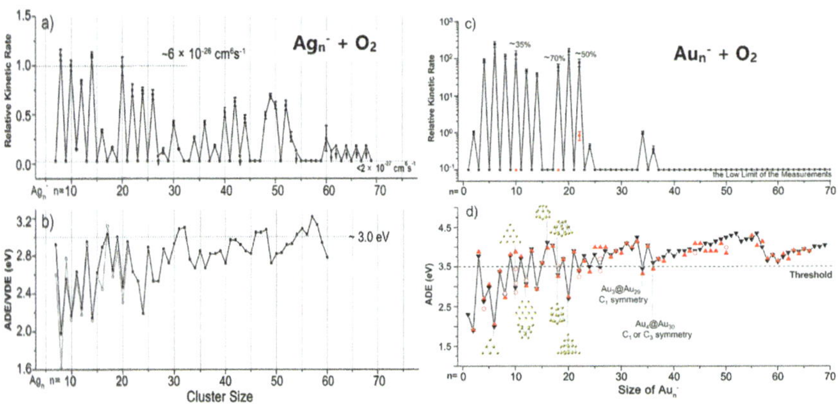

Fig. 12.10 **a** Relative kinetic rates for the reactions between Ag_n^- (n = 6–69) and O_2 at 120 K. The calibrated kinetic rate for Ag_{10}^- and the low limit of this measurement is indicated. **b** Adiabatic detachment energies (ADEs) of Ag_n^- from Ref. [141] (■) and vertical detachment energies (VDEs) from Ref. [142] (□). Reproduced with permission from Ref. [133]. Copyright 2016 Royal Society of Chemistry. **c** Relative kinetic rates of the reactions between Au_n^- (n = 1–70) and O_2 at 150 K. The red data points showed the rates of their isomers and the percentages represented the proportions of their active isomers. **d** The black data points showed the ADEs of Au_n^- from Ref. [141]; the red data points and the listed structures were from Refs. [143–152] The structure isomers were marked in red solid triangle and the red empty cycles, respectively. Reproduced with permission from Ref. [132]. Copyright 2018 American Chemical Society

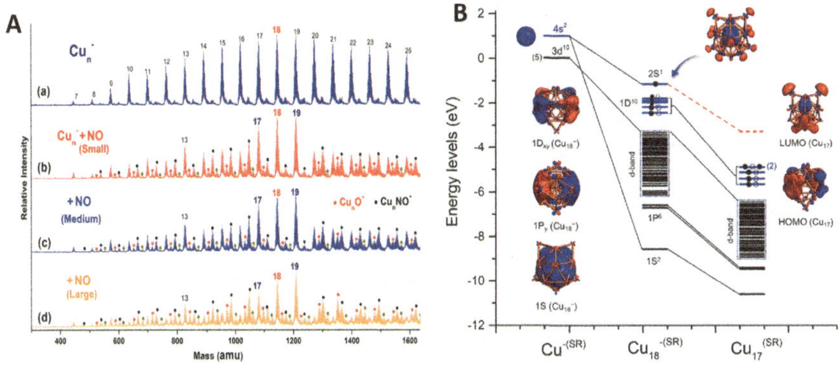

Fig. 12.11 **A** Typical TOF mass spectra of copper cluster anions produced via the laser ablation source (**a**) and after exposure to different quantity of NO gas with a partial pressure at 36 mPa (**b**), 39 mPa (**c**) and 62 mPa (**d**) respectively. The numbers of atoms in copper cluster anions are labelled on top of panel (a). **B** Kohn-sham energy level correlation between Cu_{18}^- cluster with a scalar relativistic Cu^- ($3d^{10}4s^2$), and a Cu_{17} hollow cage at a removal of the interior copper atom

Cu_{17-19}^- are inert even in the presence of sufficient reactants. While the stabilities of Cu_{17}^- and Cu_{19}^- are due to their closed electronic shell structures, it is intriguing to observe abnormal stability of an open-shell cluster Cu_{18}^-. The ab initio calculations revealed that Cu_{18}^- bears a Cu@Cu_{17}^- core-shell structure, and the unique electronic configuration allows the unpaired SOMO electron to be mainly contributed by the central copper atom (Fig. 12.11B); meanwhile the other 18 delocalized valence electrons occupy the lower-energy orbitals of the superatomic cluster. The unique electronic and geometric structures of Cu@Cu_{17}^- result in a large HOMO-LUMO gap, a large VDE value, as well as a large Cu-Cu binding energy but small NO-binding energy; meanwhile, there is -1.06 |e| negative charge distributed on the interior copper core, giving rise to tight electromagnetic shielding thus prohibiting electron transfer from the Cu_{18}^- cluster.

References

1. J. Sauer, B. Schroder, Angew. Chem. Int. Ed. **4**, 711–714 (1965)
2. M. Ichikawa, M. Soma, T. Onishi, K. Tamaru, J. Phys. Chem. **70**, 2069–2070 (1966)
3. M.S. Wrighton, H.B. Abrahamson, D.L. Morse, J. Am. Chem. Soc. **98**, 4105–4109 (1976)
4. S.J. Formosinho, Molecular Photochemistry **8**, 459–475 (1977)
5. H.B. Abrahamson, M.S. Wrighton, Inorg. Chem. **17**, 3385–3388 (1978)
6. L.N. Domelsmith, P.D. Mollere, K.N. Houk, R.C. Hahn, R.P. Johnson, J. Am. Chem. Soc. **100**, 2959–2965 (1978)
7. D. Ghesquiere, R. Arnaud, C. Caze, J. Phys. Chem. **83**, 2029–2034 (1979)
8. G. Mauclaire, R. Derai, S. Fenistein, R. Marx, R. Johnsen, J. Chem. Phys. **70**, 4023–4026 (1979)
9. K. Fujimori, N.A. Wickramasinghe, Aust. J. Chem. **33**, 189–193 (1980)
10. J.S. Keute, D.R. Anderson, T.H. Koch, J. Am. Chem. Soc. **103**, 5434–5439 (1981)

11. Y.C. Jean, C. Yu, Y.Y. Wang, Y.Y. Yeh, J. Chem. Phys. **80**, 1671–1676 (1984)
12. J. Vicens, Tetrahedron **43**, 1361–1369 (1987)
13. P.J. Kulesza, L.R. Faulkner, Colloids Surf. **41**, 123–134 (1989)
14. C. Hidalgoluangdilok, A.B. Bocarsly, Inorg. Chem. **29**, 2894–2900 (1990)
15. D. Weir, J. Phys. Chem. **94**, 5870–5875 (1990)
16. S.K. Weit, C. Kutal, Inorg. Chem. **29**, 1455–1456 (1990)
17. T. Klitsner, J.S. Nelson, Phys. Rev. Lett. **67**, 3800–3803 (1991)
18. D. Bethell, V.D. Parker, J. Phys. Org. Chem. **5**, 317–321 (1992)
19. I. Loeff, J. Rabani, A. Treinin, H. Linschitz, J. Am. Chem. Soc. **115**, 8933–8942 (1993)
20. E.R. Reddy, P.R. Rao, J. Polym. Mater. **11**, 279–283 (1994)
21. Y. Wu, C. Cohran, A.B. Bocarsly, Inorg. Chim. Acta **226**, 251–258 (1994)
22. D.C. Duncan, T.L. Netzel, C.L. Hill, Inorg. Chem. **34**, 4640–4646 (1995)
23. J.L.H. Jiwan, J.P. Soumillion, J. Phys. Chem. **99**, 14223–14230 (1995)
24. N. Haga, H. Nakajima, H. Takayanagi, K. Tokumaru, J. Org. Chem. **63**, 5372–5384 (1998)
25. M. Khajehpour, J.F. Kauffman, Chem. Phys. Lett. **297**, 141–146 (1998)
26. R. Kosloff, G. Katz, Y. Zeiri, Faraday Discuss. **117**, 291–301 (2000)
27. I. Texier, J.A. Delaire, C. Giannotti, Phys. Chem. Chem. Phys. **2**, 1205–1212 (2000)
28. F. Roncaroli, L.M. Baraldo, L.D. Slep, J.A. Olabe, Inorg. Chem. **41**, 1930–1939 (2002)
29. H. Kunkely, A. Vogler, Inorg. Chem. Commun. **6**, 1335–1337 (2003)
30. G. Ramakrishna, H.N. Ghosh, Langmuir **19**, 3006–3012 (2003)
31. A.G. Agrios, K.A. Gray, E. Weitz, Langmuir **20**, 5911–5917 (2004)
32. M. Feliz, R. Llusar, S. Uriel, C. Vicent, E. Coronado, C.I. Gomez-Garcia, Chem. Eur. J. **10**, 4308–4314 (2004)
33. H. Kunkely, A. Vogler, Inorg. Chim. Acta **357**, 888–890 (2004)
34. J. Nowotny, T. Bak, C.C. Sorrell, Adv. Appl. Ceram. **104**, 181–187 (2005)
35. T. Bak, J. Nowotny, M.K. Nowotny, L.R. Sheppard, Ionics **12**, 247–251 (2006)
36. S. Zilberg, Y. Haas, Mol. Phys. **104**, 1061–1068 (2006)
37. J. Nowotny, T. Bak, L.R. Sheppard, M.K. Nowotny, J. Am. Chem. Soc. **130**, 9984–9993 (2008)
38. T. Bak, J. Nowotny, J. Phys. Chem. C **115**, 15345–15354 (2011)
39. A.D. McNaught, A. Wilkinson, *IUPAC. Compendium of Chemical Terminology, 2nd ed. (the "Gold Book")*., Blackwell Scientific Publications, Oxford, 1997
40. Nobelprize.org, *Journal* (1986)
41. Z.X. Luo, C. Berkdemir, J.C. Smith, A.W. Castleman Jr., Chem. Phys. Lett. **582**, 24–30 (2013)
42. B.H. Bransden, M.R.C. McDowell, *Charge Exchange and the Theory of Ion—Atom Collisions* (Oxford University Press, Clarendon, 1992)
43. H. Shen, P. Hvelplund, D. Mathur, A. Bárány, H. Cederquist, N. Selberg, D.C. Lorents, Phys. Rev. A **52**, 3847–3851 (1995)
44. S. Pollack, D. Cameron, M. Rokni, W. Hill, J.H. Parks, Chem. Phys. Lett. **256**, 101–108 (1996)
45. F. Chandezon, C. Guet, B.A. Huber, D. Jalabert, M. Maurel, E. Monnand, C. Ristori, J.C. Rocco, Phys. Rev. Lett. **74**, 3784–3787 (1995)
46. C. Bréchignac, P. Cahuzac, J. Leygnier, R. Pflaum, J. Weiner, Phys. Rev. Lett. **61**, 314–317 (1988)
47. C. Bréchignac, P. Cahuzac, F. Carlier, J. Leygnier, I.V. Hertel, Z. Phys. D: At., Mol. Clusters **17**, 61–67 (1990)
48. N.D. Bhaskar, R.P. Frueholz, C.M. Klimcak, R.A. Cook, Chem. Phys. Lett. **154**, 175–180 (1989)
49. M. Abshagen, J. Kowalski, M. Meyberg, G. zu Putlitz, J. Slaby and F. Träger, Chem. Phys. Lett. **174**, 455–460 (1990)
50. C. Brechignac, P. Cahuzac, B. Concina, J. Leygnier, I. Tigneres, Eur. Phys. J. D **12**, 185–192 (2000)
51. M. Guissani, V. Sidis, J. Chem. Phys. **102**, 1288–1294 (1995)
52. W.A. Saunders, Phys. Rev. Lett. **62**, 1037 (1989)

53. M. Abshagen, J. Kowalski, M. Meyberg, G.zuPutlitz, F. Tr¨ager, J. Well, Europhys. Lett. **5**, 13 (1988)
54. M. Abshagen, J. Kowalski, M. Meyberg, G. z. Putlitz, J. Slaby and F. Tr¨ager, Chem. Phys. Lett. **174**, 455 (1990)
55. S.B.H. Bach, J.R. Eyler, J. Chem. Phys. **92**, 358–363 (1990)
56. M. Polanyi (ed.), *Atomic Reactions* (Williams & Northgate, London, 1932)
57. J.L. Magee, J. Chem. Phys. **8**, 687–698 (1940)
58. J.H. Birely, R.R. Herm, K.R. Wilson, D.R. Herschbach, J. Chem. Phys. **47**, 993–1004 (1967)
59. P.S. Weiss, J.M. Mestdagh, H. Schmidt, M.H. Covinsky, Y.T. Lee, J. Phys. Chem. **95**, 3005–3011 (1991)
60. P.Y. Cheng, D. Zhong, A.H. Zewail, Chem. Phys. Lett. **242**, 369–379 (1995)
61. C. Desfrancois, H. AbdoulCarime and J.P. Schermann, in *Physics of Electronic and Atomic Collisions*, eds. L. J. Dube, J. B. A. Mitchell, J. W. McConkey and C. E. Brion, pp 599–608 (1995)
62. M.J. Pilling, P.W. Seakins (eds.), *Reaction Kinetics* (Oxford University Press Inc., New York, 1995)
63. T. Ruchti, K. Forde, B.E. Callicoatt, H. Ludwigs, K.C. Janda, J. Chem. Phys. **109**, 10679–10687 (1998)
64. T.G. Spence, B.T. Trotter, T.D. Burns, L.A. Posey, J. Phys. Chem. A **102**, 6101–6106 (1998)
65. C.E.H. Dessent, M.A. Johnson, I. Becker, O. Cheshnovsky, in *Electron Transfer-from Isolated Molecules to Biomolecules, Pt 1*, eds. J. Jortner and M. Bixon, 1999, vol. 106, pp. 265–302
66. O. Knospe, J. Jellinek, U. Saalmann, R. Schmidt, Eur. Phys. J. D **5**, 1–4 (1999)
67. P.E. Barran, N.R. Walker, A.J. Stace, J. Chem. Phys. **112**, 6173–6177 (2000)
68. S. Bililign, B.C. Hattaway, N. Geum, G.-H. Jeung, J. Phys. Chem. A **104**, 9454–9458 (2000)
69. B. Zarour, J. Hanssen, P.A. Hervieux, M.F. Politis and F. Martin, J. Phys. B: At., Mol. Opt. Phys., **33**, L707–L714 (2000)
70. C. Brechignac, P. Cahuzac, B. Concina, J. Leygnier, L.F. Ruiz, B. Zarour, P.A. Hervieux, J. Hanssen, M.F. Politis, F. Martin, Phys. Rev. Lett. **89**, 183402 (2002)
71. T. Nagano, K. Tsumurya, H. Eguchi, D.J. Singh, Phys. Rev. B **66**, 199902 (2002)
72. C. Brechignac, P. Cahuzac, B. Concina, J. Leygnier, L.F. Ruiz, B. Zarour, P.A. Hervieux, J. Hanssen, M.F. Politis, F. Martin, Phys. Rev. A **68**, 063202 (2003)
73. V. Torma, O. Vidoni, U. Simon, G. Schmid, Eur. J. Inorg. Chem. 1121–1127 (2003)
74. D.Y. Wu, M. Hayashi, C.H. Chang, K.K. Liang, S.H. Lin, J. Chem. Phys. **118**, 4073–4085 (2003)
75. A. Datta, S.K. Pati, J. Phys. Chem. A **108**, 9527–9530 (2004)
76. C.P. Berlinguette, A. Dragulescu-Andrasi, A. Sieber, H.U. Gudel, C. Achim, K.R. Dunbar, J. Am. Chem. Soc. **127**, 6766–6779 (2005)
77. N.V. Kryzhevoi, L.S. Cederbaum, *J. Chem. Phys.* **123** (2005)
78. N.V. Kryzhevoi, N.V. Dobrodey, L.S. Cederbaum, J. Chem. Phys. **122** (2005)
79. E. Cho, S. Shin, J. Chem. Phys. **127** (2007)
80. K. Nagaya, H. Murakami, H. Iwayama, Y. Ohmasa, M. Yao, Eur. Phys. J. D **43**, 69–72 (2007)
81. C. Sporea, F. Rabilloud, M. Aubert-Frecon, J. Mol. Struct.-Theochem **802**, 85–90 (2007)
82. R.B. Gerber, Y. Miller, Abstr. Pap. Am. Chem. Soc. **236** (2008)
83. S.G. Ramesh, S. Re, J.T. Hynes, J. Phys. Chem. A **112**, 3391–3398 (2008)
84. F.F. de Biani, G. Manca, L. Marchetti, P. Leoni, S. Bruzzone, C. Guidotti, A. Atrei, A. Albinati, S. Rizzato, Inorg. Chem. **48**, 10126–10137 (2009)
85. C.L. Moss, C.M. Isborn, X. Li, Phys. Rev. A **80**, 024503 (2009)
86. M.G. Hilfiger, M. Chen, T.V. Brinzari, T.M. Nocera, M. Shatruk, D.T. Petasis, J.L. Musfeldt, C. Achim, K.R. Dunbar, Angew. Chem. Int. Ed. **49**, 1410–1413 (2010)
87. M. Jadraque, M. Martin, J. Comput. Chem. **32**, 3497–3504 (2011)
88. A.S. Moskvin, Opt. Spectrosc. **111**, 403–410 (2011)
89. T. Okubo, H. Kuwamoto, K.H. Kim, S. Hayami, A. Yamano, M. Shiro, M. Maekawa, T. Kuroda-Sowa, Inorg. Chem. **50**, 2708–2710 (2011)

90. Q. Wang, S.-W. Yang, Y. Yang, M.B. Chan-Park, Y. Chen, J. Phys. Chem. Lett. **2**, 1009–1014 (2011)
91. Z. Luo, J.C. Smith, W.H. Woodward, A.W. Castleman Jr., J. Phys. Chem. A **116**, 2012–2017 (2012)
92. M.A. Roman, O.S. Reu, S.I. Klokishner, J. Phys. Chem. A **116**, 9534–9544 (2012)
93. G. Saito, H. Hosoda, Y. Yoshida, J. Hagiwara, K. Nishimura, H. Yamochi, A. Otsuka, T. Hiramatsu, Y. Shimazaki, K. Kirakci, S. Cordier, C. Perrin, J. Mater. Chem. **22**, 19774–19791 (2012)
94. G. Pacchioni, Phys. Chem. Chem. Phys. **15**, 1737–1757 (2013)
95. O. Knospe, J. Jellinek, U. Saalmann, R. Schmidt, Phys. Rev. A **61**, 022715 (2000)
96. P. Atkins, J. d. Paul, eds., *Physcial Chemistry, Eighth Edition*, Oxford University Press, New York (2006)
97. J. Maya, P. Davidovits, J. Chem. Phys. **59**, 3143–3152 (1973)
98. E.M. Goldfield, A.M. Kosmas, E.A. Gislason, J. Chem. Phys. **82**, 3191–3197 (1985)
99. V.A. Kartoshkin, Opt. Spektrosk. **79**, 26–31 (1995)
100. C.K. Kwan, W.E. Kauppila, S.P. Parikh, T.S. Stein, S. Zhou, Hyperfine Interact. **89**, 33–45 (1994)
101. T.T. Gien, J. Phys. B: At. Mol. Opt. Phys. **24**, 2871–2876 (1991)
102. P. Davidovits, D.L. McFadden (eds.), *Alkali Halide Vapours: Structure* (Academic Press Inc, New York, Spectra and Reaction Dynamics, 1979)
103. H.H. Cornehl, G. Hornung, H. Schwarz, J. Am. Chem. Soc. **118**, 9960–9965 (1996)
104. A. Matsuda, H. Mori, Chem. Phys. **380**, 48–53 (2011)
105. Y.-P. Chang, M.-K. Hsiao, D.-K. Liu, K.-C. Lin, J. Chem. Phys. **128**, 234309 (2008)
106. H.C. Chang, Y.L. Luo, K.C. Lin, J. Chem. Phys. **94**, 3529–3536 (1991)
107. D.K. Liu, K.C. Lin, J.J. Chen, J. Chem. Phys. **113**, 5302–5310 (2000)
108. D.E. Bergeron, A.W. Castleman Jr., T. Morisato, S.N. Khanna, Science **304**, 84–87 (2004)
109. D.E. Bergeron, P.J. Roach, A.W. Castleman Jr., N. Jones, S.N. Khanna, Science **307**, 231–235 (2005)
110. S.M. Hurley, T.E. Dermota, D.P. Hydutsky, A.W. Castleman Jr., Science **298**, 202–204 (2002)
111. P.J. Roach, W.H. Woodward, A.W. Castleman Jr., A.C. Reber, S.N. Khanna, Science **323**, 492–495 (2009)
112. Z. Luo, W.H. Woodward, J.C. Smith, A.W. Castleman Jr., Int. J. Mass Spectrom. **309**, 176–181 (2012)
113. P.J. Hay, W.R. Wadt, J. Chem. Phys. **82**, 270–283 (1985)
114. Z.-Y. Jiang, K.-H. Lee, S.-T. Li, S.-Y. Chu, Phys. Rev. B **73**, 235423 (2006)
115. M.J. Frisch et al., *Gaussian 03, Revision B.05*, Gaussian Inc., Pittsburgh, PA (2003)
116. Z. Luo, G.U. Gamboa, J.C. Smith, A.C. Reber, J.U. Reveles, S.N. Khanna, A.W. Castleman Jr., J. Am. Chem. Soc. **134**, 18973–18978 (2012)
117. W. Demtröder, ed., *Molecular Physics. Theoretical Principles and Experimental Methods*, Wliey-VCH Verlag GmbH & Co. KGaG, Weinheim (2005)
118. D.R. Lide, ed., *Handbook of Chemistry and Physics, 81st Edition*, CRC Press (2000)
119. J.V. Dugan, J.L. Magee, J. Chem. Phys. **47**, 3103–3112 (1967)
120. R.D. Levine, *Molecular Reaction Dynamics* (The Press Syndicate of the University of Cambridge, Cambridge, 2005)
121. G. Herzberg (ed.), *Molecular Spectra and Molecular Structure: Spectra of Diatomic Molecules* (Prentice-Hall, New York, 1989)
122. N. Shafizadeh, B. Soep, J.M. Mestdagh, W.H. Breckenridge, Int. Rev. Phys. Chem. **28**, 359–406 (2009)
123. J. Hernandez-Trujillo, R.F.W. Bader, J. Phys. Chem. A **104**, 1779–1794 (2000)
124. P. Calaminici, A.M. Koster, A. Vela, K. Jug, J. Chem. Phys. **113**, 2199–2202 (2000)
125. K. Jug, B. Zimmermann, P. Calaminici, A.M. Koster, J. Chem. Phys. **116**, 4497–4507 (2002)
126. J.C. Polanyi, J. Chem. Phys. **31**, 1338–1351 (1959)
127. Q. Wang, Q. Sun, P. Jena, J. Chem. Phys. **131**, 124301 (2009)
128. J. Laskin, T.H. Bailey, J.H. Futrell, J. Am. Chem. Soc. **125**, 1625–1632 (2003)

129. D.M. Cox, R. Brickman, K. Creegan, A. Kaldor, Z. Phys, D—Atoms. Molecules and Clusters **19**, 353–355 (1991)
130. Y. Zhao, K.R. Yang, Z. Wang, X. Yan, S. Cao, Y. Ye, Q. Dong, X. Zhang, J.E. Thorne, L. Jin, K.L. Materna, A. Trimpalis, H. Bai, S.C. Fakra, X. Zhong, P. Wang, X. Pan, J. Guo, M. Flytzani-Stephanopoulos, G.W. Brudvig, V.S. Batista, D. Wang, Proc. Natl. Acad. Sci. U. S. A. **115**, 2902–2907 (2018)
131. J. Ma, X. Cao, X. Xing, X. Wang, J.H. Parks, Phys. Chem. Chem. Phys. **18**, 743–748 (2016)
132. J. Ma, X. Cao, H. Liu, B. Yin, X. Xing, Phys. Chem. Chem. Phys. **18**, 12819–12827 (2016)
133. S. Muramatsu, K. Koyasu, T. Tsukuda, ACS Omega **3**, 16874–16881 (2018)
134. S. Muramatsu, K. Koyasu, T. Tsukuda, J. Phys. Chem. A **120**, 957–963 (2016)
135. S. Muramatsu, K. Koyasu, T. Tsukuda, Chem. Lett. **46**, 676–679 (2017)
136. S. Muramatsu, T. Tsukuda, Chem. Asian J. **14**, 3763–3772 (2019)
137. R.L. Whetten, D.M. Cox, D.J. Trevor, A. Kaldor, Phys. Rev. Lett. **54**, 1494–1497 (1985)
138. R.L. Whetten, M.R. Zakin, D.M. Cox, D.J. Trevor, A. Kaldor, J. Chem. Phys. **85**, 1697–1698 (1986)
139. K.J. Taylor, C.L. Pettiette-Hall, O. Cheshnovsky, R.E. Smalley, J. Chem. Phys. **96**, 3319–3329 (1992)
140. H. Handschuh, C.-Y. Cha, P.S. Bechthold, G. Ganteför, W. Eberhardt, J. Chem. Phys. **102**, 6406 (1995)
141. S. Bulusu, X. Li, L.-S. Wang, X.C. Zeng, Proc. Natl. Acad. Sci. U. S. A. **103**, 8326–8330 (2006)
142. W. Huang, S. Bulusu, R. Pal, X.C. Zeng, L.S. Wang, ACS Nano **3**, 1225–1230 (2009)
143. S. Bulusu, X. Li, L.-S. Wang, X.C. Zeng, J. Phys. Chem. C **111**, 4190–4198 (2007)
144. J. Li, X. Li, H.-J. Zhai, L.-S. Wang, Science **299**, 864–867 (2003)
145. W. Huang, M. Ji, C.-D. Dong, X. Gu, L.-M. Wang, X.G. Gong, L.-S. Wang, ACS Nano **2**, 897–904 (2008)
146. N. Shao, W. Huang, Y. Gao, L.-M. Wang, X. Li, L.-S. Wang, X.C. Zeng, J. Am. Chem. Soc. **132**, 6596–6605 (2010)
147. N. Shao, W. Huang, W.-N. Mei, L.S. Wang, Q. Wu, X.C. Zeng, J. Phys. Chem. C **118**, 6887–6892 (2014)
148. S. Pande, W. Huang, N. Shao, L.M. Wang, N. Khetrapal, W.N. Mei, T. Jian, L.S. Wang, X. C. Zeng, ACS Nano (2016)
149. B. Yoon, P. Koskinen, B. Huber, O. Kostko, B. von Issendorff, H. Häkkinen, M. Moseler, U. Landman, ChemPhysChem **8**, 157–161 (2007)
150. H. Häkkinen, B. Yoon, U. Landman, X. Li, H.-J. Zhai, L.-S. Wang, J. Phys. Chem. A **107**, 6168–6175 (2003)
151. B. Yin, Q. Du, L. Geng, H. Zhang, Z. Luo, S. Zhou, J. Zhao, J. Phys. Chem. Lett. **11**, 5807–5814 (2020)
152. B. Yin, T. Wang, Y. Chen, J. Yang, G. Wang, X. Xing, Int. J. Mass Spectrom. **451**, 116312 (2020)

Chapter 13
Metal Cluster Catalysis

13.1 Introduction

To develop metal catalysts for the selective activation of certain chemical bonds (e.g., C–H) is attractive and challenging research topics in chemistry [1–8]. Noble metals and heavy metals dominant the studies and practical use of this field, because main group metals and early transition metals are too reactive to be effective catalysts. Among others, unremitting efforts have also been paid to study catalysis without precious metals [9]. For example, it has been recognized that by incorporating non-metallic elements (e.g., carbon or oxygen) into the early transition metal, the surface reactivity of the metal can be moderated to produce an effective catalyst. Extensive studies have demonstrated that oxygen-centered radicals can serve as active sites to activate C–H bonds of alkane molecules (such as methane, ethane, *and* butane etc.) under thermal collision conditions [10–29]. The activation of alkane molecules by oxygen-centered radicals over atomic clusters usually lead to single hydrogen atom abstraction (HAA) [17, 30], forming products of alkyl radicals.

Catalysis is generally caused by the ensemble operation of the intermediates on active sites and rationalized by correlative elementary reactions, along with reaction kinetics simulation based on first-principles calculations. A joint experimental and theoretical studies have revealed the catalytic activity of abundant metal clusters, and interpreted the catalysis mechanism on a basis of bond activation and reconstruction, active sites and reactive intermediates, role of promoting additives, and site-dependence of turnover frequency which evaluates the mole number of yields regarding to a mole catalysts per hour [31]. While it has been attainable to detect the adsorbed intermediates of clusters or molecules by gas-phase mass spectrometry or to observe single adsorbed atoms by STM in situ analysis [32], it is also expected to direct and control a catalytical reaction process in gas phase and on solid surfaces [33]. In particular, in recent years there is a new concept known as single-atom catalysis which stimulates reasonable research interest with vivid catalysis for reactions, such as platinum-based heterogeneous catalysis for CO oxidation [34].

Z. Luo and S. N. Khanna, *Metal Clusters and Their Reactivity*,
https://doi.org/10.1007/978-981-15-9704-6_13

On the other hand, to employ the superatom concepts to quantitatively identify electronic energy states for promising catalysts to investigate iso-valent species is also important to determine if the superatomic concepts carry over to catalytic reaction behavior. This is also an important class of catalysts likely without precious metals. Presented in this chapter are several classes of studies using clusters to unravel fundamental catalytic reaction mechanisms, including a few of those using identified superatoms and the concepts of element mimics to tailor catalysts with desired functionality [35].

13.2 Gold Cluster Catalysis

Although bulk gold is known to be chemically inert metal, gold nanoparticles have demonstrated powerful catalytic capability in various chemical reactions, such as in CO oxidation [36], epoxidation [37, 38], selective hydrogenation/reduction [39], C–C bond formation [40], and water-gas shift [41]. There are also many examples of catalysis in solution by cationic complexes of gold, in particular, gold catalysts such as Au^{I} and Au^{III} salts are powerful for C–H activation [42], in which the Au^{I}/Au^{III} catalytic cycle is very likely to be involved [43]. Cationic, neutral, and anionic gold clusters and complexes have also been identified in supported gold catalysts [23, 44, 45]. Among these catalytic reactions, it was suggested that gold may switch its role between electron donator and electron acceptor; for instance, doping atomic clusters (e.g., metal oxide clusters) with gold atoms may cause charge redistribution within the clusters during reactions. Thereby, the correlative reactivity such as the C–H activation which depends heavily on the effects of local charges can be effectively tuned by the gold-based catalysis (Fig. 13.1) [1, 23, 27, 42, 44–46]. Gold is known as a very versatile redox catalyst [36, 47–57]. It also shows potential for both selective and nonselective oxidation of hydrocarbons, for methanol synthesis by hydrogenation of carbon monoxide or dioxide, and for the reduction of nitric oxide by hydrogen, propene, or carbon monoxide, etc. [55]. Besides, the supported gold chloride was found to be the most active catalyst for the hydrochlorination of ethyne, and the hydrogenation of unsaturated hydrocarbons also occurs on highly dispersed gold catalysts [55].

13.2.1 Catalysis of Supported Gold Clusters

Together with the bonding nature of gold, the Au clusters and Au-contained heteroatomic systems have been extensively studied [58–76]. As well as the interesting relativistic effect involved in gold clusters [77–79], extensive investigations have been reported on the catalysis of gold clusters, including those (both homogeneous and heterogeneous) with gold as a key component [80–88]. The catalysis

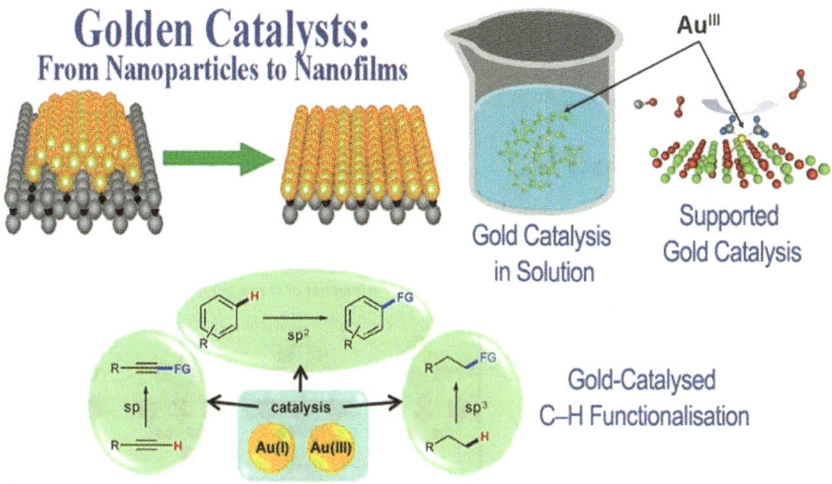

Fig. 13.1 A sketch showing the versatile gold-mediated catalysis. Reproduced with permission from Refs. [23, 42, 44]. Copyright 2008 and 2011 Royal Society of Chemistry

of gold clusters is retrospect to two significant observations in the 1980s that highlighted the special attributes of gold as a heterogeneous catalyst including: (i) the discovery that supported Au catalysts are active for CO oxidation [89], (ii) catalysis for ethyne hydrochlorination [90].

Heiz et al. [91–94] showed several studies on the catalytic activity of MgO-supported gold clusters for CO oxidation, and found strongly size dependence of the Au_n (e.g., $n = 1 - 20$). Figure 13.2 displays the temperature programmed reaction curves and the number of CO_2 molecules produced per cluster as a function of the number of atoms in the cluster. It is notable that there is no CO_2 being detected in case of $n = 1$ and $n = 2$ of such catalytic Au_n system, while little CO_2 for $n = 3 - 6$, no CO_2 for $n = 7$, and around one CO_2 per cluster for $n = 8$. For larger Au_n ($n > 8$), the yield has irregular oscillations with an ascending trend with the value of n [91]. Among $n = 1 - 22$, the most active is Au_{18}, which produces ~$2CO_2$ molecules per cluster, as seen in Fig. 13.2B. Optimized cluster structures of model catalysts comprising different number of gold atoms adsorbed at a F-center defect on MgO(100) are displayed in Fig. 13.2C.

Among the interesting Au_n clusters, Au_{25} and a magic cluster $Au_{25}(SR)_{18}$ has something unique and attracted reasonable interest [96–98]. For example, Liu et al. [99] reported the Au_{25} clusters supported on hydroxyapatite oxidized styrene in toluene, and found 100% conversion and 92% selectivity to the epoxide, under optimum conditions and using anhydrous tert-butyl hydroperoxide (TBHP) as an oxidant. Significant progress has also been made in Argonne group, where high-pressure kinetics on mass-selected Au_6–Au_{10} clusters deposited on alumina were studied, and the gold clusters were stabilized by depositing a protective layer of

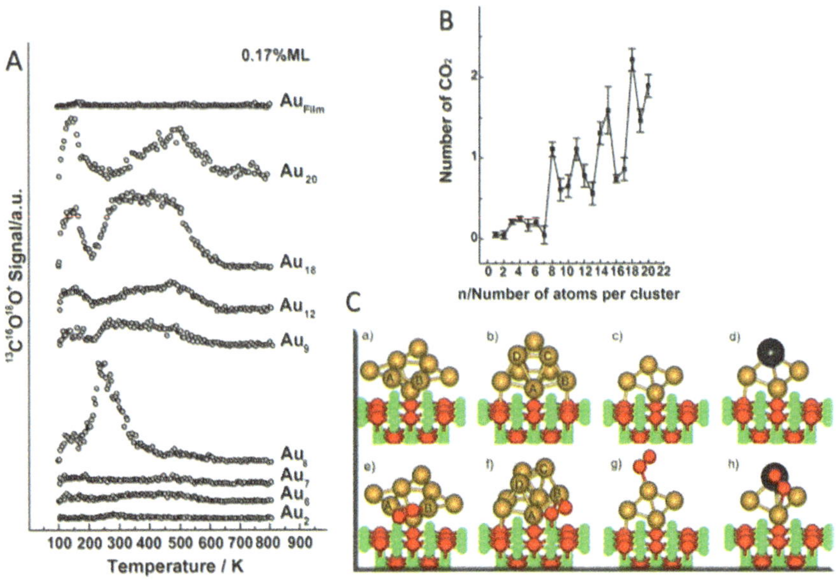

Fig. 13.2 **A** Temperature programmed reaction experiments for the CO-oxidation on selected Au_n clusters on defect-rich MgO(100) films. The model catalysts are saturated at 90 K with ^{13}CO and $^{18}O_2$ and the isotopomer $^{13}C^{18}O^{16}O$ is detected with a mass spectrometer, as a function of temperature; **B** The reactivities for Au_n expressed as the number of formed CO_2 per cluster. **C** The optimized atomic structures of model catalysts comprising **a**, **b** Au_8, **c** Au_4, and **d** Au_3Sr clusters adsorbed at a F-center defect on MgO(100). Reproduced with permission from Refs. [93, 95]. Copyright 2006 and 2003 John Wiley and Sons

alumina on them. Such cluster system was found to have high activity and selectivity for propylene epoxidation [51, 100]. Xie et al. [82] synthesized Au clusters with well-defined sizes (e.g., Au_{10}, Au_{11}, Au_{18}, Au_{25}, and Au_{39}) on solid supports, such as mesoporous silica and hydroxyapatite by using ligand-protected and size-selected Au clusters as precursors [99, 101, 102]. Moreover, they have extended such approach to precisely controlled composition of bimetallic clusters such as $Pd_1Au_{24}(SR)_{18}$ system [103, 104]. They immobilized Au_{25} and Pd_1Au_{24} on multi-walled carbon nanotubes (CNTs), as shown in Fig. 13.3 (upper). The Au_{25} and Pd_1Au_{24} clusters on multiwalled carbon nanotubes were developed via adsorption of $Au_{25}(SC_{12}H_{25})_{18}$ and $Pd_1Au_{24}(SC_{12}H_{25})_{18}$, respectively, on the nanotubes, followed by calcination. When comparing their catalysis for the aerobic oxidation of benzyl alcohol, it was noted that a single Pd atom doping (Pd_1Au_{24}/CNT) significantly improved the catalytic performance of Au_{25}/CNT [82].

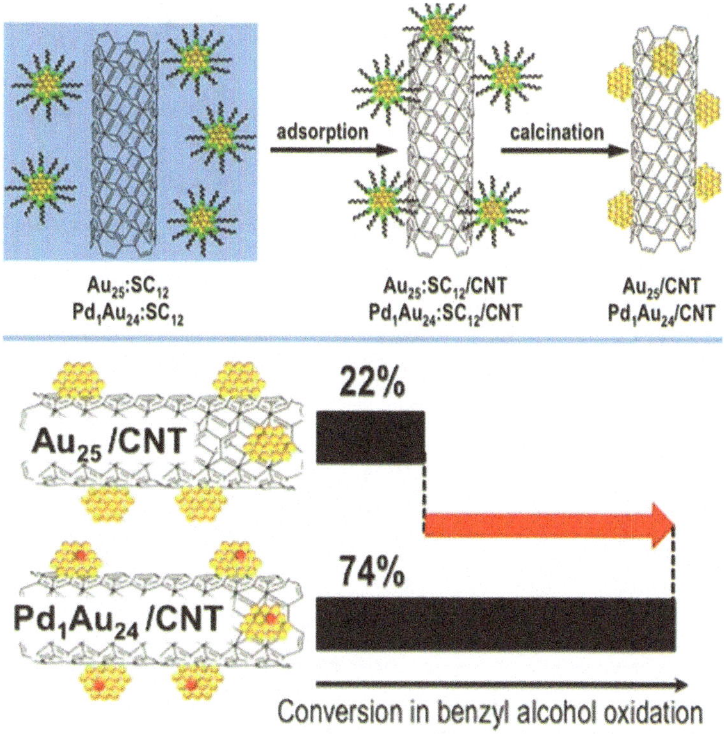

Fig. 13.3 (Upper) The synthetic scheme of Au_{25}/CNT and Pd_1Au_{24}/CNT; (Bottom) catalytic performance of Au_{25}/CNT and Pd_1Au_{24}/CNT for conversion in benzyl alcohol oxidation. Reproduced with permission from Ref. [82]. Copyright 2012 American Chemical Society

13.2.2 Catalysis of Gold Oxides

Abundant results have demonstrated that cationic gold species on oxide (Au_nO_m clusters) and zeolite supports are catalytically active especially for reactions including ethylene hydrogenation and CO oxidation [61, 105–109]. For example, Johnson et al. [107] studied the reactivity of CO with gold oxide cluster cations containing one to four gold atoms and between one and five oxygen atoms. Clusters of AuO^+, Au_2O^+, and Au_3O^+ were found to be reactive toward a CO molecule and facilitated by adsorption of a second CO. Furthermore, AuO^+ and Au_3O^+ with an odd number of gold atoms were found to be more reactive than Au_2O^+, as shown in Fig. 13.4. Species with higher oxygen content, such as $AuO_3{}^+$, $AuO_4{}^+$, $AuO_5{}^+$, $Au_2O_2{}^+$, and $Au_3O_3{}^+$ (Fig. 13.5), favor an adsorption of one CO molecule onto the cluster, accompanying with a loss of one or two O_2 molecules. This reactivity indicates that the $Au_nO_x^+$ clusters allows chemisorption of CO. Note that the $Au_nO_x^+$ clusters may be broken apart by the exothermic adsorption of CO, resulting products corresponding to a loss of

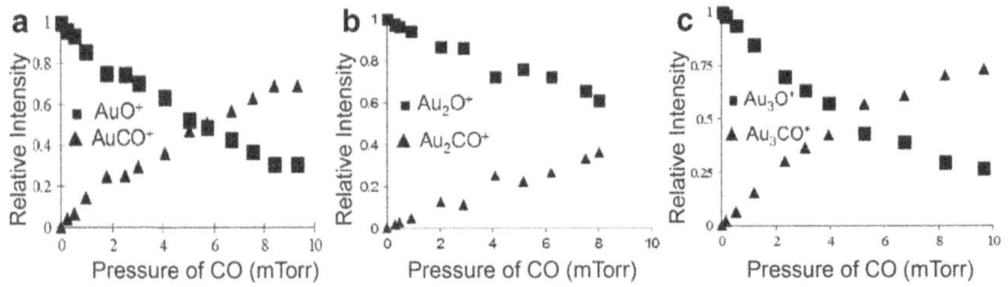

Fig. 13.4 Relative ion intensity of **a** AuO^+, **b** Au_2O^+, and **c** Au_3O^+ with increasing pressure of CO. Note the decrease in the reactant ion intensity and increase in the product corresponding to oxidation of one CO molecule and adsorption of a second CO. The relative ion intensity is plotted on the y axis. Reproduced with permission from Ref. [107]. Copyright 2008 American Chemical Society

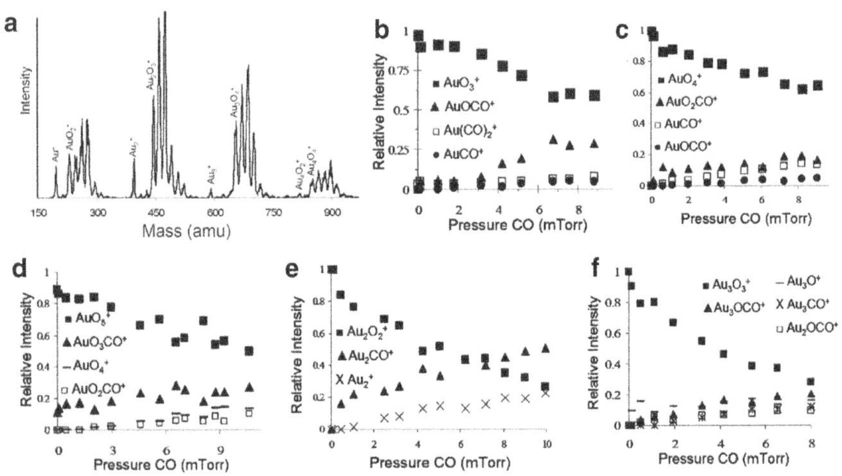

Fig. 13.5 Typical mass distribution of gold oxide cation clusters obtained through laser vaporization (**a**); relative ion intensity of $AuO_3{}^+$ (**b**), $AuO_4{}^+$ (**c**), $AuO_5{}^+$ (**d**), $Au_2O_2{}^+$ (**e**), and $Au_3O_3{}^+$ (**f**) with increasing pressure of CO. Note the decrease in the reactant ion intensity and increase in the product corresponding to adsorption of CO and loss of O_2. The relative ion intensity is plotted on the y axis. Reproduced with permission from Ref. [107]. Copyright 2008 American Chemical Society

AuO_2. In addition, products with a loss of O_2 and AuO_2 undertook further reactivity with CO.

13.2.3 Catalysis of Gold Complex

In addition, there are also investigations emphasized on gold-containing clusters of mixed oxides [105, 110]. Such systems help further understand the mechanistic details of reactions catalyzed by oxide-supported gold, which have been under debate for many years [23, 44, 45, 106]. By taking into account that niobium oxides have extraordinary catalytic properties in selective oxidation reactions, Wu et al. [110] reported such an investigation on $Au_xNb_yO_z^+$ reacting with methane, ethane, and *n*-butane respectively, as shown in Fig. 13.6. The mass spectra shown in Fig. 13.6 indicate that $AuNbO_3^+$ cluster can abstract one, two, and three H atoms from methane, ethane, and *n*-butane, respectively:

$$AuNbO_3^+ + CH_4 \rightarrow AuNbO_3H^+ + CH_3 \tag{13.1}$$

$$AuNbO_3^+ + C_2H_6 \rightarrow AuNbO_3H_2^+ + C_2H_4 \tag{13.2}$$

$$AuNbO_3^+ + C_4H_{10} \rightarrow AuNbO_3H_3^+ + C_4H_7 \tag{13.3}$$

It was found that, when the oxygen-containing clusters are doped with gold atoms, the activation of multiple C–H bonds of one alkane molecule with high selectivity is enabled. The activation of multiple C–H bonds is important as it directly generates alkenes, which are value-added products from alkanes or alkenyl radicals [110, 111].

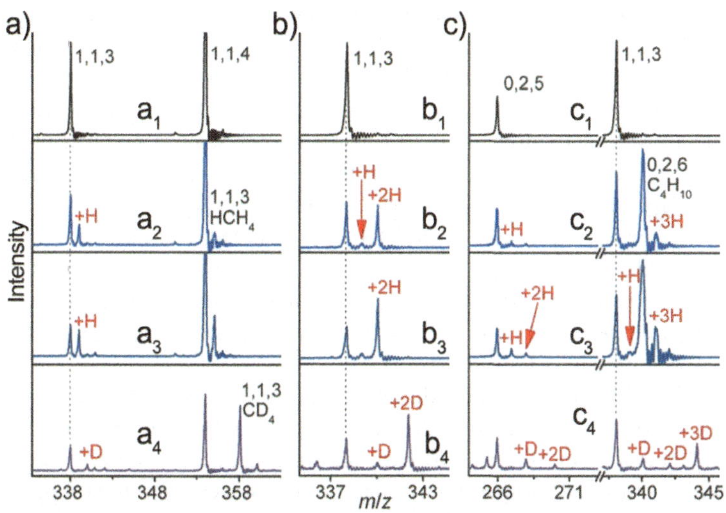

Fig. 13.6 Selected time-of-flight mass spectra for interactions of $Au_xNb_yO_z^+$ with **a** methane, **b** ethane, and **c** *n*-butane. Reference spectra without hydrocarbons in the reaction cell are shown in a_1, b_1, and c_1. The reactant gases in the cell are: a_2) CH_4 (0.25 Pa), a_3) CH_4 (0.35 Pa), a_4) CD_4 (0.35 Pa); b_2) C_2H_6 (0.17 Pa), b_3) C_2H_6 (0.28 Pa), b_4) C_2D_6 (0.28 Pa); and c_2) *n*-C_4H_{10} (0.014 Pa), c_3) *n*-C_4H_{10} (0.017 Pa), c_4) *n*-C_4D_{10} (0.017 Pa). "*x, y, z*" denotes $Au_xNb_yO_z^+$. The "+H", "+D", etc. mark the product signals with respect to $AuNbO_3^+$ or $Nb_2O_5^+$. Most of the $Au_xNb_yO_z^+$ clusters pick up the hydrocarbon molecules in the reaction cell. The $Nb_2O_6C_4H_{10}^+$ signal overlaps with $AuNbO_3H_2^+$ (c_2 and c_3). Reproduced with permission from Ref. [110]. Copyright 2013 John Wiley and Sons

13.3 Catalysis of Pt Clusters

Platinum-based heterogeneous catalysts are critical to many important commercial and industrial chemical processes. One of the aims in heterogeneous catalysis is to gain a better understanding of the catalytic behaviors of the supported metal clusters hence to optimize the efficiency and selectivity of industrial catalysts. However, there still exists rare direct experimental proof that model catalysts consisted of such small clusters indeed reveal such variations as size-dependence in the catalytic activity. Promising investigations toward such a capability have been attained by noting the size-dependent adsorption of small molecules on gas-phase clusters [112–114]. In particular, extensive results shed light on the sound catalysis of platinum clusters on various reactions [115–129], among which the catalytic oxidation of CO is known as one of the most important catalytic reactions. It was notable that, although the conversion of CO and O_2 into CO_2 in the gas phase is thermodynamically allowed and has a free enthalpy of −280 kJ/mol [130], the activation energy for the dissociation of O_2 has to be overcome, which calls on necessary catalysts to initiate such reactions.

Previous investigations have shown that, on highly coordinated Pt(111) single-crystal surfaces CO can be oxidized with the transfer of oxygen atoms produced during the dissociation process of O_2 molecules on the platinum surface and hence no additional activation for the actual oxidation step is needed [131–133]. Note that this mechanism is sensitive to the temperature and dependent on the character of the reactive sites in such systems [130, 134, 135]. Further studies on Pt(100) and Pt(110) surfaces revealed a complicated modification of the overall CO oxidation ascribed to the adsorption-induced changes of the surface structure, leading to oscillations of the steady-state rate of catalytic CO oxidation on clean Pt surfaces [136]. The strong dependence on surface structure of Pt is well interpreted through the studies on a size dependence of free Pt_n clusters [130, 137]. Figure 13.7 presents such an investigation in *Heiz* group with a focus on the catalytic oxidation of CO on monodispersed platinum clusters. The very small clusters consisting only of a few atoms show pronounced size effects in their catalytic behavior, due to the changing coordination number in different geometric structures and/or the altered electronic structures as a function of cluster size [130].

In detail, to obtain identical conditions for the study of the catalytic reactivity of the different Pt clusters, they first exposed the prepared model catalysts by a calibrated molecular beam doser at 90 K to an average of 20 $^{18}O_2$ molecules per Pt atom, which indicated saturation coverage on the clusters; and then they exposed the system to the same amount of $^{12}C^{16}O$ to exclude possible influence on the reactivity by different ratio of the reactant molecules [130, 131, 134, 135, 138]. In the temperature-programmed reaction (TPR) experiment, they detected the isotopically labelled CO_2 molecules which are catalytically produced on the cluster surfaces. As results, the catalytic action was given by integrating the TPR signal of the CO_2 molecules and normalizing to the number of Pt clusters. Figure 13.7A shows the TPR spectra for the CO oxidation on supported Pt_n ($8 \leq n \leq 20$) clusters, where each cluster size reveals different oxidation temperatures (peaks labelled with α, β_1, and

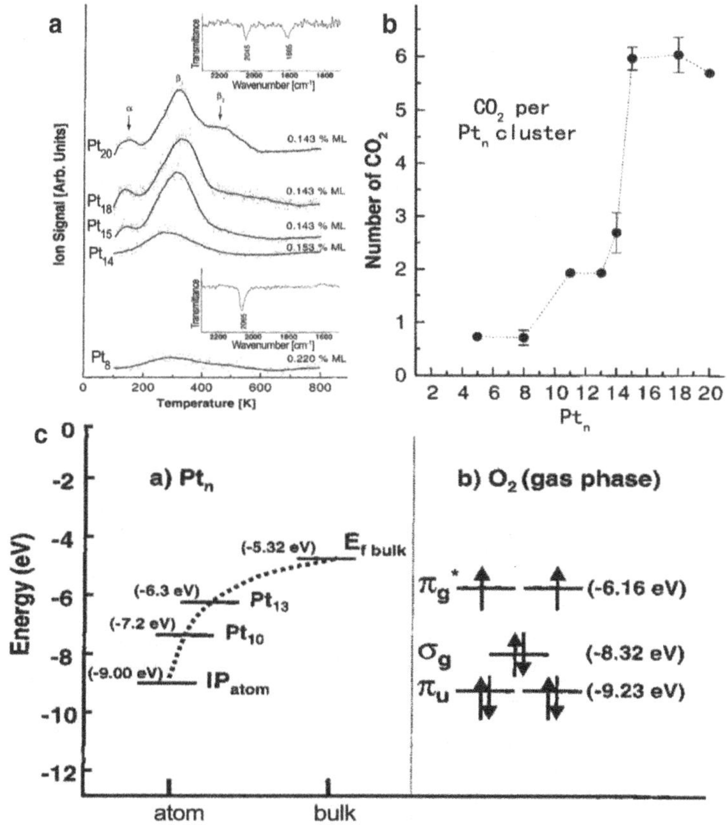

Fig. 13.7 **A** Catalytic CO_2 formation for different cluster sizes from temperature-programmed reaction (TPR) experiments. Cluster coverage is expressed in percent of a monolayer, where one monolayer corresponds to 2.2×10^{15} clusters/cm^2. Cluster coverage is scaled with an estimated area covered by each cluster size. Different CO_2 formation mechanisms (α, β_1, β_2) are labeled according to single-crystal studies. The inset shows measured vibrational frequencies of CO adsorbed on clean deposited Pt_{20} (above) and Pt_8 (below) clusters. **B** Total number of catalytically produced CO_2 molecules as a function of cluster size. **C** Energy diagram of the relevant electronic states for the oxygen dissociation for gas-phase clusters and free oxygen. The clusters' HOMOs are between the atomic limit (−9.00 eV) and the bulk limit (−5.32 eV). Included are the calculated positions of the centers of the d-bands for Pt_{10} (−7.2 eV) and Pt_{13} (−6.3 eV). Dotted line: classical conductive droplet model (the expected fine structure of the ionization potentials observed for small metal clusters is not considered). Reproduced with permission from Ref. [130]. Copyright 1999 American Chemical Society

β_2) in addition to a variety of the different ion signal intensities. Correlated FTIR experiments (insets) showed that only one CO absorption frequency (2065 cm^{-1}) for Pt_8 but two absorption frequencies (2045 and 1805 cm^{-1}) for Pt_{20}, which suggests that different oxidation temperatures correspond to different catalytic processes occurring on different active sites on the Pt clusters [130]. A further expressing on the catalytic activity of the monodispersed Pt clusters as the number of catalyzed CO_2 molecules per cluster is displayed in Fig. 13.7B. Note that the catalytic reactivity shows a local decrease for the cluster Pt_{13} which bears unique electronic structures and symmetry [139].

Through a comparison with the catalytic oxidation of CO on Pt single-crystal surfaces, the observed overall size-dependent reactivity in such system was rationalized with changes of the cluster structure, together with simple frontier orbital considerations. Considering the oxidation temperatures on size-selected Pt clusters are in a similar range as on Pt single-crystal surfaces, the reaction sites were identified accordingly based on the following two main mechanisms:

i. reaction α on Pt(111) terrace sites,

$$O_2 + 2CO_{terrace} \rightarrow 2CO_2 \tag{13.4}$$

ii. reaction β on Pt(355) stepped sites, including

$$\beta_1, O_{terrace} + CO_{terrace} \rightarrow CO_2 \tag{13.5}$$

$$\beta_2, O_{step} + CO_{step} \rightarrow CO_2 \tag{13.6}$$

$$\beta_3, O_{step} + CO_{terrace} \rightarrow CO_2 \tag{13.7}$$

where the reaction α was also found at 160 K involving hot oxygen atoms [131, 140]; while for the stepped Pt(355) surface, reactions β at 290, 350, and 200 K have also been identified, respectively [135]. The throughout presence of the β-mechanism (Fig. 13.7A) indicates dissociation of O_2 on all cluster sizes, but the α-mechanism was only observed for larger clusters, Pt_{15-20}, indicating that oxygen is adsorbed molecularly in an ionic state only on these finite species.

To further understand this size-dependent catalytic reactivity, the geometric and electronic structures of each Pt cluster must be taken into consideration. For the oxygen molecule, (Fig. 13.7C(b)), the electronic states of interest include π_u and σ_g (bonding, fully occupied) and π_g* (antibonding, half-occupied) orbitals. The adsorption and dissociation of O_2 likely occur providing there exists a resonance (energetically and symmetrically) between one of these orbitals and the cluster's density of state [141]. Specifically, fragmentation occurs when there is enough back-donation from the clusters into the antibonding π_g* state, or a sufficient donation from the π_u or σ_g orbital into the cluster. The energies of the fragment orbitals were calculated to be −9.23 eV, −8.32 eV, and −6.1 eV for π_u, σ_g, and π_g* respectively [130, 141]. On the cluster side (Fig. 13.7C(a)), the energy of the highest occupied molecular orbitals (HOMO) of the clusters changes with the cluster size and is generally associated with the energy of the d-band center [44, 45]. This shows a range from −9.0 eV (the ionization potential of a Pt atom) to −5.32 eV (the Fermi energy of bulk platinum) [130]. Based on this, the increase in the catalytic activity from Pt_8 to Pt_{15} can be rationalized with the decrease of the ionization potential and the change in the central position of the d-band, as well as the concomitantly enhanced resonance with the antibonding $\pi_g{}^*$ state of O_2 [130]. The maximum catalysis for Pt_{15} suggests its largest back-donation. Further increasing the cluster size lowers the cluster's HOMO and hence results in a weaker resonance with the antibonding $\pi_g{}^*$ state of O_2; while for the very small clusters (including Pt atom) the back-donation is small because the HOMO energy of the cluster mismatches with $\pi_g{}^*$ of oxygen (Fig. 13.7C).

These studies have made the catalysis of Pt_n clusters being widely studied; however, the efficiency is relatively low considering a per-metal atom basis as only the surface active-site atoms are used. As a precious and expensive metal, maximum atom efficiency of Pt catalysts with single-atom dispersions is highly desired. However, the challenge involves not only the limited catalytic efficiency of single Pt atoms as demonstrated above, but also technical difficulty to make it in practical use. Recently researchers have found a new solution to this challenge and synthesized a single-atom catalyst that consists of only isolated single Pt atoms anchored to the surfaces of iron oxide nanocrystallites (Fig. 13.8) [34]. This single-atom catalyst shows excellent stability and high activity for both CO oxidation and preferential oxidation of CO in H_2; also it has extremely high atom efficiency of Pt. Density functional theory (DFT) calculations showed that the high catalytic activity correlates with the partially vacant 5*d* orbitals of the positively charged, high-valent Pt atoms, which helps to reduce both the CO adsorption energy and the activation barrier for CO oxidation in forming CO_2 [34].

13.4 Catalysis of Copper-Related Systems

Some other catalysts such as copper, Cu/ZnO (/Al_2O_3) binary system and even a ternary system Cu/ZnO/Al_2O_3 were widely used in industrial production of methanol [142–146], with a worldwide demand of ~50 Mtons per year, from gas mixtures (H_2/CO_2/CO) at elevated pressures and temperatures. Binary systems (Cu/ZnO and Cu/Al_2O_3) and ternary Cu/ZnO/Al_2O_3 catalysts were important in methanol synthesis with respect to their catalytic activity and stability within the reactions. While the industrial catalysts containing low amounts of a refractory oxide displayed improved catalysis [146], the key to high performance is a largely accessible Cu surface area [142, 147]. Similar catalytic systems also attract interest for potential use of methanol as a sustainable synthetic fuel obtained by hydrogenation of captured CO_2 [148]. While the phenomenological optimization of the preparation of active catalyst on these systems is well improved, the fundamental understanding of its catalytic activity is still illusive to be further explored [149–151].

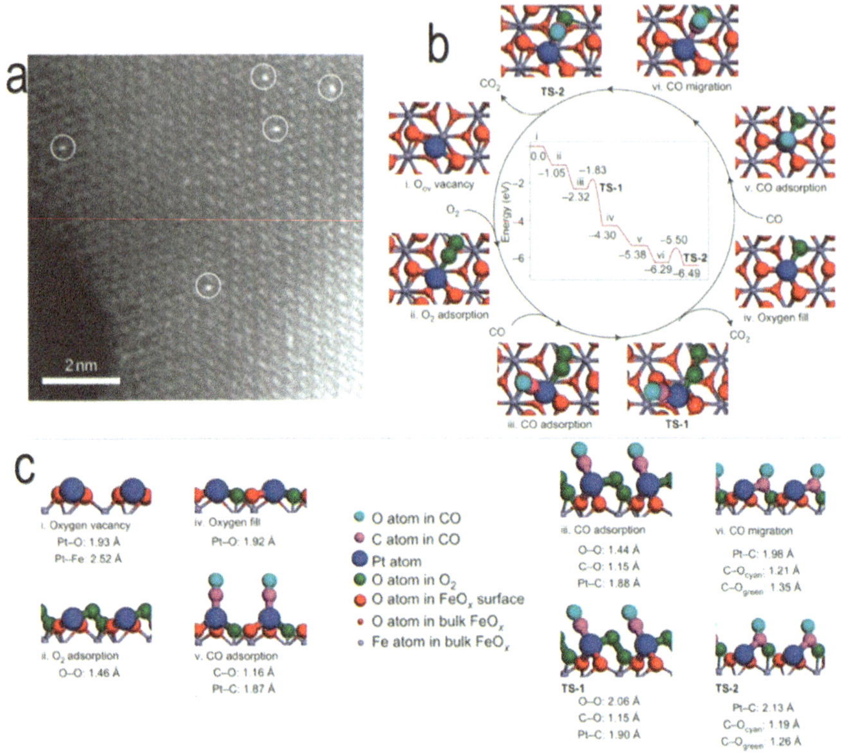

Fig. 13.8 **a** Aberration-corrected scanning transmission electron microscopy (STEM) images of Pt single atoms (white circles) which were uniformly dispersed on the FeO_x support and occupy exactly the positions of the Fe atoms. **b/c** The proposed reaction pathways for CO oxidation on the Pt_1/FeO_x catalyst, top view (**b**) and side view (**c**). After pre-treatment by H_2, the stoichiometric haematite surfaces near the Pt atoms were reduced partially to form an O_{vac} (step i) that can adsorb the O_2 reactants (step ii) as CO is adsorbed on the single Pt atoms (step iii). Through an activation barrier of 0.49 eV (TS-1), the first CO_2 molecule is released and the surface oxygen vacancy is healed by the remaining O_{ad} atom of the O_2 reactant (step iv). When the second CO molecule is adsorbed at the Pt atom (step v), it migrates to a neighboring oxygen atom (step vi) to form a transition state with a barrier of 0.79 eV (TS-2), which leads to a new CO oxidation. By releasing the second CO_2, the Pt-embedded stoichiometric surface is reduced again to form a new O_{vac} (step i). The inset in the cycle (**a**) shows the calculated energy profile, with the partially reduced sample system as the reference for the energies (in eV). After one catalytic cycle, the catalyst is recovered and releases two CO_2 molecules. Reproduced with permission from Ref. [34]. Copyright 2011 Springer Nature

For the Cu–ZnO catalysts, the synergy effect [143, 152, 153] was proposed to interpret why the presence of ZnO increases the intrinsic activity of Cu-based catalysts for methanol synthesis [154–160]. To study the role of defects in the real $Cu/ZnO/(Al_2O_3)$ composite system, Behrens [142] showed a systematic study on how to identify the crucial atomic structure motif for the industrial methanol synthesis catalyst. They developed a series of functional catalysts and compared them to a pure

Cu metal reference sample. While the ZnO-free Cu reference exhibited little activity, the catalysts that performed best were prepared by following the industrial synthesis method. The performance by the Cu surface areas resulting in the intrinsic activities, normalized to the intrinsically most active catalyst for each temperature, is shown in Fig. 13.9. The scatter of the data showed that the Cu surface area alone can do not explain the differences in performance. The neutron scattering data permitted for a sufficiently reliable fitting of the peak position 400 of the nanostructured Cu phase (Fig. 13.9c). For the inactive pure Cu sample, both ratios fell near the expected ideal value, while the catalytic materials showed a lower value for $h = 1$ and a higher one for $h = 2$, which is consistent with the presence of stacking faults in the Cu clusters (Fig. 13.9d) [142].

Utilizing aberration-corrected high-resolution transition electron microscopy (HRTEM), Behrens et al. [142] examined the relation of bulk defects and surface steps in the most active catalysts, as displayed in Fig. 13.10, which accords with the model situations of stepped Cu(211) surface or the stacking fault-created step shown in the inset of Fig. 13.9d. A vast majority of Cu nanoparticles was faulted and exhibited planar extended defects, stacking faults, and twin boundaries. The curvature of the spherical particle causes the surface to intrinsically contain a set of steps. The sample related to the HRTEM image of Fig. 13.10a showed a pattern of stepped surface facets like (211) and (522) being responsible for the curvature at the lower exposed side of the Cu nanoparticle, which is associated with an inward curvature of the surface and does not occur on regular spherical or ellipsoidal fcc particles or Wulff polyhedra. The other sample (Fig. 13.10b) showed planar bulk defects, seen as the twin boundaries, reflected in changes of the surface faceting creating a local inward curvature of the nanoparticle. Another sample (Fig. 13.10c, d) showed that twin boundaries could create distinctive surface ensembles even if the Cu surface of a larger nanoparticle appears essentially flat or the position stuck out of the regular surface. These arrangements were described as a high-energy site created by the termination of a planar defect at the surface of the Cu nanoparticles [142].

The undistorted pure Cu was quite inactive in the methanol synthesis experiment, which was also confirmed by DFT calculations for the flat Cu(111) surface. Figure 13.11 presents the DFT-calculated energy diagrams for CO_2 and CO hydrogenation on close-packed (black curve), stepped (blue curve), and Zn substituted steps (red curve) respectively. Essentially all the intermediates are thermodynamically less stable than CO_2 and H_2 in the gas phase (Fig. 13.11B), but is clearly shown that the flat Cu(111) surface bound intermediates more weakly than did Cu(211). Both the energies of the intermediates and the transition-states were stabilized considerably for the (211) surface compared with the (111) surface, rendering the steps more vigorous than the terraces. Similarly, CO hydrogenation proceeded via an initial hydrogenation of the carbon atom of CO, through a few intermediates such as HCO, H_2CO, H_3CO and then the aim product methanol (H_4CO). Both for the hydrogenation of CO_2 and CO, the last two intermediates are the same, and the order of activity is CuZn(211) > Cu(211) > Cu(111) as the steps lower the adsorption energies of the intermediates substantially compared with the flat surface [142].

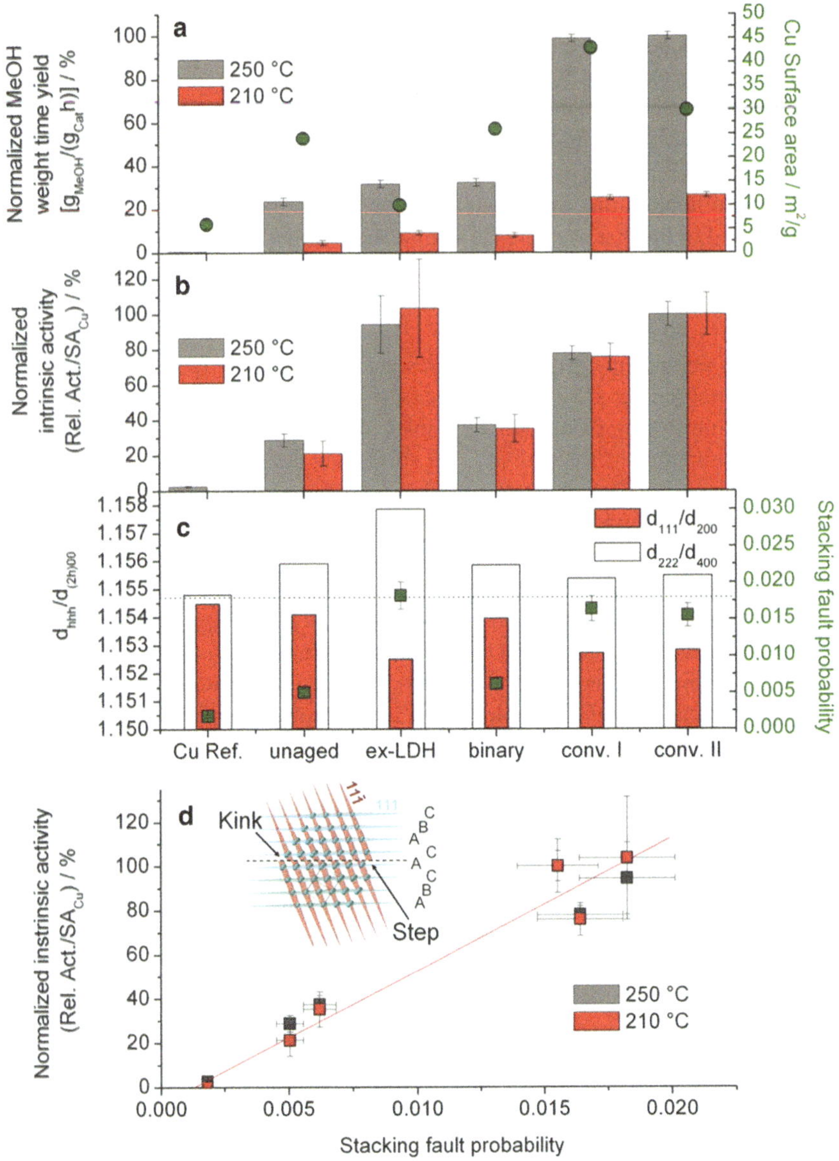

Fig. 13.9 **a** Catalytic activities and Cu surface areas of the Cu reference material and the Cu/ZnO/Al_2O_3 catalysts in methanol synthesis (P = 60 bar, T = 210°, 250°C, normalized to the most active sample). **b** Intrinsic activities per Cu surface area obtained after dividing by the Cu surface area (normalized: most active sample = 100% at each temperature). **c** Deviation of d_{111}/d_{200} and d_{222}/d_{400} observed in the neutron diffraction patterns and resulting stacking fault probabilities of the Cu particles. The dashed line refers to the ideal fcc structure. **D** Relation of the intrinsic activity of Cu to the concentration of stacking faults. (Inset) Schematic of how a stacking fault in 111 can generate kinks and surface steps in the 111 facet. Error bars indicate uncertainties determined on basis of replicate measurements (catalytic activity and copper surface area). Reproduced with permission from Ref. [142]

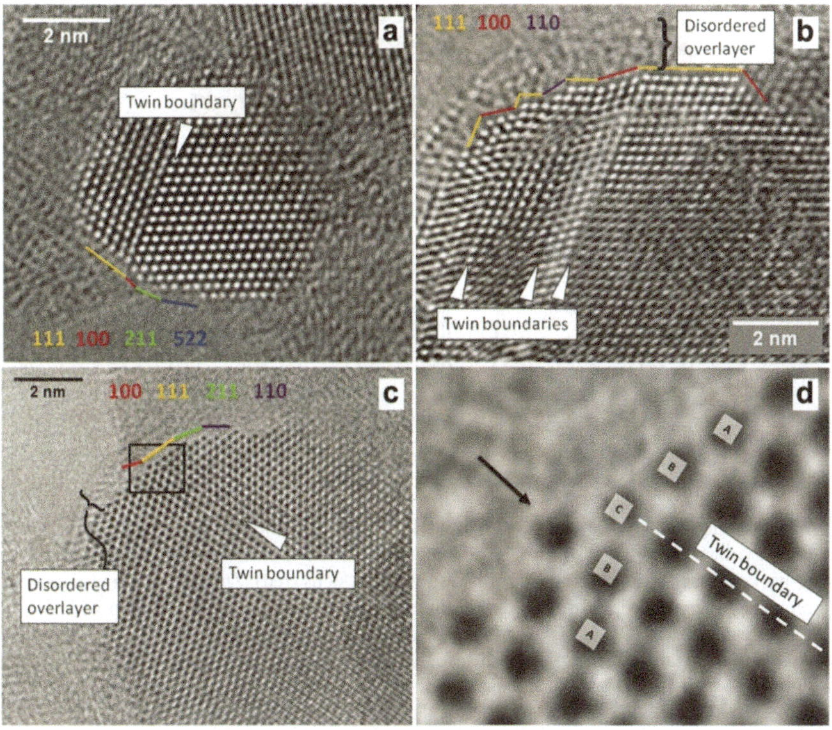

Fig. 13.10 Aberration-corrected HRTEM images of Cu particles in the conventionally prepared, most-active $Cu/ZnO/Al_2O_3$ catalysts (**a**, **b**, **c**), while **d** is a close-up of the marked area in (**c**). Reproduced with permission from Ref. [142]

With a combination analysis on the experimental and theoretical results, Behrens et al. [142] demonstrated a model for the active site of methanol synthesis over industrial catalysts, involving two aspects: (1) the presence of steps at the Cu surface, which can be stabilized by bulk defects like stacking faults or twin boundaries terminating at the surface; (2) the requirement of $Zn^{\delta+}$ at the defective (stepped) Cu surface, which is a result of a dynamically strong metal support interaction effect (in the high-performance catalyst) leading to partial coverage of the metal particles with the other oxides such as ZnO. Therefore, the increase in catalytic activity can be attributed to a stronger binding of the intermediates on stepped sites and lower energy barriers between them. Apparently, substitution of Zn into the Cu steps will strengthen the binding of the intermediates and hence enable an increase of the catalytic activity [142].

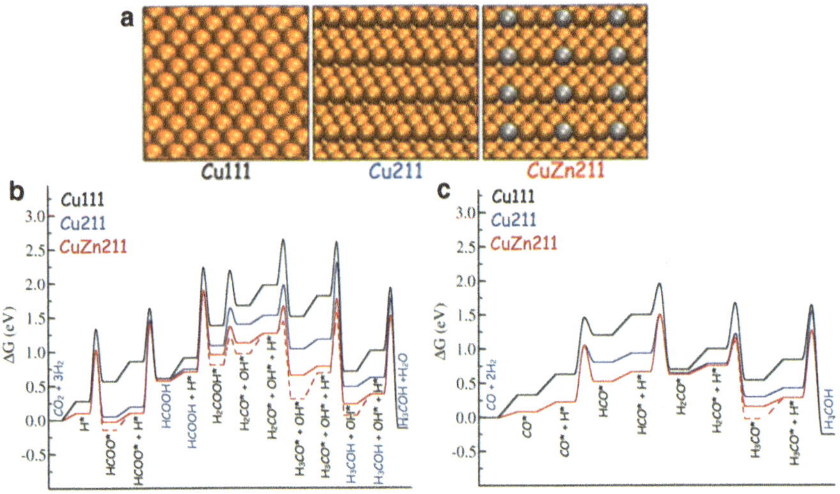

Fig. 13.11 The Cu(111), Cu(211), and CuZn(211) facets as viewed from perspective (**a**). Gibbs free energy diagram obtained from DFT calculations for CO_2 (**b**) and CO (**c**) hydrogenation on close-packed (black), stepped (blue), and Zn substituted steps (red). Zn substitution was modeled by replacing one (solid line) or two (dashed line) of the three Cu atoms of the step with Zn. All energies are relative to $CO_2 + 3H_2$ ($CO + 2H_2$) in the gas phase and the clean surfaces. Intermediates marked with a star are adsorbed on the surface. Gibbs free energies were calculated at $T = 500$ K and P of 40 bar of H_2, 10 bar of CO, and 10 bar of CO_2, respectively, and 1 bar of methanol and H_2O (corresponding to low conversion). Reproduced with permission from Ref. [142]

13.5 Catalysis of Titanium and Vanadium Oxides

Since Fujishima and Honda [161] reported the photosensitized decomposition of water into H_2 and O_2 using an electrochemical cell consisting of a Pt electrode and a TiO_2 semiconductor electrode in early 1970s, photocatalysis has attracted extensive research interest due to its potential in the conversion of light energy into useful chemical energy [162–168]. Photocatalysis, with a focus on TiO_2 and also involving many other semiconductor materials, has been applied to a variety of reactions to address the reduction and/or elimination of environmental pollutants in water and air. Extensive investigations have demonstrated the useful application, such as the decomposition of micro-organisms like bacteria and viruses [169, 170], the deactivation of cancer cells [171, 172], the degradation and elimination of offensive odors [173, 174], the photo-splitting of water to produce hydrogen [175–179], the fixation of nitrogen [180–183], and the clean-up of oil spills [184–186], etc. For these photocatalytic sensitizers (such as TiO_2, ZnO, and Fe_2O_3), light-induced redox processes occur due to the unique electronic structures of them, that is, a filled valence band and an empty conduction band. When the energy of a photon corresponds to or exceeds the band gap energy of such semiconductors, an electron is promoted into the conduction band, leaving a hole in the valence band. The electrons in the conduction band and holes in the valence band can recombine and dissipate the input energy as heat and become trapped in metastable surface states, or they can react with electron acceptors and electron donors adsorbed on the semiconductor surface or within the surrounding electrical double layer of the charged particles. Moreover, the illuminated semiconductors have been successfully applied in the remediation of contaminants for a wide variety of compounds [187–194], and recently the discovery of photo-induced superhydrophilicity and the self-cleaning effect of TiO_2 thin films has led to even wider applications for TiO_2 photocatalysts [195]. Several recent investigations have also enlightened the studies of preparing titanium oxide photocatalysts loaded on activated supports by an ionized cluster beam (ICB) method, based on which the transparent TiO_2 thin film photocatalysts showed specific interference fringes and enhanced photocatalytic reactivity [196–205].

In various studies as partly have presented above, it was found that clusters of selected sizes can serve as surface sites, where their structures may have geometries akin to steps, ledges, or corners, with characteristic accompanying charge densities [206–209]. Comparing with single-crystal surfaces, certain cluster structures can serve as ideal model surface sites and hence can be readily studied using standard methods from cluster science [210, 211]. The *Castleman* group has had a long-standing interest in cluster science applied for unraveling certain catalytic mechanisms such as oxygen transfer utilizing specific clusters as model surface sites [206, 207]. Figure 13.12 displays such a model showing the catalysis of vanadium oxides in reacting with hydrocarbons [10, 212].

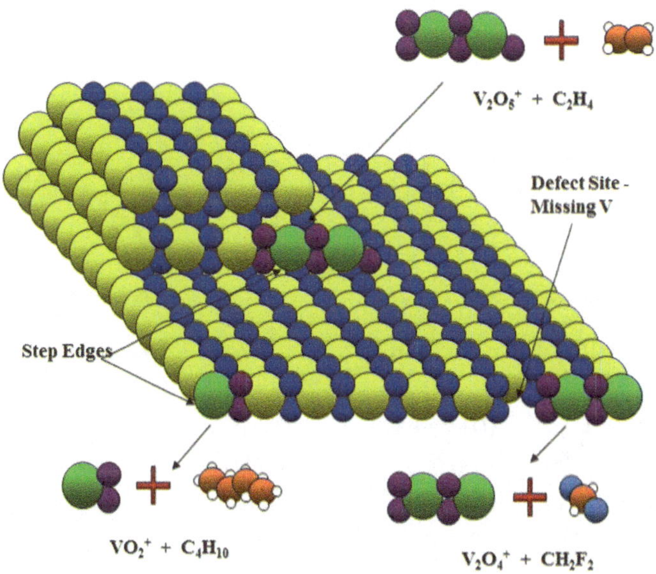

Fig. 13.12 Steps/ledges/corners: reactive centers mimicked by clusters. Reproduced with permission from Ref. [10]

One example in identifying a catalytic mechanism for oxygen transfer reactions involved the formation of acetaldehyde was smoothly made from ethylene interacting with vanadium oxides at the reaction sites. It was found that, among a wide range of $V_xO_y^+$ clusters studied (e.g., $V_2O_{4-6}^+$, $V_3O_{6-8}^+$, $V_4O_{9-11}^+$), the oxygen-rich species such as $V_2O_5{}^+$ and $V_4O_{10}{}^+$ (interesting both a 2:5 ratio of the metal-oxygen atomic composition) were found to prefer the reactivity of oxygen transfer, as displayed in Fig. 13.13.

This is also consistent with bulk catalysts that yield a similar reaction product, validating the proposal that gas-phase cluster experiments provide insight into the influence of composition, geometry, and size on catalytic reactivity. In combination with theory, the enhanced reactivity of these cluster species was traced to the presence of an oxygen-centered radical on a metal atom, and its impact on the energy barrier. See calculated reaction profile in Fig. 13.14. Specifically, for V_2O_5 the main reaction channels with energy differences were demonstrated as follows.

$$V_2O_5^+ + C_2H_4 \rightarrow V_2O_4^+ + O + C_2H_4 \Delta E = +2.24\text{ eV} \tag{13.8}$$

$$V_2O_5^+ + C_2H_4 \rightarrow V_2O_4^+ + C_2H_4O\ \Delta E = -2.53\text{ eV} \tag{13.9}$$

Equation 13.9 apparently commits the experimentally observed reactivity of catalytic oxidation. Note that the energy calculated to form the association product is -3.53 eV for the ethylene bound to one terminal oxygen atom while -4.44 eV for the ethylene bound to the two terminal oxygen atoms in the $V_2O_5{}^+$ structure (Fig. 13.14, bottom) [213].

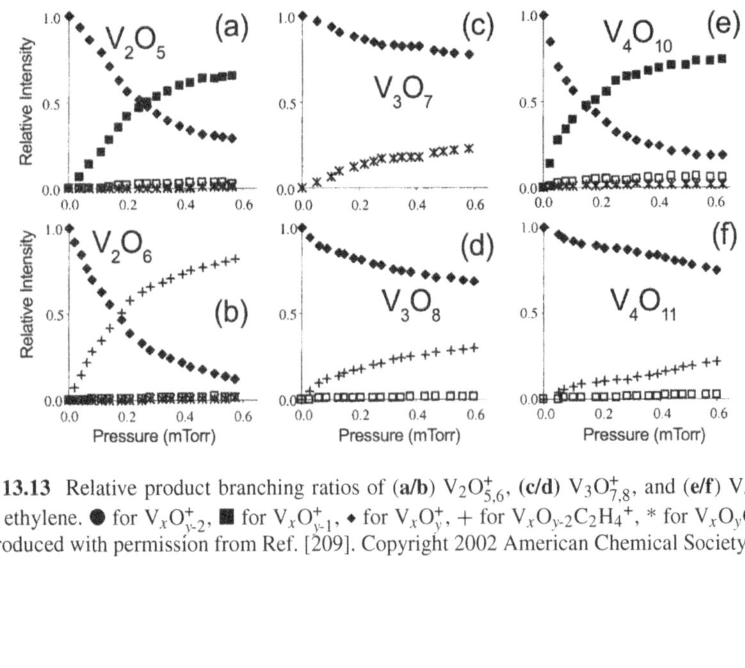

Fig. 13.13 Relative product branching ratios of **(a/b)** $V_2O_{5,6}^+$, **(c/d)** $V_3O_{7,8}^+$, and **(e/f)** $V_4O_{10,11}^+$ with ethylene. ● for $V_xO_{y-2}^+$, ■ for $V_xO_{y-1}^+$, ◆ for $V_xO_y^+$, + for $V_xO_{y-2}C_2H_4{}^+$, * for $V_xO_yC_2H_4{}^+$. Reproduced with permission from Ref. [209]. Copyright 2002 American Chemical Society

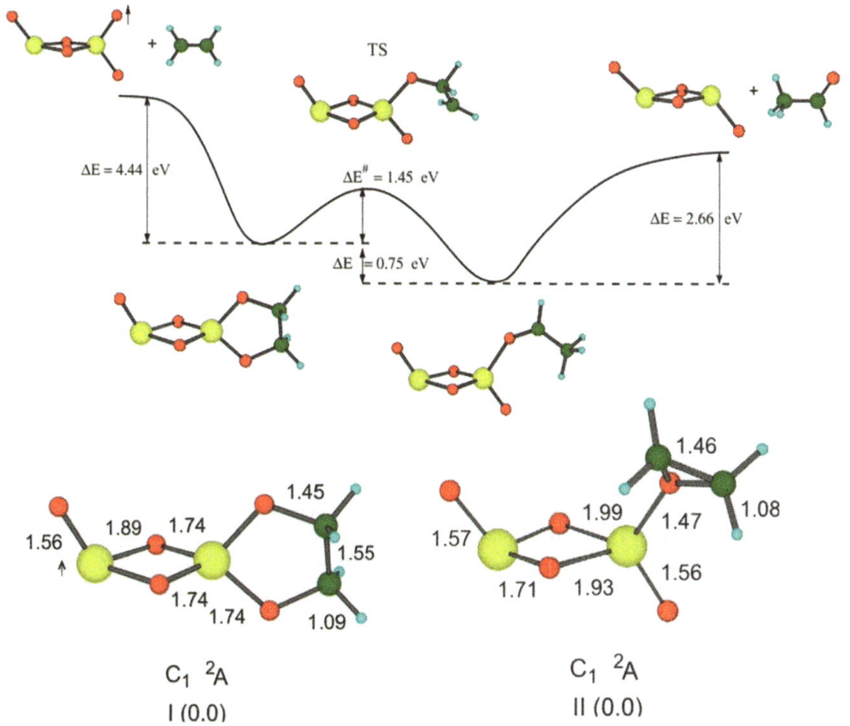

Fig. 13.14 Energetic profile for the oxygen transfer reaction of $V_2O_5^+$ with ethylene, where the structures of the association product of $V_2O_5^+$ with ethylene were displayed with the ethylene is bound to two oxygen atoms or one oxygen atom. Reproduced with permission from Ref. [213]. Copyright 2003 American Chemical Society

After the above description, it is no surprise now even to researchers outside the field of cluster science to recognize that much can be learned of its relevance to catalysis. The objective of metal cluster catalysis is to present how fundamental insights into reaction mechanisms can be acquired through a joint and compatible experimental and theoretical approach. Comparing with crystallographic planes, certain cluster structures can serve as ideal model surface sites and hence can be easily studied using standard methods from cluster science. Researches have made advances in terms of identifying suitable catalysts in place of palladium and demonstrating that a study of cluster ions is a viable alternative method to study mechanisms of catalysis, design new catalysts, and determine reactive sites.

References

1. A.E. Shilov, G.B. Shul'pin, Chem. Rev. **97**, 2879–2932 (1997)
2. J.A. Labinger, J.E. Bercaw, Nature **417**, 507–514 (2002)
3. D. Schröder, H. Schwarz, Angew. Chem. Int. Ed. **34**, 1973–1995 (1995)
4. D. Balcells, E. Clot, O. Eisenstein, Chem. Rev. **110**, 749–823 (2010)
5. W. Lai, C. Li, H. Chen, S. Shaik, Angew. Chem. Int. Ed. **51**, 5556–5578 (2012)
6. N. Dietl, M. Schlangen, H. Schwarz, Angew. Chem. Int. Ed. **51**, 5544–5555 (2012)
7. J. Roithová, D. Schröder, Chem. Rev. **110**, 1170–1211 (2009)
8. Y. Gong, M. Zhou, L. Andrews, Chem. Rev. **109**, 6765–6808 (2009)
9. R.M. Bullock, ed., in *Catalysis Without Precious Metals* (Wiley-VCH Verlag & Co. KGaA, Weinheim, 2010)
10. A.W. Castleman, Catal. Lett. **141**, 1243–1253 (2011)
11. H.-J. Zhai, L.-S. Wang, Chem. Phys. Lett. **500**, 185–195 (2010)
12. S. Feyel, J. Döbler, D. Schröder, J. Sauer, H. Schwarz, Angew. Chem. Int. Ed. **118**, 4797–4801 (2006)
13. S. Feyel, J. Döbler, D. Schröder, J. Sauer, H. Schwarz, Angew. Chem. Int. Ed. **45**, 4681–4685 (2006)
14. S. Feyel, J. Döbler, R. Höckendorf, M. K. Beyer , J. Sauer and H. Schwarz, Angew. Chem. Int. Ed. **120**, 1972–1976 (2008)
15. N. Dietl, M. Engeser, H. Schwarz, Angew. Chem. Int. Ed. **121**, 4955–4957 (2009)
16. G. de Petris, A. Troiani, M. Rosi, G. Angelini, O. Ursini, *Chem. Eur. J*. **15**, 4248–4252 (2009)
17. Y.-X. Zhao, X.-N. Wu, Z.-C. Wang, S.-G. He, X.-L. Ding, Chem. Commun. **46**, 1736–1738 (2010)
18. J.-B. Ma, X.-N. Wu, X.-X. Zhao, X.-L. Ding, S.-G. He, Phys. Chem. Chem. Phys. **12**, 12223–12228 (2010)
19. Z.-C. Wang, X.-N. Wu, Y.-X. Zhao, J.-B. Ma, X.-L. Ding and S.-G. He, Chem. Phys. Lett. **489**, 25–29 (2010)
20. Z.-C. Wang, T. Weiske, R. Kretschmer, M. Schlangen, M. Kaupp, H. Schwarz, J. Am. Chem. Soc. **133**, 16930–16937 (2011)
21. N. Dietl, R. F. Höckendorf, M. Schlangen, M. Lerch, M. K. Beyer, H. Schwarz, Angew. Chem. Int. Ed. **123**, 1466–1470 (2011)
22. J.-B. Ma, Z.-C. Wang, M. Schlangen, S.-G. He, H. Schwarz, Angew. Chem. Int. Ed. **51**, 5991–5994 (2012)
23. M. Chen, D.W. Goodman, Chem. Soc. Rev. **37**, 1860–1870 (2008)
24. X.-L. Ding, X.-N. Wu, Y.-X. Zhao, S.-G. He, Acc. Chem. Res. **45**, 382–390 (2012)
25. H.F. Liu, R.S. Liu, K.Y. Liew, R.E. Johnson, J.H. Lunsford, J. Am. Chem. Soc. **106**, 4117–4121 (1984)
26. S. Arndt, G. Laugel, S. Levchenko, R. Horn, M. Baerns, M. Scheffler, R. Schlögl, R. Schomäcker, Catal. Rev. **53**, 424–514 (2011)
27. H. Schwarz, Angew. Chem. Int. Ed. **50**, 10096–10115 (2011)
28. Y.-X. Zhao, X.-N. Wu, J.-B. Ma, S.-G. He, X.-L. Ding, Phys. Chem. Chem. Phys. **13**, 1925–1938 (2011)
29. M. Nößler, R. Mitrić, V. Bonačić-Koutecký, G.E. Johnson, E.C. Tyo, A.W. Castleman, Angew. Chem. Int. Ed. **49**, 407–410 (2010)
30. X.-L. Ding, Y.-X. Zhao, X.-N. Wu, Z.-C. Wang, J.-B. Ma, S.-G. He, Chem. - Eur. J. **16**, 11463–11470 (2010)
31. K. Tanaka, Catal. Today **154**, 105–112 (2010)
32. F. Besenbacher, J.V. Lauritsen, S. Wendt, Nano Today **2**, 30–39 (2007)
33. C. Harding, V. Habibpour, S. Kunz, A.N.S. Farnbacher, U. Heiz, B. Yoon, U. Landman, J. Am. Chem. Soc. **131**, 538–548 (2009)
34. B. Qiao, A. Wang, X. Yang, L.F. Allard, Z. Jiang, Y. Cui, J. Liu, J. Li, T. Zhang, Nat. Chem. **3**, 634–641 (2011)

35. A.W. Castleman Jr., S. Wei, Annu. Rev. Phys. Chem. **45**, 685–719 (1994)
36. M. Haruta, Catal. Today **36**, 153–166 (1997)
37. T.A. Nijhuis, T. Visser, B.M. Weckhuysen, J. Phys. Chem. B **109**, 19309–19319 (2005)
38. A.K. Sinha, S. Seelan, S. Tsubota, M. Haruta, Top. Catal. **29**, 95–102 (2004)
39. M. Juliusa, S. Robertsa, J.Q. Fletchera, Gold Bull. **43**, 298–306 (2010)
40. H. Tsunoyama, H. Sakurai, N. Ichikuni, Y. Negishi, T. Tsukuda, Langmuir **20**, 11293–11296 (2004)
41. G. Bond, Gold Bull. **42**, 337–342 (2009)
42. T.C. Boorman, I. Larrosa, Chem. Soc. Rev. **40**, 1910–1925 (2011)
43. G. Zhang, Y. Peng, L. Cui, L. Zhang, Angew. Chem. Int. Ed. **48**, 3112–3115 (2009)
44. J.C. Fierro-Gonzalez, B.C. Gates, Chem. Soc. Rev. **37**, 2127–2134 (2008)
45. R. Coquet, K.L. Howard, D.J. Willock, Chem. Soc. Rev. **37**, 2046–2076 (2008)
46. Z.-Y. Li, Y.-X. Zhao, X.-N. Wu, X.-L. Ding, S.-G. He, Chem. Eur. J. **17**, 11728–11733 (2011)
47. G.J. Hutchings, Catal. Today **100**, 55–61 (2005)
48. Y. Zhu, H. Qian, R. Jin, J. Mater. Chem. **21**, 6793–6799 (2011)
49. S.M. Lang, T.M. Bernhardt, Faraday Discuss. **152**, 337–351 (2011)
50. H.-L. Jiang, T. Akita, T. Ishida, M. Haruta, Q. Xu, J. Am. Chem. Soc. **133**, 1304–1306 (2011)
51. S. Chrétien, S.K. Buratto, H. Metiu, Curr. Opin. Solid St. M. **11**, 62–75 (2007)
52. G.J. Hutchings, M. Haruta, Appl. Catal. A: General **291**, 2–5 (2005)
53. M. Haruta, M. Date, Appl. Catal. A: General **222**, 427–437 (2001)
54. M. Haruta, Size- and support-dependency in the catalysis of noble metals: gold as a typical example (2001)
55. G.C. Bond, D.T. Thompson, Catal. Rev. Sci. Eng. **41**, 319–388 (1999)
56. R. Meyer, C. Lemire, S.K. Shaikhutdinov, H.J. Freund, Gold Bull. **37**, 72–124 (2004)
57. G.C. Bond, Catal. Today **72**, 5–9 (2002)
58. H. Hakkinen, Chem. Soc. Rev. **37**, 1847–1859 (2008)
59. J. Li, X. Li, H.-J. Zhai, L.-S. Wang, Science **299**, 864–867 (2003)
60. P. Gruene, D.M. Rayner, B. Redlich, A.F.G. van der Meer, J.T. Lyon, G. Meijer, A. Fielicke, Science **321**, 674–676 (2008)
61. C. Bürgel, N.M. Reilly, G.E. Johnson, R. Mitrić, M.L. Kimble, A.W. Castleman, V. Bonačić-Koutecký, J. Am. Chem. Soc. **130**, 1694–1698 (2008)
62. S.M. Lang, T.M. Bernhardt, R.N. Barnett, U. Landman, Angew. Chem. Int. Ed. **49**, 980–983 (2010)
63. L. Jašíková, J. Roithová, Organometallics **31**, 1935–1942 (2012)
64. X.-B. Wang, Y.-L. Wang, J. Yang, X.-P. Xing, J. Li, L.-S. Wang, J. Am. Chem. Soc. **131**, 16368–16370 (2009)
65. Y. Shichibu, K. Suzuki, K. Konishi, Nanoscale **4**, 4125–4129 (2012)
66. D.-E. Jiang, M. Walter, Nanoscale **4**, 4234–4239 (2012)
67. P.R. Nimmala, A. Dass, J. Am. Chem. Soc. **133**, 9175–9177 (2011)
68. M. Moreno, F.J. Ibañez, J.B. Jasinski, F.P. Zamborini, J. Am. Chem. Soc. **133**, 4389–4397 (2011)
69. P.A. Clayborne, O. Lopez-Acevedo, R.L. Whetten, H. Gronbeck, H. Hakkinen, J. Chem. Phys. **135** (2011)
70. J.M. Pettibone, J.W. Hudgens, J. Phys. Chem. Lett. **1**, 2536–2540 (2010)
71. O. Lopez-Acevedo, H. Tsunoyama, T. Tsukuda, H. Häkkinen, C.M. Aikens, J. Am. Chem. Soc. **132**, 8210–8218 (2010)
72. J. Akola, K.A. Kacprzak, O. Lopez-Acevedo, M. Walter, H. Grönbeck, H. Häkkinen, J. Phys. Chem. C **114**, 15986–15994 (2010)
73. G. Shafai, S. Hong, M. Bertino, T.S. Rahman, J. Phys. Chem. C **113**, 12072–12078 (2009)
74. M. Walter, J. Akola, O. Lopez-Acevedo, P.D. Jadzinsky, G. Calero, C.J. Ackerson, R.L. Whetten, H. Grönbeck, H. Häkkinen, Proc. Natl. Acad. Sci. U. S. A. **105**, 9157–9162 (2008)
75. Y. Shichibu, Y. Negishi, T. Watanabe, N.K. Chaki, H. Kawaguchi, T. Tsukuda, J. Phys. Chem. C **111**, 7845–7847 (2007)

76. C.E. Briant, B.R.C. Theobald, J.W. White, L.K. Bell, D.M.P. Mingos, A.J. Welch, J. Chem. Soc., Chem. Commun. **0**, 201–202 (1981)
77. P. Pyykkö, Angew. Chem. Int. Ed. **43**, 4412–4456 (2004)
78. H. Schwarz, Angew. Chem. Int. Ed. **42**, 4442–4454 (2003)
79. P. Pyykkö, Annu. Rev. Phys. Chem. **63**, 45–64 (2012)
80. F. Mancin, L.J. Prins, P. Scrimin, Curr. Opin. Colloid Interface Sci. **18**, 61–69 (2013)
81. L. D'Souza, M. Noeske, R.M. Richards, U. Kortz, J. Colloid Interface Sci. **394**, 157–165 (2013)
82. S. Xie, H. Tsunoyama, W. Kurashige, Y. Negishi, T. Tsukuda, Acs Catal. **2**, 1519–1523 (2012)
83. A. Tesfamichael, K. Suggs, Z. Felfli, X.-Q. Wang, A.Z. Msezane, J. Phys. Chem. C **116**, 18698–18705 (2012)
84. M. Schlangen, H. Schwarz, Catal. Lett. **142**, 1265–1278 (2012)
85. S. Yin, E.R. Bernstein, Int. J. Mass Spectrom. **321–322**, 49–65 (2012)
86. T. Takei, T. Akita, I. Nakamura, T. Fujitani, M. Okumura, K. Okazaki, J. Huang, T. Ishida, M. Haruta, in *Advances in Catalysis*, vol. 55, ed. by B.C. Gates, F. C. Jentoft (2012), pp. 1–126
87. A.Z. Msezane, Z. Felfli, K. Suggs, A. Tesfamichael, X.-Q. Wang, Gold Bull. **45**, 127–135 (2012)
88. S. Antonello, M. Hesari, F. Polo, F. Maran, Nanoscale **4**, 5333–5342 (2012)
89. M. Haruta, T. Kobayashi, H. Sano, N. Yamada, Chem. Lett. **4**, 405 (1987)
90. G.J. Hutchings, J. Catal. **96**, 292 (1985)
91. A. Sanchez, S. Abbet, U. Heiz, W.D. Schneider, H. Häkkinen, R.N. Barnett, U. Landman, J. Phys. Chem. A **103**, 9573–9578 (1999)
92. U. Heiz, A. Sanchez, S. Abbet, W.D. Schneider, Chem. Phys. **262**, 189–200 (2000)
93. H. Häkkinen, S. Abbet, A. Sanchez, U. Heiz, U. Landman, Angew. Chem. Int. Ed. **42**, 1297–1300 (2003)
94. U. Heiz, E.L. Bullock, J. Mater. Chem. **14**, 564–577 (2004)
95. M. Arenz, U. Landman, U. Heiz, ChemPhysChem **7**, 1871–1879 (2006)
96. K.A. Kacprzak, L. Lehtovaara, J. Akola, O. Lopez-Acevedo, H. Hakkinen, Phys. Chem. Chem. Phys. **11**, 7123–7129 (2009)
97. D.-E. Jiang, S. Dai, Inorg. Chem. **48**, 2720–2722 (2009)
98. M. Walter, M. Moseler, J. Phys. Chem. C **113**, 15834–15837 (2009)
99. Y. Liu, H. Tsunoyama, T. Akita, T. Tsukuda, Chem. Commun. **46**, 550–552 (2010)
100. M. Haruta, Gold Bull. **37**, 27–36 (2004)
101. Y. Liu, H. Tsunoyama, T. Akita, T. Tsukuda, J. Phys. Chem. C **113**, 13457–13461 (2009)
102. Y. Liu, H. Tsunoyama, T. Akita, S. Xie, T. Tsukuda, Acs Catal. **1**, 2–6 (2010)
103. Y. Negishi, W. Kurashige, Y. Niihori, T. Iwasa, K. Nobusada, Phys. Chem. Chem. Phys. **12**, 6219–6225 (2010)
104. C.A. Fields-Zinna, M.C. Crowe, A. Dass, J.E.F. Weaver, R.W. Murray, Langmuir **25**, 7704–7710 (2009)
105. H. Himeno, K. Miyajima, T. Yasuike, F. Mafuné, J. Phys. Chem. A **115**, 11479–11485 (2011)
106. J. Gong, Chem. Rev. **112**, 2987–3054 (2011)
107. G.E. Johnson, N.M. Reilly, E.C. Tyo, A.W. Castleman Jr., J. Phys. Chem. C **112**, 9730–9736 (2008)
108. M.L. Kimble, A.W. Castleman Jr., Int. J. Mass Spectrom. **233**, 99–101 (2004)
109. M.L. Kimble, A.W. Castleman Jr., R. Mitrić, C. Bürgel, V. Bonačić-Koutecký, J. Am. Chem. Soc. **126**, 2526–2535 (2004)
110. X.-N. Wu, X.-N. Li, X.-L. Ding, S.-G. He, Angew. Chem. Int. Ed. **52**, 2444–2448 (2013)
111. V.D. Knyazev, I.R. Slagle, J. Phys. Chem. A **102**, 8932–8940 (1998)
112. E.K. Parks, G.C. Nieman, K.P. Kerns, S.J. Riley, J. Chem. Phys. **108**, 3731–3739 (1998)
113. A. Berces, P.A. Hackett, L. Lian, S.A. Mitchell, D.M. Rayner, J. Chem. Phys. **108**, 5476–5490 (1998)
114. R.L. Whetten, D.M. Cox, D.J. Trevor, A. Kaldor, Phys. Rev. Lett. **54**, 1494–1497 (1985)
115. Y. Xu, W.A. Shelton, W.F. Schneider, Abstr. Pap. Am. Chem. Soc. **232** (2006)
116. O.S. Alexeev, F. Li, M.D. Amiridis, B.C. Gates, J. Phys. Chem. B **109**, 2338–2349 (2005)

117. S. Bhaduri, Curr. Sci. **78**, 1318–1324 (2000)
118. A. Fukuoka, M. Osada, T. Shido, S. Inagaki, Y. Fukushima, M. Ichikawa, Inorg. Chim. Acta **294**, 281–284 (1999)
119. N. Toshima, K. Nakata, H. Kitoh, Inorg. Chim. Acta **265**, 149–153 (1997)
120. Y.Z. Yuan, K. Asakura, H.L. Wan, K. Tsai, Y. Iwasawa, Chem. Lett. 129–130 (1996)
121. L.H. Pignolet, M.A. Aubart, K.L. Craighead, R.A.T. Gould, D.A. Krogstad, J.S. Wiley, Coord. Chem. Rev. **143**, 219–263 (1995)
122. M.E. Curley, J.W. Bacon, I.V. Gubkina, L.H. Pignolet, Abstr. Pap. Am. Chem. Soc. **209**, 334-CHED (1995)
123. N. Toshima, T. Takahashi, Bull. Chem. Soc. Jpn. **65**, 400–409 (1992)
124. M.A. Aubart, L.H. Pignolet, Abstr. Pap. Am. Chem. Soc. **203**, 141-CHED (1992)
125. M.A. Aubart, L.H. Pignolet, J. Am. Chem. Soc. **114**, 7901–7903 (1992)
126. G.J. Li, T. Fujimoto, A. Fukuoka, M. Ichikawa, J. Chem. Soc., Chem. Commun. 1337–1339 (1991)
127. N. Toshima, K. Kushihashi, T. Yonezawa, H. Hirai, Chem. Lett. 1769–1772 (1989)
128. A. Fusi, R. Ugo, R. Psaro, P. Braunstein, J. Dehand, J. Mol. Catal. **16**, 217–230 (1982)
129. M. Ichikawa, Bull. Chem. Soc. Jpn. **51**, 2268–2272 (1978)
130. U. Heiz, A. Sanchez, S. Abbet, W.D. Schneider, J. Am. Chem. Soc. **121**, 3214–3217 (1999)
131. K.H. Allers, H. Pfnur, P. Feulner, D. Menzel, J. Chem. Phys. **100**, 3985–3998 (1994)
132. T. Matsushima, Surf. Sci. **127**, 403–423 (1983)
133. Y. Ohno, T. Matsushima, Surf. Sci. **241**, 47–53 (1991)
134. A. Szabo, M.A. Henderson, J.J.T. Yates, J. Chem. Phys. **96**, 6191–6202 (1992)
135. J. Xu, J.J.T. Yates, J. Chem. Phys. **99**, 725–732 (1993)
136. G. Ertl, P.R. Norton, J. Rüstig, Phys. Rev. Lett. **49**, 177–180 (1982)
137. Y. Shi, K.M. Ervin, J. Chem. Phys. **108**, 1757–1760 (1998)
138. A. Rar, T. Matsushima, Surf. Sci. **318**, 89–96 (1994)
139. N. Watari, S. Ohnishi, Phys. Rev. B **58**, 1665–1677 (1998)
140. T. Zambelli, J.V. Barth, J. Wintterlin, G. Ertl, Nature **390**, 495–497 (1997)
141. A.W.E. Chan, R. Hoffmann, W. Ho, Langmuir **8**, 1111–1119 (1992)
142. M. Behrens, F. Studt, I. Kasatkin, S. Kühl, M. Hävecker, F. Abild-Pedersen, S. Zander, F. Girgsdies, P. Kurr, B.-L. Kniep, M. Tovar, R.W. Fischer, J.K. Nørskov, R. Schlögl, Science **336**, 893–897 (2012)
143. M.S. Spencer, Top. Catal. **8**, 259–266 (1999)
144. I. Kasatkin, P. Kurr, B. Kniep, A. Trunschke, R. Schlögl, Angew. Chem. Int. Ed. **119**, 7465–7468 (2007)
145. M. Behrens, J. Catal. **267**, 24–29 (2009)
146. M. Kurtz, H. Wilmer, T. Genger, O. Hinrichsen, M. Muhler, Catal. Lett. **86**, 77–80 (2003)
147. M. Kurtz, N. Bauer, C. Büscher, H. Wilmer, O. Hinrichsen, R. Becker, S. Rabe, K. Merz, M. Driess, R. Fischer, M. Muhler, Catal. Lett. **92**, 49–52 (2004)
148. G.A. Olah, A. Goeppert, G.K.S. Prakash, *Beyond Oil and Gas: The Methanol Economy* (Wiley-VCH, Weinheim, Germany, 2006)
149. J. Yoshihara, C.T. Campbell, J. Catal. **161**, 776–782 (1996)
150. J. Szanyi, D.W. Goodman, Catal. Lett. **10**, 383–390 (1991)
151. P.B. Rasmussen, P.M. Holmblad, T. Askgaard, C.V. Ovesen, P. Stoltze, J.K. Nørskov, I. Chorkendorff, Catal. Lett. **26**, 373–381 (1994)
152. R. Burch, S.E. Golunski, M.S. Spencer, J. Chem. Soc., Faraday Trans. **86**, 2683–2691 (1990)
153. J. Nakamura, T. Uchijima, Y. Kanai, T. Fujitani, Catal. Today **28**, 223–230 (1996)
154. K. Klier, in *Advances in Catalysis*, vol. 31, ed. by H.P.D.D. Eley, B.W. Paul (Academic Press, 1982), pp. 243–313
155. V. Ponec, Surf. Sci. **272**, 111–117 (1992)
156. W.P.A. Jansen, J. Beckers, J.C.v.d. Heuvel, A.W. Denier v.d. Gon, A. Bliek, H.H. Brongersma, J. Catal. **210**, 229–236 (2002)
157. J.C. Frost, Nature **334**, 577–580 (1988)
158. J. Nakamura, Y. Choi, T. Fujitani, Top. Catal. **22**, 277–285 (2003)

159. K.C. Waugh, Catal. Today **15**, 51–75 (1992)
160. P.L. Hansen, J.B. Wagner, S. Helveg, J.R. Rostrup-Nielsen, B.S. Clausen, H. Topsøe, Science **295**, 2053–2055 (2002)
161. A. Fujishima, K. Honda, Nature **238**, 37–38 (1972)
162. M. Anpo, in *Green Chemistry: Challenging Perspectives*, ed. by P. Tundo, P. Anastas (Oxford University Press, New York, 2000), p. 269
163. M. Anpo, M. Takeuchi, H. Yamashita, S. Kishiguchi, in *Molecular and Supramolecular Photochemistry: Semiconductor Photochemistry and Photophysics*, vol. 10, ed. by V. Ramamurthy, K. Schanze (Dekker, New York, 2003), p. 361
164. M. Anpo, M. Takeuchi, K. Ikeue, S. Dohshi, Current Opinion Solid State Mater. Sci. **6**, 381–388 (2002)
165. M. Anpo, S. Dohshi, M. Kitano, Y. Hu, M. Takeuchi, M. Matsuoka, Annu. Rev. Mater. Res. **35**, 1–27 (2005)
166. H. Si-Ahmed, M. Calatayud, C. Minot, E.L. Diz, A.E. Lewandowska, M.A. Banares, Catal. Today **126**, 96–102 (2007)
167. J.-B. Ma, B. Xu, J.-H. Meng, X.-N. Wu, X.-L. Ding, X.-N. Li, S.-G. He, J. Am. Chem. Soc. **135**, 2991–2998 (2013)
168. M. Anpo, M. Takeuchi, J. Catal. **216**, 505–516 (2003)
169. J.C. Ireland, P. Klostermann, E.W. Rice, R.M. Clark, Appl. Environ. Microbiol. **59**, 1668–1670 (1993)
170. J.C. Sjogren, R.A. Sierka, Appl. Environ. Microbiol. **60**, 344–347 (1994)
171. R.X. Cai, Y. Kubota, T. Shuin, H. Sakai, K. Hashimoto, A. Fujishima, Cancer Res. **52**, 2346–2348 (1992)
172. R. Cai, K. Hashimoto, Y. Kubota and A. Fujishima, Chem. Lett. 427–430 (1992)
173. D. Ollis, H. Al-Ekabi, *Photocatalytic Purification and Treatment of Water and Air* (Elsevier, Tokyo/Amsterdam, 1993)
174. K.E. Karakitsou, X.E. Verykios, J. Phys. Chem. **97**, 1184–1189 (1993)
175. M. Gratzel, Acc. Chem. Res. **14**, 376–384 (1981)
176. D. Duonghong, E. Borgarello, M. Gratzel, J. Am. Chem. Soc. **103**, 4685–4690 (1981)
177. E. Borgarello, J. Kiwi, E. Pelizzetti, M. Visca, M. Gratzel, Nature **289**, 158–160 (1981)
178. A. Wold, Chem. Mater. **5**, 280–283 (1993)
179. K. Kalyanasundaram, E. Borgarello, D. Duonghong, M. Gratzel, Angew. Chem. Int. Ed. **20**, 987–988 (1981)
180. M.M.T. Khan, N.N. Rao, J. Photochem. Photobiol. A **56**, 101–111 (1991)
181. M. Schiavello, Electrochim. Acta **38**, 11–14 (1993)
182. M.M.T. Khan, D. Chatterjee, M. Krishnaratnam, M. Bala, J. Mol. Catal. **72**, 13–18 (1992)
183. M.M.T. Khan, D. Chatterjee, M. Bala, J. Photochem. Photobiol. A **67**, 349–352 (1992)
184. H. Gerischer, A. Heller, J. Electrochem. Soc. **139**, 113–118 (1992)
185. N.B. Jackson, C.M. Wang, Z. Luo, J. Schwitzgebel, J.G. Ekerdt, J.R. Brock, A. Heller, J. Electrochem. Soc. **138**, 3660–3664 (1991)
186. M. Nair, Z.H. Luo, A. Heller, Ind. Eng. Chem. Res. **32**, 2318–2323 (1993)
187. E. Pelizzetti, C. Minero, P. Piccinini, M. Vincenti, Coord. Chem. Rev. **125**, 183–193 (1993)
188. J.C. Doliveira, G. Alsayyed, P. Pichat, Environ. Sci. Technol. **24**, 990–996 (1990)
189. H. Hidaka, J. Zhao, E. Pelizzetti, N. Serpone, J. Phys. Chem. **96**, 2226–2230 (1992)
190. E.R. Carraway, A.J. Hoffman, M.R. Hoffmann, Environ. Sci. Technol. **28**, 786–793 (1994)
191. G. Mills, M.R. Hoffmann, Environ. Sci. Technol. **27**, 1681–1689 (1993)
192. C. Kormann, D.W. Bahnemann, M.R. Hoffmann, Environ. Sci. Technol. **25**, 494–500 (1991)
193. A. Chemseddine, H.P. Boehm, J. Mol. Catal. **60**, 295–311 (1990)
194. J.C. Doliveira, C. Minero, E. Pelizzetti, P. Pichat, J. Photochem. Photobiol. A **72**, 261–267 (1993)
195. K.S. Liu, L. Jiang, in *Annual Review of Materials Research*, vol. 42, ed. by D.R. Clarke (Annual Reviews, Palo Alto, 2012), pp. 231–263
196. M. Nolan, A. Iwaszuk, H. Tada, Aust. J. Chem. **65**, 624–632 (2012)

197. M. Takeuchi, M. Anpo, Preparation of highly transparent TiO2-based thin film photocatalysts by an ion engineering method: ionized cluster beam deposition (2010)
198. M. Lee, P. Amaratunga, J. Kim, D. Lee, J. Phys. Chem. C **114**, 18366–18371 (2010)
199. J.K. Zhou, M. Takeuchi, X.S. Zhao, A.K. Ray, M. Anpo, Catal. Lett. **106**, 67–70 (2006)
200. H. Yamashita, M. Anpo, Catal. Surv. Asia **8**, 35–45 (2004)
201. M. Takeuchi, S. Dohshi, T. Eura, M. Anpo, J. Phys. Chem. B **107**, 14278–14282 (2003)
202. M. Takeuchi, M. Matsuoka, H. Yamashita, M. Anpo, J. Synchrotron Radiat. **8**, 643–644 (2001)
203. H. Yamashita, M. Harada, A. Tanii, M. Honda, M. Takeuchi, Y. Ichihashi, M. Anpo, N. Iwamoto, N. Itoh, T. Hirao, Catal. Today **63**, 63–69 (2000)
204. M. Takeuchi, H. Yamashita, M. Matsuoka, M. Anpo, T. Hirao, N. Itoh, N. Iwamoto, Catal. Lett. **66**, 185–187 (2000)
205. M. Harada, A. Tanii, H. Yamashita, M. Anpo, Z. Phys, Chem. **213**, 59–65 (1999)
206. N. Reilly, G. Johnson, A.W. Castleman Jr., in *Model Systems in Catalysis*, ed. by R. Rioux (Springer New York, 2010), pp. 293–317
207. G.E. Johnson, R. Mitrić, V. Bonačić-Koutecký, A.W. Castleman Jr., Chem. Phys. Lett. **475**, 1–9 (2009)
208. G.E. Johnson, E.C. Tyo, A.W. Castleman, Proc. Natl. Acad. Sci. U. S. A. **105**, 18108–18113 (2008)
209. K.A. Zemski, D.R. Justes, A.W. Castleman Jr., J. Phys. Chem. B **106**, 6136–6148 (2002)
210. R.C. Bell, K.A. Zemski, D.R. Justes, A.W. Castleman Jr., J. Chem. Phys. **114**, 798–811 (2001)
211. R.C. Bell, K.A. Zemski, K.P. Kerns, H.T. Deng, A.W. Castleman Jr., J. Phys. Chem. A **102**, 1733–1742 (1998)
212. A.E. Lewandowska, M. Calatayud, E. Lozano-Diz, C. Minot, M.A. Banares, Catal. Today **139**, 209–213 (2008)
213. D.R. Justes, R. Mitric, N.A. Moore, V. Bonacic-Koutecky, A.W. Castleman Jr., J. Am. Chem. Soc. **125**, 6289–6299 (2003)

Chapter 14
Creating Genetic Materials of Metal Clusters

14.1 Introduction

While the preceding chapters present the reactivity of metal clusters, elucidating the chemistry of condensed matter, studies of cluster reactivity also serve to reveal the microscopic aspects such as nucleation phenomena, formation of highly dispersed media like ultrafine particles and nanoscale materials/surfaces. It has been outlined in aforementioned chapters how the potential use of metal clusters, particularly the reaction products in initiating cluster-assembled materials, has generated reasonable research interest in the activity of clusters on surfaces [1–4]. Unique properties of abundant metal clusters were found to give rise to promising potential use, such as the catalytic properties of noble metal clusters [5, 6], the oxygen-etching resistance of certain aluminum clusters and the energetic materials production by cluster reactions [7–9], or the selectivity of band gap and optical properties based on specific precise clusters [10].

However, due to the free-electron characteristics of metal clusters and thus their corresponding reactivity, precisely controlled deposition onto surfaces proves to be difficult. Even assuming a cluster can be successfully soft-landed without implanting, embedding or fragmenting, in many cases the cluster on surfaces may deform thus losing its desirable electronic and geometric structure [11, 12]. Furthermore, even though a successful soft-landing onto a surface, clusters may still diffuse unless they bind to defect sites, step edges, or reactive sites; and depending on the abundance, they can agglomerate into large islands that no longer exhibit the original properties regarding the strongly size-dependent properties of small individuals [13]. In the gas phase, small clusters have been shown to exhibit substantially different properties compared with bulk materials, and also differ from each other even if just a single atom or electron is added or removed [14, 15].

Even the all-known challenges, extensive investigations have been undertaken on studying the deposited cluster systems which display so interesting characteristics that they have prompted the growth of an entire sub-field of cluster science [16–28].

Z. Luo and S. N. Khanna, *Metal Clusters and Their Reactivity*,
https://doi.org/10.1007/978-981-15-9704-6_14

The electronic and geometric structures of the deposited clusters were investigated although a notable difference from the nascent clusters in gas phase [17, 19, 29, 30]. The deposition clusters can be a size distribution or a mass-selected species having an exact number of atoms, feasibly grown on diverse surfaces composed of metals, metal oxides, graphite and silicon etc. generally with controlled defect sites [31–33]. Because of the small sizes, high specific surface area, likely fluorescence and flexible chemistry of the cluster assembly, the properties of as-prepared materials can be tuned through quantum confinement by adsorbing surface-active species, or by altering the elemental composition of the clusters building blocks. It is anticipated that superatom clusters will be a significant aim in creating materials of metal clusters via assembly.

14.2 Building Blocks Identified from Gas Phase

One of the most exciting developments in cluster science is the realization that chosen stable clusters can mimic the chemical behavior of an atom or group of the periodic table, known as superatoms [34]. The major finding in the field of metal cluster science pertaining to the development of the superatom concept, originally termed a unified atom [35], came from study of aluminum cluster reactivity conducted in the *Castleman* group in 1989 [7], where they found a dramatic size-dependence of reactivity that cluster anions containing 13, 23, and 37 atoms were unreactive toward oxygen although the other species were etched away. This observation was accounted for by shell closings at 40, 70, and 112 electrons predicted by the well-established jellium shell model [36–38]. Replacement of the term "superatom" for unified atom, and the initial conceptual framework behind the idea that clusters mimicking different elements of the periodic table could be designed by changing size, composition, and the charged state was introduced by Khanna and Jena in a series of pioneering papers starting in 1990s [39–50]. Among them, the recognition of superatoms with one less electron than a full shell has been ascertained as superhalogens, such as Al_{13} which has an adiabatic electron affinity of 3.40 eV exhibiting behaviour reminiscent to a halogen atom [34]. Extensive studies in this field have made the stability of such superatomic clusters being understood within a few models, including the jellium model, aromaticity (e.g., a planar-structured boron cluster B_{19}^- [51]), and Wade-Mingos rules (e.g., organometallic clusters) depending on the geometry and metallicity of the cluster [1, 40, 52, 53]. The cluster sizes and electron counts of these clusters are quite stable relative to others of similar size, and hence their physical and chemical properties are dominated as expectation to reach a certain valence state [54].

While the behavior of metals is determined by the electronic levels near the Fermi energy, the entire valence electronic structure in simple metal clusters is made of discrete electronic shells which are multiple highly degenerate states and much like atoms. For such clusters a simplified framework to explain their behavior is the jellium model which was first proposed by Martins et al. in 1981 [14] and verified experimentally by Knight et al. in 1984 [6]. Sizes and electron counts of superatoms interpreted by the jellium model are quite stable compared with others of similar sizes; also their properties are dominated as expectation to reach a certain valence state. Typical superatom species having a spherical-like structure are Al_{13} and Al_{13}^- which share an icosahedral structure of 13 aluminum atoms. The stability of a cluster with 13 atoms is not limited to aluminum and allowing for modifications of the theory to account for non-spherical symmetries. For example, some other metal clusters such as Ag_{13}^- [15], Au_{13} [16], Pt_{13} and Pd_{13} [17], also exhibit enhanced stability although their lowest energy structures are not icosahedral. Among them, for instance, the stability of Ag_{13}^- is assigned to the large spin excitation energy and a big HOMO-LUMO gap associated with its bi-layer triangular cluster structure.

According to the metallic near-free electron gas (NFEG) theory, special superatoms can be simply classified into groups based on their valence electron counts, e.g., superatomic noble gas (with a closed shell), superhalogens (one less electron than a closed shell), superalkalis (one more electron than a closed shell) and superatomic alkaline-earth metals, also magnetic superatoms indicating importance in spin electronics [9]. The emergence of the periodic table of elements in the 1800s allowed chemists to predict properties and structures of elements and molecules. Hopefully the interesting researches of superatoms will enable the development of a 3D periodic table of elements (Fig. 14.1) hence leading to better predictability and design of related materials with tunable properties in nanoscale [55].

Several nonmetal-doped metal clusters have also been demonstrated as superatoms, such as a few magic species of Al_nC^-, Al_nN^- and Al_nH^- clusters (typically $Al_4H_7^-$, Al_7H_7 and Al_7H_3 etc.) [36, 37]. These clusters display enhanced stability due to large HOMO–LUMO gaps, low-electron affinities and/or high ionization potentials; however, these superatoms do not strictly follow a jellium model on the basis of NFEG theory, and the electronic/geometric structure of the core metal cluster may alter when brought into contact with doped atoms or molecules [38, 39]. A better

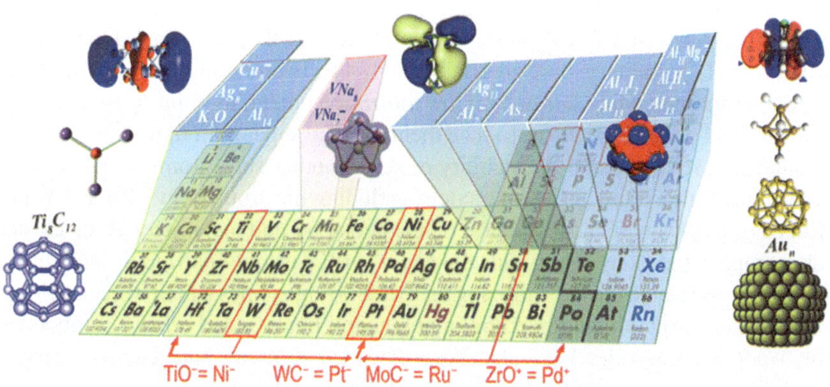

Fig. 14.1 Special and general superatoms within a 3D periodic table of elements

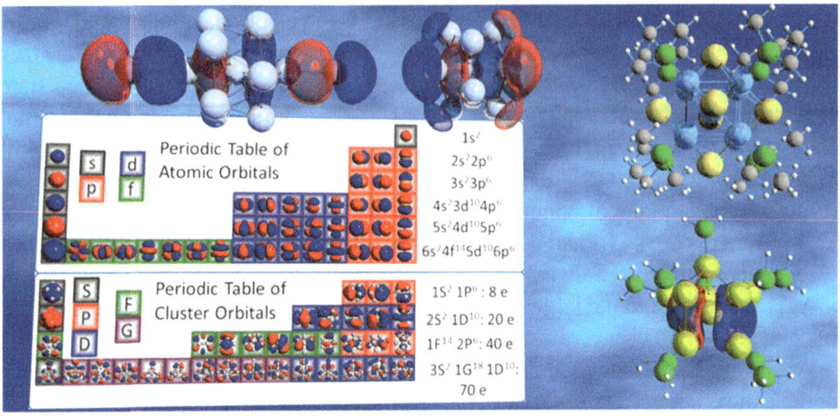

Fig. 14.2 The role of electronic and geometric effects on the stability of clusters, shedding light on the superatom characteristics [56]

interpretation on their stability and reactivity (Fig. 14.2) could be the use of molecular orbital analysis through which successful cases have been addressed for alkali metals coordinated with ammonia or ethylenediamine dopings [40], shedding light on the superatom characteristics which is applicable to metal clusters including both naked ones and ligand-protected clusters [56].

Further insight into the general superatom concept has made us recall investigations of several years ago when Metallo-Carbohedrenes ("Met-Cars" for short) were discovered in *Castleman* group [41, 42]. It was more or less serendipity when they used the laser to probe various titanium reactions and noticed a species showing a very strong peak at 528 mass units referring to Ti_8C_{12}. Subsequently, extensive studies were performed for the reaction with a number of other hydrocarbon gases (including those composed of deuterium and a use of ^{13}C in the hydrocarbon raw material) and also with a different transition metal to substitute titanium, until finally the stoichiometry M_8C_{12} (M = Ti, V, Zr and Hf) and its pentagonal dodecahedral structure of T_h symmetry were ascertained [41–44]. In addition to the unique geometry and properties, Met-Cars exhibit low ionization potentials pointing to alkali-like character and hence here we introduce them into the superatom family [45].

Studies have also made the stability of several other superatomic clusters being understood within a few different models, including the aromaticity [46] and Wade-Mingos rules [47] with a dependence on the geometry and metallicity of the cluster. For example, studies by Wang and coworkeers [48] have rationalized the stability of several boron-related clusters utilizing the Wade-Mingos rules which were developed to understand structure and bonding in polyhedral boranes and related compounds. The Wade-Mingos rules have also been extended to many bare element clusters which bear structures similar to polyhedral boranes, providing a basis to extend the concept of aromaticity from 2D planar hydrocarbons to 3D polyhedral clusters.

Through the point of view of material science, researchers in *Khanna* group [64] proposed cluster complexes and assemblies on a basis of the phosphorus-like superatoms. For example, the cluster As_7^{3-} was used as a stable building block and linked together with multiple lithium, potassium, rubidium or caesium atoms, in which they demonstrated the assemblies of a new class of semiconductor cluster materials for potential electronics [64]. Further, interesting cluster assemblies were postulated with a focus on superatom units of both Al_{13} and K_3O [29]. Among various $(Al_{13}K_3O)_n$ superatom compounds that they studied, $(Al_{13}K_3O)_3$ has its first two ionization potentials lower than any other atom in the periodic table and was described as an ultra-alkali motif (3.17 eV for K_3O). Calculation results on assembly structures of three typical species, $(Al_{13}K_3O)_{4,5,6}$.

14.3 Nanoclusters Synthesized via Wet Chemistry

Metal nanoclusters (NCs) consisting of a metallic core and protection ligands have attracted extensive research interest indicating potential use in various fields [57–60]. The reported Au/Ag nanoclusters with precise formulas are mostly protected by thiolate ligands, such as Au_{102} [52], Au_{92} [61], Au_{60} [62], Au_{55} [63], Au_{52} [64], Au_{44} [65], Au_{40} [64], Au_{38} [66], Au_{37} [67], Au_{36} [68, 69], Au_{30} [69, 70], Au_{28} [71–73], Au_{25} [74–76], Au_{24} [77–79], Au_{23} [80], Au_{22} [81], Au_{21} [82], Au_{20} [83, 84], Au_{18} [85], and Au_{13} [86, 87]; also $Ag_{14}(SR)_{12}$ [88], $Ag_{16}(SR)_{14}$ [89], $Ag_{20}/Ag_{21}((Se/S)_2R)_{12}$ [90, 91], $Ag_{25}(SR)_{18}$ [92], $Ag_{29}(S_2R)_{12}$ [93], $Ag_{32}(SR)_{24}$ [89], Ag_{33} [94], Ag_{34} [95], $Ag_{38}(SR)_{26}$ [96], $Ag_{44}(SR)_{30}$ [97, 98], $Ag_{48}(SR)_{42}$ [99], $Ag_{50}(SR)_{30}$ [100], $Ag_{63}(SR)_{36}$ [96], $Ag_{67}(SR)_{32}$ [101], etc. Besides, a few silylated chalcogenide sources [102], and electron-deficient alkynyl ligands were also applicable to successful synthesis procedures, such as $Ag_{19}(C \equiv CR)_{14}$ [103], $Ag_{25}(C \equiv CR)_{20}$ [103], $Ag_{74}(C \equiv CPh)_{44}$ [104]. On the other hand, water-soluble ligands such as 4-mercaptobenzoic acid (MBA), mercaptosuccinic acid (MSA), dimercaptosuccinic acid (DMSA), glutathione (GSH), and D-penicillamine (DPA) have also been utilized to synthesize water-soluble silver nanoclusters. Utilizing these interesting ligands, various monolayer protected water-soluble silver clusters have been synthesized, such as $Ag_{5,6}(DDT)_{4,2}$ [105], $Ag_7(DMSA)_4$ [106], $Ag_{7,8}(MSA)_{7,8,12}$ [107], $Ag_9(MSA)_7$ [108], $Ag_9(SG)_6$ [109], $Ag_{11}(SG)_7$ [110], $[Ag_{12}(HSMA)_6Na_6]^{2+}$ [111], $Ag_{14}(SG)_{11,12}$ [112, 113], $Ag_{15}(SG)_{11}$ [114], $Ag_{16}(SG)_9$ [109], $Ag_{20}(DPA)_{18}$ [115], $Ag_{31}(SG)_{19}$ [114], $Ag_{32}(SG)_{19}$ [116], $Ag_{44}(SR)_{30}$ [117], $Ag_{75}(SG)_{40}$ [118], etc. There are similar systems leading to the isolation and determination of several gold NCs, such as $Au_{10}(SG)_{10}$, $Au_{15}(SG)_{13}$, $Au_{18}(SG)_{14}$, $Au_{22}(SG)_{16}$, $Au_{22}(SG)_{17}$, $Au_{25}(SG)_{18}$, $Au_{29}(SG)_{20}$, $Au_{33}(SG)_{22}$, and $Au_{39}(SG)_{24}$, etc., mainly synthesized in *Tsukuda* group [119]. The assembly of these coordination clusters could exhibit unique characteristics, e.g., high stability, low cytotoxicity, and decent fluorescence in the solid state and in solution, giving rise to promising application as a fluorescent probe in vivo and in vitro [120]. An example shown in Fig. 14.3 illustrates the superatomic Ag_{14} NCs stabilized by face-capping ligands [57].

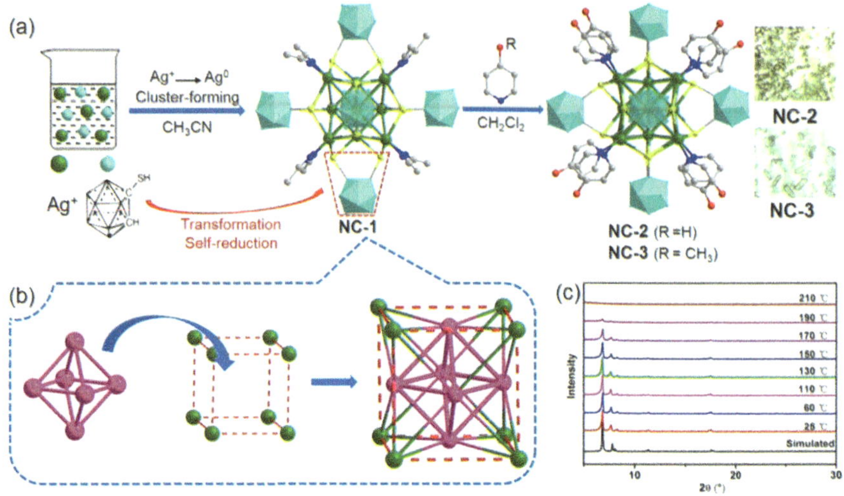

Fig. 14.3 **a** Synthesis of desired Ag_{14} NCs; note that the (fcc) array (represented by green spheres) of Ag_{14} superatoms are stabilized by face-capping 1,2-dithiolate-*o*-carborane ($C_2B_{10}H_{10}S_2$) ligands. **b** Structural dissection of **NC-1**. **c** Variable-temperature PXRD patterns of **NC-2**. Color codes: green and pink = silver; yellow = sulfur; gray = carbon; blue = nitrogen; turquoise = carborane. Reproduced with permission from Ref. [57]. Copyright 2018 American Chemical Society

Table 14.1 provides a list of the crystal information for a few ligand-protected gold clusters. While the rapid development in synthesizing various clusters via wet chemistry, there remains a challenge to synthesize superatom clusters with atomically precise structures as explored in gas phase. Nevertheless, it is notable that, the stable NCs were often found to have a "magic" 13-atom metal kernel [121, 122], with unique bonding nature [123], degenerated electronic state and geometrical symmetry [124, 125], likely relativistic effect [126], and crystal-field-like splitting of orbitals [127, 128]. The advances of metal clusters not only enrich the design of cluster building blocks [1, 15, 39, 40, 129–139], but also indicate applications in a various areas including catalysis [140–142], biotechnology and medicine [143, 144], magnetic devices and functional materials [145–151].

The 13-atom clusters including the NCs consiting of 13-atom cores have received particular research interest as it could form an icosahedron geometric structure and likely closed electronic shell. In particular, as a supporting kernel of cluster stability, 13-atom motif ubiquitous presence in various clusters owing to its double magic nature. This ubiquitous stable 13-atom moiety can serve as a "gene" for a considerable volume of cluster assembled materials, and the epigrowth of this analog of gene can be used to construct functional materials; also, aggregating, extending, stacking, printing and other possible methods could be used to originate a construction of the 13-atom motifs for a variety of genetic materials (Fig. 14.4) [152]. Insights into the structural chemistry of metal clusters enable to facilitate better understanding of fundamentals

Table 14.1 Crystal information for ligand-protected gold clusters with an Au_{13} core

Cluster composition	Core		Shell		Au–Au Bond (Å)	Reference
	Composition	Symmetry	Composition	Number		
$Au_{30}S(S\text{-}t\text{-}Bu)_{18}$	$2Au_{13}$	Bi-cuboctahedral	$Au(SR)_2$	2	/	[69]
			$Au_3(SR)_4$	2		
$Au_{38}(SC_2H_4Ph)_{18}$	$2Au_{13}$	Icosahedron	$Au(SR)_2$	3	2.80	[66]
			$Au_2(SR)_3$	6		
$Au_{40}(o\text{-}MBT)_{24}$	$Au_7 + S_6$	D_3	Au_4	6	2.76	[64]
			$Au(SR)_2$	6		
			$Au_3(SR)_4$	3		
$[Au_{20}(PP_3)_4]^{4+}$	Au_{13}	Icosahedron	Au_7	1	2.57	[154]
$[Au_{23}(SC_6H_{11})_{16}]^-$	Au_{13}	Cuboctahedron	$Au_3(SR)_4$	2	2.96	[80]
			$Au(SR)_2$	1		
$[Au_{25}(SCH_2CH_2Ph)_{18}]^-$	Au_{13}	Icosahedron	$Au_2(SR)_3$	6	2.93	[155]
$Au_{28}(TBBT)_{20}$	$2Au_{13}$	FCC	$Au_2(SR)_3$	4	2.99	[80]
Au_{55}	Au_{13}	icosahedron	Au_{42}	1	/	[85]

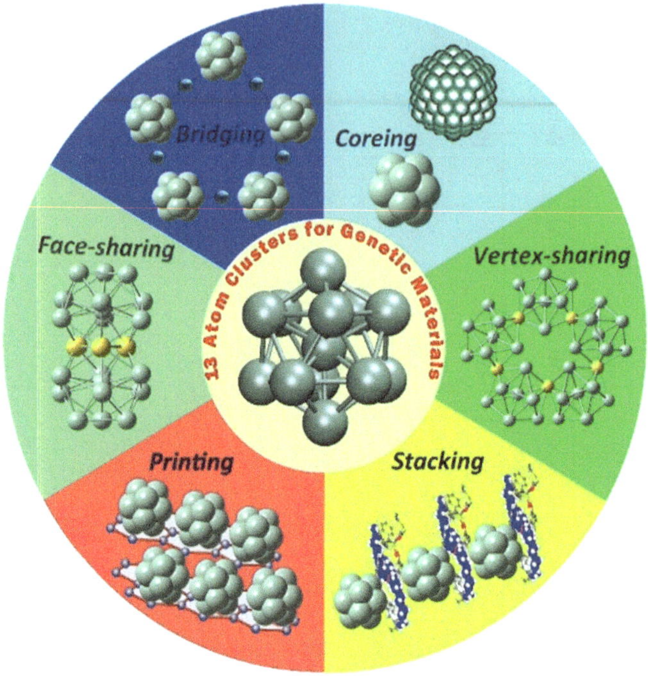

Fig. 14.4 The proposal of the thirteen-atom metal clusters for genetic materials

in condensed-phase chemistry and provide templates for the construction process of new functional materials [153].

14.4 Solid-Supported Metal Clusters

It is of interest in many areas of science to study modification of surfaces so as to control their chemical and physical properties including microelectronics, catalysis, optics, and electrochemistry, etc. [16, 31, 156–159]. For that purpose, deposition of size-controlled nanoclusters have attracted extensive research interest [5, 157, 158, 160–172]. Different approaches have been used as self-organization on suitable substrates, including direct surface reactions, surfactant templating [173], and chemical/physical deposition/coating methods (e.g., CVD, PLD, etc.) [174–176]. In addition to the extensive approaches related to nanoclusters, some other attempts have also been performed to perform the experiments through soft-landing of gas-phase clusters [177–190].

14.4.1 *Soft and Reactive Landing on Self-assembled Monolayers (SAMs)*

Soft-landing which was first introduced by Cooks and co-workers [191] refers to the deposition of intact projectile ions onto targets with or without the retention of the initial charge onto certain surfaces. In this study, they reported such a method of preparing modified surfaces, where intact polyatomic ions were deposited from the gas phase into a monolayer fluorocarbon surface at room temperature [191]. The ions were trapped in the fluorocarbon matrix for certain hours; and then they were intactly released upon sputtering at low or high energy or by thermal desorption, so that their molecular compositions were confirmed by isotopic labelling and mass analysis. Figure 14.5a illustrates the deposition of $(CH_3)_2SiNCS^+$ projectile ions into the surface of fluorinated self-assembled monolayer (F-SAM), where two $(CH_3)_2SiNCS^+$ ions penetrated in different depths, and a third approaches the F-SAM surface [191]. As the surface disorder produced by ionic collisions was not considered, this is actually a simplified representation of a soft-landing process. Note that the bulky substituent groups are important in facilitating soft-landing by steric interactions. The F-SAMs have been recognized a significant class of soft landing deposition substrates available for various metal clusters [189, 191–193].

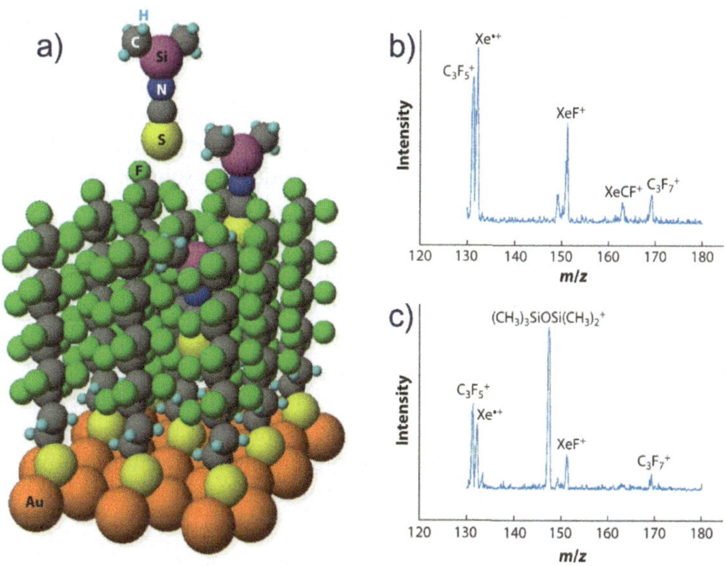

Fig. 14.5 **a** Three-dimensional molecular modeling representation of the soft-landing process for $(CH_3)_2SiNCS^+$ projectile ions impinging on a fluorinated self-assembled monolayer (FSAM) surface. **b**, **c** Mass spectra recorded by 60-eV Xe^+ sputtering of (**b**) an FSAM surface and (**c**) the same surface after treatment for 1 h at a collision energy of 5 eV, with $(CH_3)_3SiOSi(CH_3)_2^+$ ions (*m/z* 147), at a total dose corresponding to 7% of a monolayer. Reproduced with permission from Ref. [191]. Copyright 1997 American Association for the Advancement of Science

According to the procedure by Miller et al. [191], F-SAM surfaces were prepared by soaking the substrate (a glass layer, 1.6 mm thick, covered with 50 Å of Ti and 1000 Å of polycrystalline gold) in a 1-mM solution of $CF_3(CF_2)_7(CH_2)_2SH$ in ethanol for a few days [193]; and then the surfaces were then rinsed and sonicated in ethanol several times before modification by the low-energy ion beams. F-SAM surface was examined before and after the deposition of the ions with the use of $^{132}Xe^{·+}$ sputtering for surface analysis. The spectral results (Fig. 14.5b vs. c) showed that, after the deposition experiment, only a single prominent new ion species was observed with $m/z = 147$ assigned to $(CH_3)_3SiOSi\,(CH_3)_2^+$. Some other similar soft-landing experiments confirmed the availability of soft-landing of polyatomic ions at the F-SAM surfaces [191]. These results indicated successful soft-landing and retrieval of polyatomic ions favored by relatively bulky steric groups [188].

Laskin and coworkers [195] demonstrated the soft-landing of mass-selected peptide ions on SAM surfaces. Surfaces modified with peptides are commonly used in biological and medical applications, including characterization of molecular recognition events at the amino acid level, identification of biologically active motifs in proteins, as well as the development of novel biosensors and substrates for improved cell adhesion [196, 197]. For example, in one of their studies they showed the potential use to prepare conformation-specific peptide arrays by using the singly protonated Ac-A_{15}K peptide (Ac = acetyl, A = alanine, K = lysine) which was selected as a model system in their study as this peptide forms a very stable α-helical conformation due to the interaction between the protonated C-terminal lysine residue and the dipole of the helix [194]. Experiments were performed using an electrospray ionization (ESI) cluster source and an ion deposition apparatus as shown in Fig. 14.6. Based on such instrumentation, the physical and chemical properties of peptide film on substrates were determined by noticing very different FTIR spectra obtained following deposition of the AcA_{15}K peptide from solution and from the gas phase (Fig. 14.7).

The spectrum obtained by electrospray deposition (ESD) showed a dominant absorption band corresponding to a mixture of the β-sheet, α-helix and other secondary structure motifs; however, the spectrum obtained by soft-landing yields a narrow amide-I band that corresponds to the α-helical conformation [195]. In comparison, a narrow α-helical amide-I band was also obtained following reactive-landing of $Ac–A_{15}K$ on a NHS-SAM. These findings indicated that, while ESD resulted in the formation of a peptide layer dominated by the β-sheet structure, a stable α-helical peptide layer was formed by both soft-landing and reactive-landing, which enables a technique for controllable preparing of conformation-selected peptide layers [178, 195].

14.4.2 Soft-Landing onto Unreactive Solid Supports

Previous studies have revealed that, clusters with kinetic energies of less than 1 eV per atom could be nondestructively landed on bare substrates, but severe impact

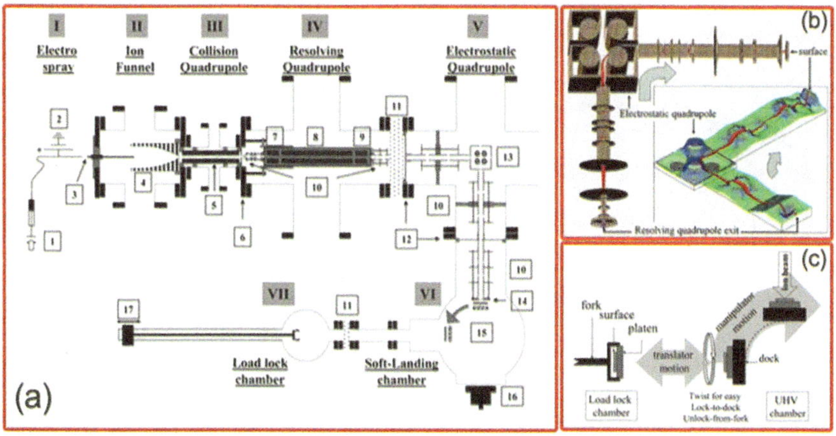

Fig. 14.6 **a** Schematic view of the ion soft-landing instrument: I, electrospray source (760 Torr). II, high-transmission ion funnel (2×10^{-1} Torr). III, ion thermalization and focusing stage (10^{-2} Torr). IV, *m*/*z* ion selection stage (4×10^{-5} Torr). V, 90° ion bending stage (10^{-7} Torr). VI, UHV chamber for ion soft landing (2×10^{-9} Torr). VII, surface introduction stage (from 760 to 2×10^{-8} Torr). (1) Syringe pump, (2) HV needle, (3) heated capillary, (4) electrodynamic ion funnel, (5) collision quadrupole (CQ), (6) 1-mm conductance limit (CL), (7) prefilter, (8) resolving quadrupole (RQ), (9) postfilter, (10) Einzel lenses, (11) gate valve, (12) 2-mm CL, (13) electrostatic quadrupole (bender), (14) deceleration area, (15) surface and phosphorus screen detector, (16) CCD camera, and (17) magnetic translator. **b** SIMION simulation showing the 3D plot of the ion optics from the resolving quadrupole to the surface, and potential surface plot showing the transport of the ion beam and the sudden deceleration stage right before the surface. **c** Schematic view of the sample-transfer system for the transfer and positioning of the surface holder inside the UHV chamber. Reproduced with permission from Ref. [194]. Copyright 2007 American Chemical Society

energies towards the substrate will lead to fragmentation of the clusters and damage of the substrate materials [158]. In this basis, controlled deposition of metal clusters have been extensively studied rendering a promising method to tailor monodispersed nanostructures at solid surfaces [198–229]. Examples of such soft-landing deposition of clusters are presented for abundant metal clusters on a variety of supports including noble metal surfaces typically Au(111) [28, 193, 230], Pt(111) [231], Si(111) [203], Ni(001) [209], Cu(001) [232], metal oxides (e.g., Al_2O_3 [233], MgO [5, 161, 234], SiO_2 [235], TiO_2 [186], etc.), molybdenum disulphide (MoS_2) [23, 236], amorphous carbon (e.g., graphene) [237] and mica [238]. For these systems, the cluster stability and morphologies, cluster-support electronic interactions, novel reactivity and catalysis [227], are extensively investigated. Also, rare-gas matrix are used for the soft-landing deposition, where the clusters are co-deposited with Ar(/kr/Xe) gas on a cooled sapphire or CaF_2 window allowing the presence of low energy electrons [239].

On the other hand, energetic cluster beams were found an efficient tool for dry etching, smoothing and cleaning of solid surfaces, and the energetic cluster–surface collisions could induce specific chemical reactions in view of the temporary build-up high particle densities, as well as the ultrafast energy dissipation and redistribution at

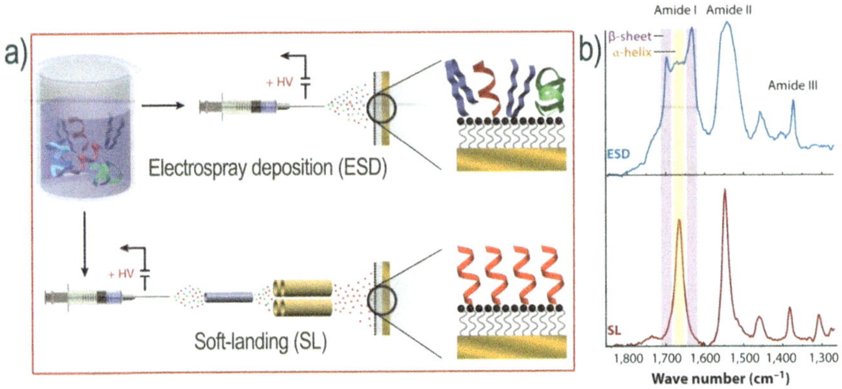

Fig. 14.7 **a** Schematic drawing of (top) electrospray deposition (ESD) and (bottom) soft landing (SL) of peptide ions on self-assembled monolayer (SAM) surfaces. ESD of $AcA_{15}K$ from solution results in the formation of a peptide layer dominated by the β-sheet structure, and a stable α-helical peptide layer on SAM surfaces is formed by SL. **b** Infrared reflection-absorption spectroscopy spectra of an $Ac\text{-}A_{15}K$ peptide layer on the hydrocarbon SAM surface prepared by (*top*) ESD and (*bottom*) SL. The purple areas correspond to characteristic absorption of the β-sheet conformation, the position of the α-helical band is highlighted in gold. Reproduced with permission from Ref. [195]. Copyright 2008 John Wiley and Sons

the localized region [240]. In specific, implantation of keV-energy clusters or MeV-energy could promote shallow junction formation and infusion doping of the shallow layers, leading to nanosized hillocks or pillars on the surfaces.

14.4.3 Factors in Affecting the Soft-Landing Deposition

Controlled deposition of peptide and protein ions onto surfaces profits a new approach for better understanding the interactions of biomolecules with various hydrophobic or hydrophilic substrates [241–246]. The factors affecting soft-landing include the effect of the primary structure of the ions, their kinetic energy and initial charge states, physical and chemical properties of SAM surfaces on the efficiency of soft-landing or reactive landing. Employing similar method as above, polydisperse diphosphine-capped gold clusters were synthesized in solution and introduced into the gas phase by electrospray ionization. Mass selection was employed to isolate a multiply charged cationic cluster species, $Au_{11}L_5^{3+}$ in which the ligand "L" refers to 1,3-bis(diphenylphosphino)propane. The $Au_{11}L_5^{3+}$ clusters were delivered onto four different self-assembled monolayers (SAMs) supported on gold with controlled coverages at 10^{11} and 10^{12} clusters. Employing time-of-flight secondary ion mass spectrometry (TOF-SIMS), they found that the coverage of cationic gold clusters on the surface is associated with the relative abundance of different charge states of the soft-landed multiply charged clusters. In the case a lower coverage of ~10^{11}

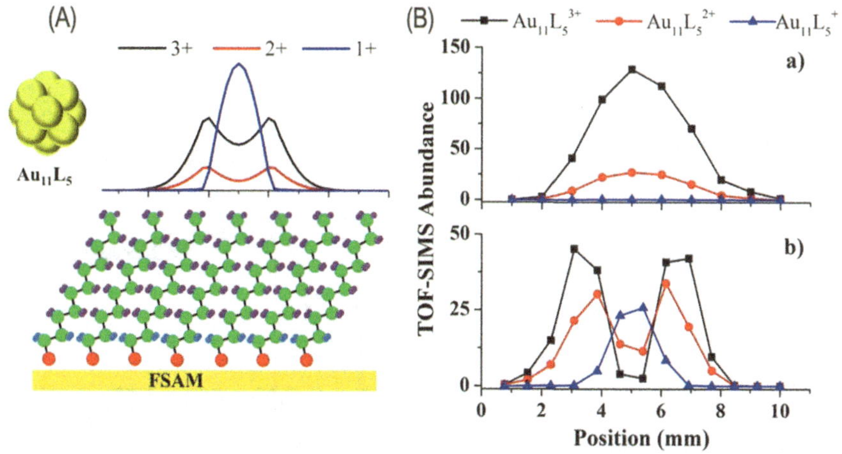

Fig. 14.8 **A** A sketch showing the soft-landing $Au_{11}L_5$ on the FSAM surface. **B** In situ TOF-SIMS abundance line profiles of $Au_{11}L_5{}^{3+}$, $Au_{11}L_5{}^{2+}$, and $Au_{11}L_5{}^{+}$ on the surface of the FSAM following deposition of **a** 1.5×10^{11} clusters and **b** 1.2×10^{12} clusters. Reproduced with permission from Ref. [185]. Copyright 2012 American Chemical Society

clusters on a fluoride monolayer (FSAM), almost complete retention of charge by the deposited $Au_{11}L_5{}^{3+}$ clusters was observed; while in sharp contrast, pronounced reduction of charge to $Au_{11}L_5{}^{2+}$ and $Au_{11}L_5{}^{+}$ was observed on the FSAM at a higher coverage of ~10^{12} clusters, as shown in Fig. 14.8 [185].

When using mercaptohexadecanoic acid surfaces on gold (COOH-SAMs) for the same soft-landing experiments, the mass-selected $Au_{11}L_5{}^{3+}$ clusters exhibited partial reduction of charge to $Au_{11}L_5{}^{2+}$ at lower coverage, while allowing additional reduction of charge to both $Au_{11}L_5{}^{2+}$ and $Au_{11}L_5{}^{+}$ at higher coverage. Repeated experiments on the surfaces of 1-dodecanethiol (HSAM) monolayers supported on gold, the most abundant charge state was $Au_{11}L_5{}^{2+}$ for lower coverage while $Au_{11}L_5{}^{+}$ for higher-coverage cases. The results demonstrated that one of the critical factors that affect the properties of supported metal clusters and their ionic charge state could be the coverage of the charged species soft landed onto SAM surfaces.

Further investigations have also found that, in addition to the overage situation, the property of SAMs themselves may bring influences. The findings through in situ TOF-SIMS approach by Laskin and coworkers [241–246] demonstrated that the $Au_{11}L_5{}^{3+}$ cluster retains its 3 + charge state when soft landed onto the surface of a FSAM on gold; however, when deposited onto COOH-SAM and HSAM surfaces, the clusters exhibit larger relative abundances of the 2+ and 1+ charge states respectively. These interesting experimental observations are addressed in Fig. 14.9. Investigations on the kinetics of charge reduction on the different surfaces through in situ Fourier transform ion cyclotron resonance (FT-ICR) SIMS have ascertained this origin. It was found that an slow interfacial charge reduction occurs on the FSAM surface while an instantaneous neutralization takes place on the HSAM surface [177].

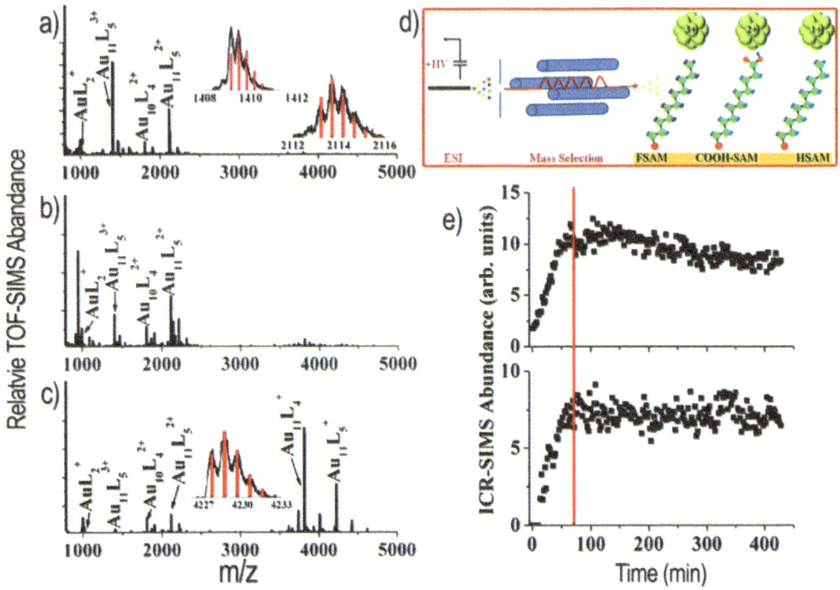

Fig. 14.9 A schematic illustration explaining the soft landing of mass-selected gold clusters onto surfaces. The clusters are introduced into the gas phase from solution using electrospray ionization, filtered according to mass-to-charge ratio using a quadrupole mass filter, and delivered to different SAM surfaces at controlled energies: **a** FSAM, **b** COOH-SAM, **c** HSAM. High-mass range of the positive mode in situ TOF-SIMS mass spectrum (m/z = 800–5000) of the SAM surfaces after soft landing of 5×10^{11} $Au_{11}L_5^{3+}$ (L = DPPP) (m/z = 1409.5) clusters for all. **d** a sketch showing the soft-landing process; **e** the relationship of the ICR-SIMS abundance with deposition time, representatively for FSAM. Reproduced with permission from Ref. [177]. Copyright 2011 American Chemical Society

As illustrated above, the interesting process of soft-landing has been widely utilized for the deposition of small molecules [192, 247–250], peptides [244, 246, 251, 252], and proteins [251–253], oligonucleotides [254] and viruses [255], and clusters [17, 18, 256–259], onto various substrates for surface chemistry, photochemistry and catalysis investigations. Besides the morphology characterization, these soft-landing modified surfaces were subsequently applied to several aspects, such as surface-enhanced Raman spectroscopy [248], deposition of low-energy ions on liquid surfaces was used to investigate transport properties of small ions through thin/thick films [260, 261]. Gas-phase ion chemistry combined with soft-landing provides a unique opportunity for preparation of novel synthetic materials, such as chiral-enriched products demonstrated by Cooks and coworkers [262], through soft-landing of protonated serine octamers. The soft-landing technique is particularly beneficial to cluster assembly as a tool for selective isolation of metal clusters especially those cannot be synthesized in condensed phase such as the Metallo-Carbohedrenes (Met-Cars) as mentioned in Chap. 10 [210].

14.4.4 Characterization of Soft-Landed Clusters

Soft-landed clusters enable morphology characterization as various nanomaterials and nanoscale surfaces, profiting from the development of nanotechnology especially scanning tunneling microscopy (STM). Combining STM and spectroscopy at cryogenic (UHV, 4 K) conditions, the cluster conductance with complete control of their chemical and physical environment can be measured, where thermal broadening of their electronic states as well as their mobility is minimized. For example, Weiss and coworkers [263] studied the diffusion in the tunneling spectra of isolated, ligand-stabilized undecagold Au_{11} clusters immobilized by attachment to α,ω-alkanedithiolate tethers inserted into alkanethiolate SAMs, as shown in Fig. 14.10. While soft-landed clusters are chemically adsorbed on the SAM surfaces, their assemblies were supposed to be sufficiently dynamic to affect their transport properties significantly. Surprising results were found that the chemically-identical individual particles produced different families of tunneling spectra, comparable to previous results for heterogeneous distributions of gold particles. It was also found that the

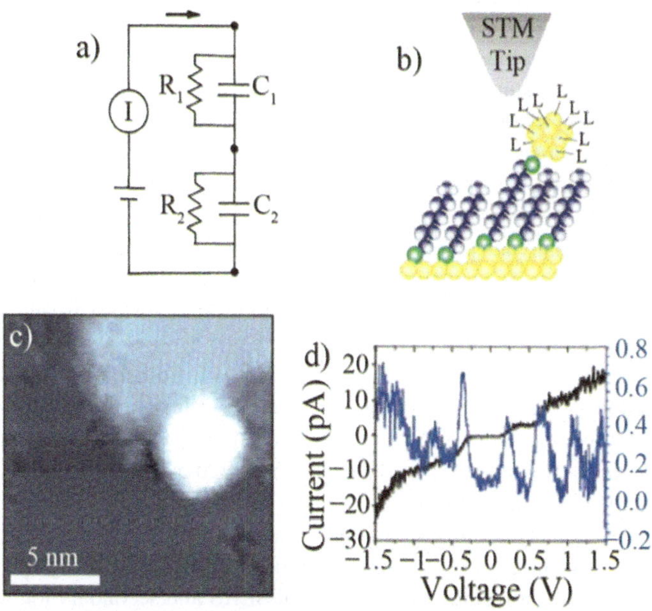

Fig. 14.10 **a/b** Schematic of STM circuit and ligand (L)-stabilized Au_{11} cluster immobilized via a 1,10-DDT tether inserted into a C_8 SAM. **c** A 157 Å × 157 Å STM image of an Au_{11} − TPP attached at an Au step edge. The location of the cluster, as well as the molecular lattice of the host SAM, can be easily resolved ($V_{sample} = +\ 1.5$ V, $i_{tunnel} = 14$ pA, $T = 4.2$ K). **d** A spectrum showing current − voltage data ($I(V)$, black), plotted with the simultaneously acquired dI/dV data (blue). $V_{sample} = +1.5$ V; $i_{tunnel} = 18$ pA (n), 9 pA (q). Reproduced with permission from Ref. [263]. Copyright 2006 American Chemical Society

Au_{11} clusters demonstrate Coulomb blockade behavior at low temperature, with zero-conductance gaps resulting from quantum size effects [263].

Cooperation work in *Castleman* and *Weiss* groups presented a study in which they softly land Al_{17}^- clusters onto hydroxyl-terminated SAMs using reactivity previously characterized in the gas phase before imaging the deposited clusters via the STM technique. Among the reactive Al clusters as discussed in Chap. 6, Al_{17}^- is unique in that it exhibits several active sites—one on each face of its structure—in reacting with water [264]. It was noted that an Al_{17}^- cluster approaching a hydroxyl group such as SAM consisting of only hydroxyl-terminated molecules have a high probability of interacting in this manner to form a chemisorbed product. Further, as a SAM presents a continuous surface of nucleophiles of which each can donate electrons to the approaching cluster, it was conjectured the SAMs also profit the link of Al clusters covalently to the substrate. This is similar to the soft-landing studies that were performed for the deposition of pre-tethered assemblies onto a surface, as well as the aforementioned peptide ions onto SAMs where the ions could bind retaining their charge state [265]. Nevertheless, as Al_{17}^- and Al_{17} neutral clusters have same structures and free-electron characteristics, the charge of such a deposited species would be irrelevant when using standard microscopic techniques [266, 267].

This soft-landing of Al clusters extends the novel deposition scheme where fragile all-metal clusters are deposited in a predictable fashion, and is recognized to display the important initial states for the bottom-up construction of substrate-supported clusters. It is worth mentioning in this study there is a well design and implementation of a versatile vacuum suitcase for use in transporting air-sensitive samples (such as Al clusters) between ultra-high vacuum (UHV) instruments, especially the portability and stability when the sample is transferred into the final vacuum chamber. This system was easily adaptable to a wide variety of applications involving sample preparation and analyses where two or more of the procedural steps occur in separate vacuum chambers. It is important to design such a system with the ease of adaptability of this transfer device to other vacuum systems.

Besides the STM characterization, there are also some other approaches that have been applied to the solid-supported metal clusters, for instance, by means of X-ray photoelectron spectroscopy (XPS). For example, an early work in *Cooks* group [183] reported the in situ Raman analysis of surfaces prepared by ion soft landing, on which surface-enhanced Raman spectroscopy (SERS) effect was noted for crystal violet, Rhodamine 6G, methyl orange and copper phthalocyanine. Furthermore, imaging of the modified surfaces was attained utilizing the 2D Raman imaging technique. The combination of molecular spectral tools of SERS and secondary ion mass spectrometer (SIMS) fitted with in-vacuum sample transport capability facilitates in situ analysis of such novel surfaces modified by cluster soft landing deposition. It is worth mentioning that, because of the fingerprint spectra of Raman spectroscopy and the availability to identify charge transfer between metals and analytes, the *in situ* Raman measurements could be applicable to judge the charge state of the soft-landed metal clusters [268].

References

1. A.W. Castleman Jr., S.N. Khanna, J. Phys. Chem. C **113**, 2664–2675 (2009)
2. S.A. Claridge, A.W. Castleman, S.N. Khanna, C.B. Murray, A. Sen, P.S. Weiss, ACS Nano **3**, 244–255 (2009)
3. D.P. Woodruff, *The Chemical Physics of Solid Surfaces* (Elsevier, Oxford, 2007)
4. W. Harbich, in *Metal Clusters at Surfaces: Structure, Quantum Properties, Physical Chemistry* (New York, 2000)
5. B. Yoon, H. Hakkinen, U. Landman, A.S. Worz, J.M. Antonietti, S. Abbet, K. Judai, U. Heiz, Science **307**, 403–407 (2005)
6. A.A. Herzing, C.J. Kiely, A.F. Carley, P. Landon, G.J. Hutchings, Science **321**, 1331–1335 (2008)
7. R.E. Leuchtner, A.C. Harms, A.W. Castleman Jr., J. Chem. Phys. **91**, 2753–2754 (1989)
8. A.C. Reber, S.N. Khanna, P.J. Roach, W.H. Woodward, A.W. Castleman Jr., J. Am. Chem. Soc. **129**, 16098–16101 (2007)
9. R. Burgert, H. Schnockel, A. Grubisic, X. Li, S.T. Stokes, K.H. Bowen, G.F. Gantefor, B. Kiran, P. Jena, Science **319**, 438–442 (2008)
10. R.E. Leuchtner, A.C. Harms, A.W. Castleman Jr., J. Chem. Phys. **94**, 1093–1101 (1991)
11. C.L. Cleveland, U. Landman, Science **257**, 355–361 (1992)
12. B. Nacer, C. Massobrio, C. Felix, Phys. Rev. B **56**, 10590–10595 (1997)
13. G.M. Francis, I.M. Goldby, L. Kuipers, B. von Issendorff, R.E. Palmer, Dalton Trans. 665–671 (1996)
14. D.E. Bergeron, P.J. Roach, A.W. Castleman Jr., N. Jones, S.N. Khanna, Science **307**, 231–235 (2005)
15. P.J. Roach, A.C. Reber, W.H. Woodward, S.N. Khanna, A.W. Castleman, Proc. Natl. Acad. Sci. U.S.A. **104**, 14565–14569 (2007)
16. P. Jensen, Rev. Mod. Phys. **71**, 1695 (1999)
17. S. Messerli, S. Schintke, K. Morgenstern, A. Sanchez, U. Heiz, W.D. Schneider, Surf. Sci. **465**, 331–338 (2000)
18. W. Yamaguchi, K. Yoshimura, Y. Tai, Y. Maruyama, K. Igarashi, S. Tanemura, J. Murakami, Chem. Phys. Lett. **311**, 341–345 (1999)
19. F. Claeyssens, S. Pratontep, C. Xirouchaki, R.E. Palmer, Nanotechnology **17**, 805–807 (2006)
20. H. Yasumatsu, T. Hayakawa, S. Koizumi, T. Kondow, J. Chem. Phys. **123**, 124709 (2005)
21. S. Lee, B. Lee, F. Mehmood, S. Seifert, J. Libera, J. Elam, J. Greeley, P. Zapol, L. Curtiss, M. Pellin, P. Stair, R. Winans, S. Vajda, J. Phys. Chem. C **114**, 10342–10348 (2010)
22. S.A. Kandel, P.S. Weiss, J. Phys. Chem. B **105**, 8102–8106 (2001)
23. J.G. Kushmerick, P.S. Weiss, J. Phys. Chem. B **102**, 10094–10097 (1998)
24. Y. Kuk, M.F. Jarrold, P.J. Silverman, J.E. Bower, W.L. Brown, Phys. Rev. B **39**, 11168–11170 (1989)
25. W.E. Kaden, T. Wu, W.A. Kunkel, S.L. Anderson, Science **326**, 826–829 (2009)
26. S. Lee, L.M. Molina, M.J. Lopez, J.A. Alonso, B. Hammer, B. Lee, S. Seifert, R.E. Winans, J.W. Elam, M.J. Pellin, S. Vajda, Angew. Chem. Int. Ed. **48**, 1467–1471 (2009)
27. S.S. Lee, C.Y. Fan, T.P. Wu, S.L. Anderson, J. Am. Chem. Soc. **126**, 5682–5683 (2004)
28. J.M. Lightstone, M.J. Patterson, P. Liu, J.C. Lofaro, M.G. White, J. Phys. Chem. C **112**, 11495–11506 (2008)
29. K. Schouteden, N. Vandamme, E. Janssens, P. Lievens, C. Van Haesendonck, Surf. Sci. **602**, 552–558 (2008)
30. S. Duffe, T. Irawan, M. Bieletzki, T. Richter, B. Sieben, C. Yin, B. Issendorff, M. Moseler, H. Hovel, Eur. Phys. J. D **45**, 401–408 (2007)
31. H. Brune, M. Giovannini, K. Bromann, K. Kern, Nature **394**, 451–453 (1998)
32. A. Sanchez, S. Abbet, U. Heiz, W.D. Schneider, H. Hakkinen, R.N. Barnett, U. Landman, J. Phys. Chem. A **103**, 9573–9578 (1999)
33. S. Chrétien, S.K. Buratto, H. Metiu, Curr. Opin. Solid St. M. **11**, 62–75 (2007)

34. D.E. Bergeron, A.W. Castleman Jr., T. Morisato, S.N. Khanna, Science **304**, 84–87 (2004)
35. R.S. Mulliken, Rev. Mod. Phys. **4**, 1–86 (1932)
36. W. Ekardt, Phys. Rev. B **29**, 1558–1564 (1984)
37. M. Brack, Rev. Mod. Phys. **65**, 677–732 (1993)
38. W. Ekardt, *Metal Clusters* (Wiley, Berlin, 1999)
39. S.N. Khanna, P. Jena, Phys. Rev. Lett. **69**, 1664–1667 (1992)
40. S.N. Khanna, P. Jena, Phys. Rev. B **51**, 13705–13716 (1995)
41. S.N. Khanna, P. Jena, Chem. Phys. Lett. **219**, 479–483 (1994)
42. F. Liu, M. Mostoller, T. Kaplan, S.N. Khanna, P. Jena, Chem. Phys. Lett. **248**, 213–217 (1996)
43. P. Jena, S.N. Khanna, B.K. Rao, Surf. Rev. Lett. **03**, 993–999 (1996)
44. C. Ashman, S.N. Khanna, F. Liu, P. Jena, T. Kaplan, M. Mostoller, Phys. Rev. B **55**, 15868–15873 (1997)
45. B.K. Rao, S.N. Khanna, P. Jena, J. Cluster Sci. **10**, 477–491 (1999)
46. B.K. Rao, S.N. Khanna, P. Jena, Phys. Rev. B **62**, 4666–4671 (2000)
47. C. Ashman, S.N. Khanna, M.R. Pederson, Chem. Phys. Lett. **324**, 137–142 (2000)
48. C. Ashman, S.N. Khanna, M.R. Pederson, J. Kortus, Phys. Rev. B **62**, 16956–16961 (2000)
49. S.N. Khanna, B.K. Rao, P. Jena, Phys. Rev. B **65**, 125105 (2002)
50. C. Ashman, S.N. Khanna, M.R. Pederson, Phys. Rev. B **66**, 193408 (2002)
51. W. Huang, A.P. Sergeeva, H.-J. Zhai, B.B. Averkiev, L.-S. Wang, A.I. Boldyrev, Nat. Chem. **2**, 202–206 (2010)
52. P.D. Jadzinsky, G. Calero, C.J. Ackerson, D.A. Bushnell, R.D. Kornberg, Science **318**, 430–433 (2007)
53. M. Walter, J. Akola, O. Lopez-Acevedo, P.D. Jadzinsky, G. Calero, C.J. Ackerson, R.L. Whetten, H. Grönbeck, H. Häkkinen, Proc. Natl. Acad. Sci. U.S.A. **105**, 9157–9162 (2008)
54. J.U. Reveles, S.N. Khanna, P.J. Roach, A.W. Castleman, Proc. Natl. Acad. Sci. U.S.A. **103**, 18405–18410 (2006)
55. Z. Luo, A.W. Castleman, Acc. Chem. Res. **47**, 2931–2940 (2014)
56. A.C. Reber, S.N. Khanna, Acc. Chem. Res. **50**, 255–263 (2017)
57. Z.-Y. Wang, M.-Q. Wang, Y.-L. Li, P. Luo, T.-T. Jia, R.-W. Huang, S.-Q. Zang, T.C.W. Mak, J. Am. Chem. Soc. **140**, 1069–1076 (2018)
58. I. Chakraborty, T. Pradeep, Chem. Rev. **117**, 8208–8271 (2017)
59. P. Liu, R. Qin, G. Fu, N. Zheng, J. Am. Chem. Soc. **139**, 2122–2131 (2017)
60. Q.-M. Wang, Y.-M. Lin, K.-G. Liu, Acc. Chem. Res. **48**, 1570–1579 (2015)
61. C. Zeng, C. Liu, Y. Chen, N.L. Rosi, R. Jin, J. Am. Chem. Soc. **138**, 8710–8713 (2016)
62. Y. Song, F. Fu, J. Zhang, J. Chai, X. Kang, P. Li, S. Li, H. Zhou, M. Zhu, Angew. Chem. Int. Ed. **54**, 8430–8434 (2015)
63. C.M. Chang, C. Cheng, C.M. Wei, J. Chem. Phys. **128**, 124710 (2008)
64. C. Zeng, Y. Chen, C. Liu, K. Nobusada, N.L. Rosi, R. Jin, Sci. Adv. **1**, e1500425 (2015)
65. L. Liao, S. Zhuang, C. Yao, N. Yan, J. Chen, C. Wang, N. Xia, X. Liu, M.-B. Li, L. Li, X. Bao, Z. Wu, J. Am. Chem. Soc. **138**, 10425–10428 (2016)
66. H. Qian, W.T. Eckenhoff, Y. Zhu, T. Pintauer, R. Jin, J. Am. Chem. Soc. **132**, 8280–8281 (2010)
67. R. Jin, C. Liu, S. Zhao, A. Das, H. Xing, C. Gayathri, Y. Xing, N.L. Rosi, R.R. Gil, R. Jin, ACS Nano **9**, 8530–8536 (2015)
68. C.J. Zeng, H.F. Qian, T. Li, G. Li, N.L. Rosi, B. Yoon, R.N. Barnett, R.L. Whetten, U. Landman, R.C. Jin, Angew. Chem. Int. Ed. **51**, 13114–13118 (2012)
69. D. Crasto, S. Malola, G. Brosofsky, A. Dass, H. Häkkinen, J. Am. Chem. Soc. **136**, 5000–5005 (2014)
70. H.Y. Yang, Y. Wang, A.J. Edwards, J.Z. Yan, N.F. Zheng, Chem. Commun. **50**, 14325–14327 (2014)
71. C. Zeng, Y. Chen, K. Iida, K. Nobusada, K. Kirschbaum, K.J. Lambright, R. Jin, J. Am. Chem. Soc. **138**, 3950–3953 (2016)
72. Y. Chen, C. Liu, Q. Tang, C. Zeng, T. Higaki, A. Das, D.-E. Jiang, N.L. Rosi, R. Jin, J. Am. Chem. Soc. **138**, 1482–1485 (2016)

73. C. Zeng, T. Li, A. Das, N.L. Rosi, R. Jin, J. Am. Chem. Soc. **135**, 10011–10013 (2013)
74. M.W. Heaven, A. Dass, P.S. White, K.M. Holt, R.W. Murray, J. Am. Chem. Soc. **130**, 3754–3755 (2008)
75. M. Zhu, C.M. Aikens, F.J. Hollander, G.C. Schatz, R. Jin, J. Am. Chem. Soc. **130**, 5883–5885 (2008)
76. M.Z. Zhu, W.T. Eckenhoff, T. Pintauer, R.C. Jin, J. Phys. Chem. C **112**, 14221–14224 (2008)
77. A. Das, T. Li, G. Li, K. Nobusada, C. Zeng, N.L. Rosi, R. Jin, Nanoscale **6**, 6458–6462 (2014)
78. Y. Song, S. Wang, J. Zhang, X. Kang, S. Chen, P. Li, H. Sheng, M. Zhu, J. Am. Chem. Soc. **136**, 2963–2965 (2014)
79. Z. Gan, Y. Lin, L. Luo, G. Han, W. Liu, Z. Liu, C. Yao, L. Weng, L. Liao, J. Chen, X. Liu, Y. Luo, C. Wang, S. Wei, Z. Wu, Angew. Chem. Int. Ed. **128**, 11739–11743 (2016)
80. A. Das, T. Li, K. Nobusada, C. Zeng, N.L. Rosi, R. Jin, J. Am. Chem. Soc. **135**, 18264–18267 (2013)
81. J. Chen, Q.-F. Zhang, T.A. Bonaccorso, P.G. Williard, L.-S. Wang, J. Am. Chem. Soc. **136**, 92–95 (2014)
82. S. Chen, L. Xiong, S. Wang, Z. Ma, S. Jin, H. Sheng, Y. Pei, M. Zhu, J. Am. Chem. Soc. **138**, 10754–10757 (2016)
83. C.J. Zeng, C. Liu, Y.X. Chen, N.L. Rosi, R.C. Jin, J. Am. Chem. Soc. **136**, 11922–11925 (2014)
84. X.K. Wan, Z.W. Lin, Q.M. Wang, J. Am. Chem. Soc. **134**, 14750–14752 (2012)
85. Y. Chen, C. Zeng, C. Liu, K. Kirschbaum, C. Gayathri, R.R. Gil, N.L. Rosi, R. Jin, J. Am. Chem. Soc. **137**, 10076–10079 (2015)
86. M. Sugiuchi, Y. Shichibu, T. Nakanishi, Y. Hasegawa, K. Konishi, Chem. Commun. **51**, 13519–13522 (2015)
87. Y. Shichibu, K. Suzuki, K. Konishi, Nanoscale **4**, 4125–4129 (2012)
88. H. Yang, J. Lei, B. Wu, Y. Wang, M. Zhou, A. Xia, L. Zheng, N. Zheng, Chem. Commun. **49**, 300–302 (2013)
89. H. Yang, Y. Wang, N. Zheng, Nanoscale **5**, 2674–2677 (2013)
90. W.-T. Chang, P.-Y. Lee, J.-H. Liao, K.K. Chakrahari, S. Kahlal, Y.-C. Liu, M.-H. Chiang, J.-Y. Saillard, C.W. Liu, Angew. Chem. Int. Ed. **56**, 10178–10182 (2017)
91. R.S. Dhayal, J.-H. Liao, Y.-C. Liu, M.-H. Chiang, S. Kahlal, J.-Y. Saillard, C.W. Liu, Angew. Chem. Int. Ed. **54**, 3702–3706 (2015)
92. C.P. Joshi, M.S. Bootharaju, M.J. Alhilaly, O.M. Bakr, J. Am. Chem. Soc. **137**, 11578–11581 (2015)
93. L.G. AbdulHalim, M.S. Bootharaju, Q. Tang, S. Del Gobbo, R.G. AbdulHalim, M. Eddaoudi, D.E. Jiang, O.M. Bakr, J. Am. Chem. Soc. **137**, 11970–11975 (2015)
94. F. Tian, R. Chen, J. Am. Chem. Soc. **141**, 7107–7114 (2019)
95. Z.-J. Guan, F. Hu, S.-F. Yuan, Z.-A. Nan, Y.-M. Lin, Q.-M. Wang, Chem. Sci. **10**, 3360–3365 (2019)
96. H. Yang, J. Yan, Y. Wang, H. Su, L. Gell, X. Zhao, C. Xu, B.K. Teo, H. Häkkinen, N. Zheng, J. Am. Chem. Soc. **139**, 31–34 (2017)
97. A. Desireddy, B.E. Conn, J. Guo, B. Yoon, R.N. Barnett, B.M. Monahan, K. Kirschbaum, W.P. Griffith, R.L. Whetten, U. Landman, T.P. Bigioni, Nature **501**, 399–402 (2013)
98. H. Yang, Y. Wang, H. Huang, L. Gell, L. Lehtovaara, S. Malola, H. Häkkinen, N. Zheng, Nat. Commun. **4**, 2422 (2013)
99. S. Ahmar, D.G. MacDonald, N. Vijayaratnam, T.L. Battista, M.S. Workentin, J.F. Corrigan, Angew. Chem. Int. Ed. **49**, 4422–4424 (2010)
100. W. Du, S. Jin, L. Xiong, M. Chen, J. Zhang, X. Zou, Y. Pei, S. Wang, M. Zhu, J. Am. Chem. Soc. **139**, 1618–1624 (2017)
101. M.J. Alhilaly, M.S. Bootharaju, C.P. Joshi, T.M. Besong, A.-H. Emwas, R. Juarez-Mosqueda, S. Kaappa, S. Malola, K. Adil, A. Shkurenko, H. Häkkinen, M. Eddaoudi, O.M. Bakr, J. Am. Chem. Soc. **138**, 14727–14732 (2016)
102. O. Fuhr, S. Dehnen, D. Fenske, Chem. Soc. Rev. **42**, 1871–1906 (2013)

103. S.-F. Yuan, P. Li, Q. Tang, X.-K. Wan, Z.-A. Nan, D.-E. Jiang, Q.-M. Wang, Nanoscale **9**, 11405–11409 (2017)
104. H. Yang, Y. Wang, X. Chen, X. Zhao, L. Gu, H. Huang, J. Yan, C. Xu, G. Li, J. Wu, A.J. Edwards, B. Dittrich, Z. Tang, D. Wang, L. Lehtovaara, H. Häkkinen, N. Zheng, Nat. Commun. **7**, 12809 (2016)
105. B.S. González, M. Blanco, M.A. López-Quintela, Nanoscale **4**, 7632–7635 (2012)
106. H. Xiang, S.-H. Wei, X. Gong, J. Am. Chem. Soc. **132**, 7355–7360 (2010)
107. A. Leelavathi, T.U. Bhaskara Rao, T. Pradeep, Int. J. Nanosci. **10**, 839–843 (2011)
108. T.U.B. Rao, B. Nataraju, T. Pradeep, J. Am. Chem. Soc. **132**, 16304–16307 (2010)
109. X. Yuan, M.I. Setyawati, A.S. Tan, C.N. Ong, D.T. Leong, J. Xie, NPG Asia Mater. **5**, e39 (2013)
110. A. Baksi, M. Bootharaju, X. Chen, H. Hakkinen, T. Pradeep, J. Phys. Chem. C **118**, 21722–21729 (2014)
111. P. An, R. Anumula, H. Wu, J. Han, Z. Luo, Nanoscale **10**, 10635–10640 (2018)
112. W. Ding, L. Guan, J. Han, R. Mangala, Z. Luo, Sensors Actuators B Chem. **250**, 364–371 (2017)
113. J. Yang, N. Xia, X. Wang, X. Liu, A. Xu, Z. Wu, Z. Luo, Nanoscale **7**, 18464–18470 (2015)
114. F. Bertorelle, R. Hamouda, D. Rayane, M. Broyer, R. Antoine, P. Dugourd, L. Gell, A. Kulesza, R. Mitric, V. Bonacic-Koutecky, Nanoscale **5**, 5637–5643 (2013)
115. X. Liu, W. Ding, Y. Wu, C. Zeng, Z. Luo, H. Fu, Nanoscale **9**(11), 3986–3994 (2017)
116. J. Guo, S. Kumar, M. Bolan, A. Desireddy, T.P. Bigioni, W.P. Griffith, Anal. Chem. **84**, 5304–5308 (2012)
117. K.M. Harkness, Y. Tang, A. Dass, J. Pan, N. Kothalawala, V.J. Reddy, D.E. Cliffel, B. Demeler, F. Stellacci, O.M. Bakr, Nanoscale **4**, 4269–4274 (2012)
118. I. Chakraborty, T. Udayabhaskararao, T. Pradeep, Chem. Commun. **48**, 6788–6790 (2012)
119. Y. Negishi, K. Nobusada, T. Tsukuda, J. Am. Chem. Soc. **127**, 5261–5270 (2005)
120. M.H. Zeng, Z. Yin, Z.H. Liu, H.B. Xu, Y.C. Feng, Y.Q. Hu, L.X. Chang, Y.X. Zhang, J. Huang, M. Kurmoo, Angew. Chem. Int. Ed. **55**, 11407–11411 (2016)
121. W.A. De Heer, Rev. Mod. Phys. **65**, 611–676 (1993)
122. V.M. Medel, J.U. Reveles, S.N. Khanna, V. Chauhan, P. Sen, A.W. Castleman, Proc. Natl. Acad. Sci. U.S.A. **108**, 10062–10066 (2011)
123. G. Frenking, N. Fröhlich, Chem. Rev. **100**, 717–774 (2000)
124. N. Haruta, T. Tsukamoto, A. Kuzume, T. Kambe, K. Yamamoto, Nat. commun. **9**, 3758 (2018)
125. B. Stephanidis, S. Adichtchev, S. Etienne, S. Migot, E. Duval, A. Mermet, Phys. Rev. B **76**, 121404 (2007)
126. H. Hakkinen, M. Moseler, U. Landman, Phys. Rev. Lett. **89**, 033401 (2002)
127. P.J. Roach, W.H. Woodward, A.C. Reber, S.N. Khanna, A.W. Castleman Jr., Phys. Rev. B **81**, 195404 (2010)
128. Z. Luo, G.U. Gamboa, J.C. Smith, A.C. Reber, J.U. Reveles, S.N. Khanna, A.W. Castleman Jr., J. Am. Chem. Soc. **134**, 18973–18978 (2012)
129. H. E. Johnson. and C. M. Aikens., J. Phys. Chem. A, 2009, **113**, 4445–4450
130. D.L. Long, R. Tsunashima, L. Cronin, Angew. Chem. Int. Ed. **49**, 1736–1758 (2010)
131. W. Song, W.-C. Lu, Q.-J. Zang, C.Z. Wang, K.M. Ho, Int. J. Quantum Chem. **112**, 1717–1724 (2012)
132. P.K. Jena, S.N. Khanna, B.K. Rao, Mater. Sci. Forum **232**, 1–26 (1996)
133. F.K. Sheong, J.X. Zhang, Z. Lin, Inorg. Chem. **55**, 11348–11353 (2016)
134. P. Alivisatos, P.F. Barbara, A.W. Castleman, J. Chang, D.A. Dixon, M.L. Klein, G.L. McLendon, J.S. Miller, M.A. Ratner, P.J. Rossky, S.I. Stupp, M.E. Thompson, Adv. Mater. **10**, 1297–1336 (1998)
135. B.D. Leskiw, A.W. Castleman Jr., Chem. Phys. Lett. **316**, 31–36 (2000)
136. V. Kumar, Comput. Mater. Sci. **36**, 1–11 (2006)
137. W.J. Zheng, O.C. Thomas, T.P. Lippa, S.J. Xu, K.H. Bowen Jr., J. Chem. Phys. **124**, 144304 (2006)
138. E.G. Mednikov, M.C. Jewell, L.F. Dahl, J. Am. Chem. Soc. **129**, 11619–11630 (2007)

139. K.H. Kuo, Struct. Chem. **13**, 221–222 (2002)
140. E.C. Tyo, S. Vajda, Nat. Nanotechnol. **10**, 577–588 (2015)
141. Y. Watanabe, Sci. Technol. Adv. Mater. **15**, 063501 (2014)
142. X. Nie, H. Qian, Q. Ge, H. Xu, R. Jin, ACS Nano **6**, 6014–6022 (2012)
143. I. Chakraborty, T. Udayabhaskararao, G.K. Deepesh, T. Pradeep, J. Mater. Chem. B **1**, 4059–4064 (2013)
144. M.A.H. Muhammed, P.K. Verma, S.K. Pal, R.C.A. Kumar, S. Paul, R.V. Omkumar, T. Pradeep, Chem. Eur. J. **15**, 10110–10120 (2009)
145. Y.S. Chen, H. Choi, P.V. Kamat, J. Am. Chem. Soc. **135**, 8822–8825 (2013)
146. T. Udaya Bhaskara Rao, T. Pradeep, Angew. Chem. Int. Ed. **49**, 3925–3929 (2010)
147. N. Satoh, T. Nakashima, K. Yamamoto, Sci. Rep. **2013**, 3 (1959)
148. X. Yuan, B. Zhang, Z. Luo, Q. Yao, D.T. Leong, N. Yan, J. Xie, Angew. Chem. Int. Ed. **53**, 4623–4627 (2014)
149. S.M. Copp, D.E. Schultz, S. Swasey, E.G. Gwinn, ACS Nano **9**, 2303–2310 (2015)
150. J.U. Reveles, P.A. Clayborne, A.C. Reber, S.N. Khanna, K. Pradhan, P. Sen, M.R. Pederson, Nat. Chem. **1**, 310–315 (2009)
151. X. Zhang, Y. Wang, H. Wang, A. Lim, G. Gantefoer, K.H. Bowen, J.U. Reveles, S.N. Khanna, J. Am. Chem. Soc. **135**, 4856–4861 (2013)
152. Y. Jia, Z. Luo, Coord. Chem. Rev. **400**, 213053 (2019)
153. P. Jena, Q. Sun, Chem. Rev. **118**, 5755–5780 (2018)
154. X.K. Wan, S.F. Yuan, Z.W. Lin, Q.M. Wang, Angew. Chem. Int. Ed. **53**, 2923–2926 (2014)
155. M.W. Heaven, A. Dass, P.S. White, K.M. Holt, R.W. Murray, J. Am. Chem. Soc. (2008)
156. K. Berggren, A. Bard, J. Wilbur, J. Gillaspy, A. Helg, J. McClelland, S. Rolston, W. Phillips, M. Prentiss, G. Whitesides, Science **269**, 1255–1257 (1995)
157. H.P. Cheng, U. Landman, Science **260**, 1304–1307 (1993)
158. K. Bromann, C. Félix, H. Brune, W. Harbich, R. Monot, J. Buttet, K. Kern, Science **274**, 956–958 (1996)
159. M.-C. Daniel, D. Astruc, Chem. Rev. **104**, 293–346 (2003)
160. Z.Y. Li, N.P. Young, M. Di Vece, S. Palomba, R.E. Palmer, A.L. Bleloch, B.C. Curley, R.L. Johnston, J. Jiang, J. Yuan, Nature **451**, 46–U42 (2008)
161. M. Moseler, H. Hakkinen, U. Landman, Phys. Rev. Lett. **89**, 176103 (2002)
162. B. Yoon, V.M. Akulin, P. Cahuzac, F. Carlier, M. de Frutos, A. Masson, C. Mory, C. Colliex, C. Brechignac, Surf. Sci. **443**, 76–88 (1999)
163. H.P. Cheng, U. Landman, J. Phys. Chem. **98**, 3527–3537 (1994)
164. W. Harbich, S. Fedrigo, F. Meyer, D.M. Lindsay, J. Lignieres, J.C. Rivoal, D. Kreisle, J. Chem. Phys. **93**, 8535–8543 (1990)
165. E.C. Honea, A. Ogura, D.R. Peale, C. Felix, C.A. Murray, K. Raghavachari, W.O. Sprenger, M.F. Jarrold, W.L. Brown, J. Chem. Phys. **110**, 12161–12172 (1999)
166. C.-C. Kuo, W.-C. Lin, H.-Y. Chang, K.-J. Song, Appl. Phys. Express **6**, 025501 (2013)
167. B. Yoon, U. Landman, V. Habibpour, C. Harding, S. Kunz, U. Heiz, M. Moseler, M. Walter, J. Phys. Chem. C **116**, 9594–9607 (2012)
168. B. Wang, B. Yoon, M. Koenig, Y. Fukamori, F. Esch, U. Heiz, U. Landman, Nano Lett. **12**, 5907–5912 (2012)
169. K.A. Wepasnick, X. Li, T. Mangler, S. Noessner, C. Wolke, M. Grossmann, G. Gantefoer, D.H. Fairbrother, K.H. Bowen, J. Phys. Chem. C **115**, 12299–12307 (2011)
170. M. Diaz, L. Martinez, M.M. Ruano, D. Llamosa P, E. Roman, M. Garcia-Hernandez, C. Ballesteros, R. Fermento, A. Cebollada, G. Armelles, Y. Huttel, J. Nanopart. Res. **13**, 5321–5333 (2011)
171. A.J. Logsdail, J. Akola, J. Phys. Chem. C **115**, 15240–15250 (2011)
172. L. Bardotti, F. Tournus, M. Pellarin, M. Broyer, P. Mélinon, V. Dupuis, Surf. Sci. **606**, 110–114 (2012)
173. D. Sun, A.E. Riley, A.J. Cadby, E.K. Richman, S.D. Korlann, S.H. Tolbert, Nature **441**, 1126–1130 (2006)
174. P. Schwerdtfeger, Angew. Chem. Int. Ed. **42**, 1892–1895 (2003)

175. M.H. Chisholm, A.M. Macintosh, Chem. Rev. **105**, 2949–2976 (2005)
176. L.M.C. Beltran, J.R. Long, Acc. Chem. Res. **38**, 325–334 (2005)
177. G.E. Johnson, C. Wang, T. Priest, J. Laskin, Anal. Chem. **83**, 8069–8072 (2011)
178. G.E. Johnson, Q. Hu, J. Laskin, in *Annual Review of Analytical Chemistry*, vol. 4, ed. by R.G. Cooks, E.S. Yeung (2011), pp. 83–104
179. S. Nagaoka, K. Ikemoto, K. Horiuchi, A. Nakajima, J. Am. Chem. Soc. **133**, 18719–18727 (2011)
180. H. Li, Y. Pei, X.C. Zeng, J. Chem. Phys. **133**, 134707–134707 (2010)
181. A.K. Badu-Tawiah, C. Wu, R.G. Cooks, Anal. Chem. **83**, 2648–2654 (2011)
182. V.N. Popok, I. Barke, E.E.B. Campbell, K.-H. Meiwes-Broer, Surf. Sci. Rep. **66**, 347–377 (2011)
183. J. Cyriac, M. Wleklinski, G. Li, L. Gao, R.G. Cooks, Analyst **137**, 1363–1369 (2012)
184. G. Verbeck, W. Hoffmann, B. Walton, Analyst **137**, 4393–4407 (2012)
185. G.E. Johnson, T. Priest, J. Laskin, J. Phys. Chem. C **116**, 24977–24986 (2012)
186. S.P. Price, X. Tong, C. Ridge, V. Shapovalov, Z. Hu, P. Kemper, H. Metiu, M.T. Bowers, S.K. Buratto, Surf. Sci. **605**, 972–976 (2011)
187. I. Rabin, W. Schulze, G. Ertl, C. Felix, C. Sieber, W. Harbich, J. Buttet, Chem. Phys. Lett. **320**, 59–64 (2000)
188. J. Laskin, P. Wang, O. Hadjar, J. Phys. Chem. C **114**, 5305–5311 (2010)
189. S. Nagaoka, K. Ikemoto, T. Matsumoto, M. Mitsui, A. Nakajima, J. Phys. Chem. C **112**, 15824–15831 (2008)
190. W.P. Peng, G.E. Johnson, I.C. Fortmeyer, P. Wang, O. Hadjar, R.G. Cooks, J. Laskin, Phys. Chem. Chem. Phys. **13**, 267–275 (2011)
191. S.A. Miller, H. Luo, S.J. Pachuta, R.G. Cooks, Science **275**, 1447–1450 (1997)
192. H. Luo, S.A. Miller, R.G. Cooks, S.J. Pachuta, Int. J. Mass Spectrom. Ion Processes **174**, 193–217 (1998)
193. G.-Y. Liu, P. Fenter, C.E.D. Chidsey, D.F. Ogletree, P. Eisenberger, M. Salmeron, J. Chem. Phys. **101**, 4301–4306 (1994)
194. O. Hadjar, P. Wang, J.H. Futrell, Y. Dessiaterik, Z. Zhu, J.P. Cowin, M.J. Iedema, J. Laskin, Anal. Chem. **79**, 6566–6574 (2007)
195. P. Wang, J. Laskin, Angew. Chem. Int. Ed. **47**, 6678–6680 (2008)
196. J.C. Love, L.A. Estroff, J.K. Kriebel, R.G. Nuzzo, G.M. Whitesides, Chem. Rev. **105**, 1103–1170 (2005)
197. U. Reimer, U. Reineke, J. Schneider-Mergener, Curr. Opin. Biotechnol. **13**, 315–320 (2002)
198. D.A. Eastham, J. Phys. D Appl. Phys. **28**, 2176–2180 (1995)
199. W. Mahoney, S.T. Lin, R.P. Andres, in *Evolution of Thin Film and Surface Structure and Morphology*, vol. 355, ed. by B.G. Demczyk, E. Garfunkel, B.M. Clemens, E.D. Williams, J.J. Cuomo (1995), pp. 83–88
200. P. Melinon, B. Prevel, V. Dupuis, A. Perez, B. Champagnon, Y. Guyot, M. Boudeulle, M. Pellarin, P. Dugourd, M. Broyer, Mater. Sci. Eng. Struct. Mater. Prop. Microstruct. Process. **217**, 69–73 (1996)
201. M. Han, Z.Y. Wang, P.P. Chen, S.W. Yu, G.H. Wang, Nucl. Instrum. Methods Phys. Res. Sect. B Beam Interact. Mater. Atoms **135**, 564–569 (1998)
202. W. Yamaguchi, K. Yoshimura, Y. Maruyama, K. Igarashi, S. Tanemura, J. Murakami, Chem. Phys. Lett. **311**, 415–420 (1999)
203. M.O. Watanabe, N. Uchida, T. Kanayama, Phys. Rev. B **61**, 7219–7222 (2000)
204. W. Yamaguchi, K. Yoshimura, Y. Tai, Y. Maruyama, K. Igarashi, S. Tanemura, J. Murakami, J. Chem. Phys. **112**, 9961–9966 (2000)
205. B. Klipp, M. Grass, J. Muller, D. Stolcic, U. Lutz, G. Gantefor, T. Schlenker, J. Boneberg, P. Leiderer, Appl. Phys. A Mater. Sci. Process. **73**, 547–554 (2001)
206. C. Brechignac, P. Cahuzac, F. Carlier, N. Kebaili, J. Le Roux, A. Masson, Surf. Sci. **518**, 192–200 (2002)
207. W.T. Nichols, D.T. O'Brien, G. Malyavanatham, M.F. Becker, J.W. Keto, in *Nanophase and Nanocomposite Materials Iv*, vol. 703, ed. by S. Komarneni, J.C. Parker, R.A. Vaia, G.Q. Lu, J.I. Matsushita (2002), pp. 209–214

208. X.Y. Wang, Z.Y. Pan, Y. Xu, Z. Huang, A.J. Du, Y.K. Ho, Surf. Coat. Technol. **158**, 263–268 (2002)
209. Y.X. Wang, Z.Y. Pan, Q. Wei, A.J. Du, Z. Huang, Y. Xu, Y.K. Ho, Surf. Sci. **512**, 128–134 (2002)
210. L. Gao, M.E. Lyn, D.E. Bergeron, A.W. Castleman, Int. J. Mass Spectrom. **229**, 11–17 (2003)
211. K. Shintani, T. Nakajima, Y. Taniguchi, in *Morphological and Compositional Evolution of Thin Films*, vol. 749, ed. by M.J. Aziz, N.C. Bartelt, I. Berbezier, J.B. Hannon, S.J. Hearne (2003), pp. 361–366
212. K. Meinander, K. Nordlund, J. Keinonen, Nucl. Instrum. Methods Phys. Res. Sect. B Beam Interact. Mater. Atoms **228**, 69–74 (2005)
213. X. Yang, H.J. Rader, A. Rouhanipour, K. Mullen, Eur. J. Mass Spectrom. **11**, 287–293 (2005)
214. A. Awasthi, S.C. Hendy, P. Zoontjens, S.A. Brown, Phys. Rev. Lett. **97**, 186103 (2006)
215. M.A. Rottgen, K. Judai, J.-M. Antonietti, U. Heiz, S. Rauschenbach, K. Kern, Rev. Sci. Instrum. **77**, 013302–013306 (2006)
216. J.-M. Antonietti, J. Gong, V. Habibpour, M.A. Roettgen, S. Abbet, C.J. Harding, M. Arenz, U. Heiz, C. Gerber, Rev. Sci. Instrum. **78**, 054101 (2007)
217. A. Awasthi, S.C. Hendy, P. Zoontjens, S.A. Brown, F. Natali, Phys. Rev. B **76**, 115437 (2007)
218. G. Schultes, M. Schmidt, M. Truar, D. Goettel, O. Freitag-Weber, U. Werner, Thin Solid Films **515**, 7790–7797 (2007)
219. T. Hayakawa, H. Yasumatsu, T. Koudow, Eur. Phys. J. D **52**, 95–98 (2009)
220. C.o.-K. Chen, S.-C. Chang, Appl. Surf. Sci. **256**, 2890–2897 (2010)
221. M.A. Gracia-Pinilla, E. Martinez, G. Silva Vidaurri, E. Perez-Tijerina, Nanoscale Res. Lett. **5**, 180–188 (2010)
222. V.N. Morozov, in *Nano/Micro Biotechnology*, vol. 119, ed. by I. Endo, T. Nagamune (2010), pp. 115–162
223. N. Ning, H. Vach, J. Phys. Chem. A **114**, 3297–3305 (2010)
224. C. Binns, P. Prieto, S. Baker, P. Howes, R. Dondi, G. Burley, L. Lari, R. Kroeger, A. Pratt, S. Aktas, J.K. Mellon, J. Nanopart. Res. **14**, 1136 (2012)
225. S. Rauschenbach, G. Rinke, N. Malinowski, R.T. Weitz, R. Dinnebier, N. Thontasen, Z. Deng, T. Lutz, P.M.d.A. Rollo, G. Costantini, L. Harnau, K. Kern, Adv. Mater. **24**, 2761–2767 (2012)
226. B.L. Walton, W.D. Hoffmann, G.F. Verbeck, Int. J. Mass Spectrom. **377**, 214–221 (2015)
227. S. Vajda, M.G. White, Acs Catal. **5**, 7152–7176 (2015)
228. Y. Duan, A.V. Teplyakov, J. Chem. Phys. **146**, 052814 (2017)
229. M. Khojasteh, S. Haghighat, J.M. Dawlaty, V.V. Kresin, Nanotechnology **29**, 215603 (2018)
230. I. Popov, T. Kunze, S. Gemming, G. Seifert, Eur. Phys. J. D **45**, 439–446 (2007)
231. K. Bromann, H. Brune, C. Felix, W. Harbich, R. Monot, J. Buttet, K. Kern, Surf. Sci. **377**, 1051–1055 (1997)
232. Y.X. Wang, Z.Y. Pan, Y.K. Ho, Z. Huang, A.J. Du, Q. Wei, Y. Xu, Surf. Coat. Technol. **158**, 258–262 (2002)
233. J. Carrey, J.L. Maurice, F. Petroff, A. Vaures, Phys. Rev. Lett. **86**, 4600–4603 (2001)
234. T. Gleitsmann, M.E. Vaida, T.M. Bernhardt, V.B. Koutecky, C. Burgel, A.E. Kuznetsov, R. Mitric, Eur. Phys. J. D **45**, 477–483 (2007)
235. S. Nagaoka, K. Horiuchi, M. Shikishima, A. Nakajima, J. Phys. Chem. C **115**, 24215–24220 (2011)
236. M.J. Cuddy, K.P. Arkill, Z.W. Wang, H.-P. Komsa, A.V. Krasheninnikov, R.E. Palmer, Nanoscale **6**, 12463–12469 (2014)
237. S. Linas, F. Jean, T. Zhou, C. Albin, G. Renaud, L. Bardotti, F. Tournus, Sci. Rep. **5**, 3053 (2015)
238. US2015045205-A1; US9574263-B2
239. W. Harbich, S. Fedrigo, F. Meyer, J. Chem. Phys. **93**, 8535–8543 (1990)
240. W. Christen, U. Even, J. Phys. Chem. A **102**, 9420–9426 (1998)
241. P. Wang, O. Hadjar, J. Laskin, J. Am. Chem. Soc. **129**, 8682–8683 (2007)
242. O. Hadjar, J.H. Futrell, J. Laskin, J. Phys. Chem. C **111**, 18220–18225 (2007)

243. J. Laskin, P. Wang, O. Hadjar, J.H. Futrell, J. Alvarez, R.G. Cooks, Int. J. Mass Spectrom. **265**, 237–243 (2007)
244. J. Alvarez, R.G. Cooks, S.E. Barlow, D.J. Gaspar, J.H. Futrell, J. Laskin, Anal. Chem. **77**, 3452–3460 (2005)
245. B. Gologan, J.R. Green, J. Alvarez, J. Laskin, R. Graham Cooks, Phys. Chem. Chem. Phys. **7**, 1490–1500 (2005)
246. J. Alvarez, J.H. Futrell, J. Laskin, J. Phys. Chem. A **110**, 1678–1687 (2005)
247. J.W. Shen, C. Evans, N. Wade, R.G. Cooks, J. Am. Chem. Soc. **121**, 9762–9763 (1999)
248. M. Volný, A. Sengupta, C.B. Wilson, B.D. Swanson, E.J. Davis, F. Tureček, Anal. Chem. **79**, 4543–4551 (2007)
249. M.B.J. Wijesundara, L. Hanley, B. Ni, S.B. Sinnott, Proc. Natl. Acad. Sci. U.S.A. **97**, 23–27 (2000)
250. M.B.J. Wijesundara, E. Fuoco, L. Hanley, Langmuir **17**, 5721–5726 (2001)
251. M. Volný, W.T. Elam, B.D. Ratner, F. Tureček, Anal. Chem. **77**, 4846–4853 (2005)
252. B. Gologan, Z. Takáts, J. Alvarez, J. Wiseman, N. Talaty, Z. Ouyang, R.G. Cooks, J. Am. Soc. Mass Spectrom. **15**, 1874–1884 (2004)
253. Z. Ouyang, Z. Takáts, T.A. Blake, B. Gologan, A.J. Guymon, J.M. Wiseman, J.C. Oliver, V.J. Davisson, R.G. Cooks, Science **301**, 1351–1354 (2003)
254. B. Feng, D.S. Wunschel, C.D. Masselon, L. Pasa-Tolic, R.D. Smith, J. Am. Chem. Soc. **121**, 8961–8962 (1999)
255. G. Siuzdak, B. Bothner, M. Yeager, C. Brugidou, C.M. Fauquet, K. Hoey, C.-M. Change, Chem. Biol. **3**, 45–48 (1996)
256. M. Mitsui, S. Nagaoka, T. Matsumoto, A. Nakajima, J. Phys. Chem. B **110**, 2968–2971 (2006)
257. R. Neuendorf, R.E. Palmer, R. Smith, Chem. Phys. Lett. **333**, 304–307 (2001)
258. B. Kaiser, T.M. Bernhardt, B. Stegemann, J. Opitz, K. Rademann, Phys. Rev. Lett. **83**, 2918–2921 (1999)
259. R.E. Palmer, S. Pratontep, H.G. Boyen, Nat. Mater. **2**, 443–448 (2003)
260. J.P. Cowin, A.A. Tsekouras, M.J. Iedema, K. Wu, G.B. Ellison, Nature **398**, 405–407 (1999)
261. A.A. Tsekouras, M.J. Iedema, J.P. Cowin, J. Chem. Phys. **111**, 2222–2234 (1999)
262. S. Nanita, Z. Takats, R.G. Cooks, S. Myung, D. Clemmer, J. Am. Soc. Mass Spectrom. **15**, 1360–1365 (2004)
263. R.K. Smith, S.U. Nanayakkara, G.H. Woehrle, T.P. Pearl, M.M. Blake, J.E. Hutchison, P.S. Weiss, J. Am. Chem. Soc. **128**, 9266–9267 (2006)
264. P.J. Roach, W.H. Woodward, A.W. Castleman Jr., A.C. Reber, S.N. Khanna, Science **323**, 492–495 (2009)
265. J. Laskin, P. Wang, O. Hadjar, Phys. Chem. Chem. Phys. **10**, 1079–1090 (2008)
266. L. Ma, B. Issendorff, A. Aguado, J. Chem. Phys. **132**, 104303 (2010)
267. F.C. Chuang, C.Z. Wang, K.H. Ho, Phys. Rev. B **73**, 125431 (2006)
268. M. Guo, B. Yin, B. Huang, H. Wu, Z. Luo, J. Mater. Chem. C (2020). https://doi.org/10.1039/D1030TC02296A

Chapter 15
Future Directions

As metal clusters provide complementary active sites for surface catalysis and reactions, there is future promising potential to create unique compounds with tailored properties where one atom site makes a difference. On this basis, one of the prime objectives of cluster science is to lay the foundation for forming sub-nano materials via assembly of stable cluster building blocks or varied size and composition. This pursuit is recognized as one of the most promising frontiers in nanoscience, and has been realized by soft-landing deposition technique. Numerous studies have been conducted on solid-supported clusters revealing valuable applications in catalysis, typically redox reactions on electrodes. Also, there are ongoing efforts devoted to elucidating interactions occurring within heterogeneous catalytic systems, providing a dearth of knowledge pertaining to structure-reactivity relationships. The investigations of solid-supported clusters are anticipated to bloom with the rapid development of electron microscope technology, especially spherical aberration electron microscopy and frozen electron microscopy, as well as scanning tunneling microscopy. In particular, extensive experimental studies of metal cluster catalysis from wet chemistry synthesis have been conducted enabling to unravel the intrinsic mechanisms of metal nanoparticles at reduced precise sizes [1, 2].

One of the areas where future catalysts are going to play an important role is the green energy. For example, Methane from shale and biomass resources is emerging as an important feedstock for the fuel, chemical, and power industries. Methane needs to be converted to high-value liquid fuels and chemicals for utilization. The traditional approach to methane conversion is based on oxidation to syngas ($CO + H_2$) followed by Fischer Tropsch synthesis to higher hydrocarbons. This two-step process has an inherent inefficiency since the breaking of all methane C–H bonds to produce syngas in the first step must be substantially reversed in the second step to produce hydrocarbons, thus resulting in low energy efficiency and high capital-cost. There is a critical need to develop direct pathways to convert methane into high-value liquid aromatics and one possibility is to use single-atom and metal cluster catalysts.

Z. Luo and S. N. Khanna, *Metal Clusters and Their Reactivity*,
https://doi.org/10.1007/978-981-15-9704-6_15

Such a development will present a transformational approach to catalysis with the possibility of significantly high activity due to a large extent of uncoordinated atoms.

Another growing area of research is the catalysis by supported clusters. The primary role of support systems in metal-catalyzed heterogeneous catalysis has historically centered on the ability to provide a high surface area to which metal clusters can be affixed, thus limiting deactivation through sintering. However, the emergence of shape selective catalysts in hydrothermal cracking applications provides an excellent example of how supports can add additional value to catalytic processes. For example, the use of titanium, cerium, and zirconium oxides as substrate promoters in cobalt-catalyzed Fischer-Tropsch reactions offers an additional example of how supports can contribute to heterogeneous catalytic activity. As we show in Chap. 9, the support can play a more active role in catalysis by acting as a ligand that can allow charge flow thus enabling one to reduce reaction barriers on supported clusters during oxidation and reduction steps. These developments that need further work can be transformational in reducing cost for pharmaceutical drugs opening the pathway to "medicine for everybody".

Most of the catalysts currently used in industry are based on precious metals and there is a tremendous interest in developing catalysts using cheaper metals. The development of superatoms [3–5], where the clusters of a given element can mimic the chemical behavior of another element provides a viable alternative. Such findings will be further extended to binary metals and compound systems, allowing a lollapalooza effect of both electronic and geometric structures identified to govern the stability and presence of reactive sites. It is anticipated of superatom building blocks for new materials with tailored properties, for instance, isovalent inexpensive clusters will lead to a new generation of catalysts, as well as magnetic superatoms for spin electronics, and superatomic semiconductive clusters for transistors. A success of future applications of metal nanoclusters will rely on the availability of low-dimensionally assembled materials. For this purpose, the genetic structures of superatom clusters are expected to open a new research areas [6].

Moreover, there could be research topics on the energetics-related metal cluster reactions with organic molecules which simulate fuels. Such reactions conducted in endothermic conductions will bring important implications to aviation and rockets, namely the ability to acquire large thrust without undue stress being placed on the aircraft engines and related components. With well-selected conditions, the engines run efficiently acquiring better combusting rates. Investigations on exothermic cluster reactions will stimulate researchers to look for promising reactant pairs which may also be undertaken with binary metal cluster systems.

While it is well recognized that clusters of selected composition, stoichiometry, and charge-state provide ideal systems for studying heterogeneous catalysis and energetics-related reactions, there could be anticipated research interest devoted to understanding the principles and measures upon anticorrosion which cost a great deal of money especially in naval vessels, cross-sea bridge, water piping system, hydroelectric-dam reservoirs, etc. A few future directions of this topic could involve the investigations to understand the physical or chemical damage of metal surfaces by environmental interactions, to unravel the principles and mechanism corresponding to varied corrosion phenomenon and processes, to propose anticorrosion approaches relative to different environmental conditions, and to establish theoretical consensus for metal materials application.

On the other hand, photochemistry, including optical absorption, luminescence, nonlinear optics and photocatalysis, are also the most attractive properties of metal nanoclusters [7, 8], indicating promising applications in chemical sensing, bio-imaging and cell labeling, phototherapy and drug delivery [2, 8–18]. It is notable that, the valence electrons of the metals often occupy the higher energy levels in the metal NCs, and metal alloys could display altered optical properties even with same ligand and similar structure, or even though the heterometal belongs to the same group of elements (that is, no changes of the total valence electrons) [19]. The electronic structure of the nanoclusters (HOMO-LUMO gap and charge density) could be meticulously adjustable pertaining to alloying techniques and ligand engineering strategy, allowing likely multicolor emissions within and against Kasha's Rule. The high adjustability, reasonable rigidity and photostability of metal cluster luminance enable to develop high-performance microdevices for light-emitting diodes and chemo-sensors [20].

References

1. Y. Du, H. Sheng, D. Astruc, M. Zhu, Chem. Rev. **120**, 526–622 (2020)
2. I. Chakraborty, T. Pradeep, Chem. Rev. **117**, 8208–8271 (2017)
3. Z. Luo, A.W. Castleman, Acc. Chem. Res. **47**, 2931–2940 (2014)
4. A.C. Reber, S.N. Khanna, Acc. Chem. Res. **50**, 255–263 (2017)
5. P. Jena, Q. Sun, Chem. Rev. **118**, 5755–5780 (2018)
6. Y. Jia, Z. Luo, Coord. Chem. Rev. **400**, 213053 (2019)
7. K.L.D.M. Weerawardene, C.M. Aikens, J. Am. Chem. Soc. **138**, 11202–11210 (2016)
8. X. Kang, M. Zhu, Chem. Soc. Rev. **48**, 2422–2457 (2019)
9. M. Wang, W. Wang, C. Liu, J. Liu, T.-S. Kang, C.-H. Leung, D.-L. Ma, Materials Chemistry Frontiers **1**, 128–131 (2017)
10. Y. Tao, M. Li, J. Ren, X. Qu, Chem. Soc. Rev. **44**, 8636–8663 (2015)
11. X. Hu, Y. Zheng, J. Zhou, D. Fang, H. Jiang, X. Wang, Chem. Mater. **30**, 1947–1955 (2018)
12. N. Goswami, K. Zheng, J. Xie, Nanoscale **6**, 13328–13347 (2014)
13. L. Nie, X. Xiao, H. Yang, J. Nanosci. Nanotechnol. **16**, 8164–8175 (2016)
14. B. Du, X. Jiang, A. Das, Q. Zhou, M. Yu, R. Jin, J. Zheng, Nat. Nanotechnol. **12**, 1096–1102 (2017)
15. T. Udayabhaskararao, T. Pradeep, J. Phys. Chem. Lett. **4**, 1553–1564 (2013)
16. T.-A.D. Nguyen, Z.R. Jones, B.R. Goldsmith, W.R. Buratto, G. Wu, S.L. Scott, T.W. Hayton, J. Am. Chem. Soc. **137**, 13319–13324 (2015)
17. K.K. Chakrahari, R.P.B. Silalahi, J.-H. Liao, S. Kahlal, Y.-C. Liu, J.-F. Lee, M.-H. Chiang, J.-Y. Saillard, C.W. Liu, Chem. Sci. **9**, 6785–6795 (2018)
18. S. Sharma, K.K. Chakrahari, J.-Y. Saillard, C.W. Liu, Acc. Chem. Res. **51**, 2475–2483 (2018)
19. S. Wang, Q. Li, X. Kang, M. Zhu, Acc. Chem. Res. **51**, 2784–2792 (2018)
20. M. Xie, C. Han, Q. Liang, J. Zhang, G. Xie, H. Xu, Sci. Adv. **5**, eaav9857 (2019)

The manufacturer's authorised representative in the EU is Springer
Nature Customer Service Centre GmbH, Europaplatz 3, 69115 Heidelberg,
Germany. If you have any concerns regarding our products, please
contact ProductSafety@springernature.com

Printed and bound by CPI Group (UK) Ltd, Croydon, CR0 4YY
30/09/2026
02225329-0010